T0296085

CAMBRIDGE GEOLOGICAL SERIES

THE GEOLOGY OF THE
METALLIFEROUS DEPOSITS

THE
GEOLOGY OF THE
METALLIFEROUS DEPOSITS

BY

R. H. RASTALL, Sc.D., M.Inst.M.M.

*University Lecturer in Economic Geology, Cambridge,
and formerly Fellow of Christ's College*

CAMBRIDGE
AT THE UNIVERSITY PRESS
1923

CAMBRIDGE
UNIVERSITY PRESS

University Printing House, Cambridge CB2 8BS, United Kingdom

Published in the United States of America by Cambridge University Press, New York

Cambridge University Press is part of the University of Cambridge.

It furthers the University's mission by disseminating knowledge in the pursuit of
education, learning and research at the highest international levels of excellence.

www.cambridge.org
Information on this title: www.cambridge.org/9781107659575

© Cambridge University Press 1923

First published 1923
First paperback edition 2014

A catalogue record for this publication is available from the British Library

ISBN 978-1-107-65957-5 Paperback

PREFACE

IN writing this book the author's chief objects have been two-fold; first to give a clear account, founded on basal theories of petrology and mineralogy, of the general principles that underlie the formation of ore-deposits and the changes that they subsequently undergo; secondly, to describe and illustrate a certain number of typical examples of mineral deposits, selected so far as possible to bring out the geological relations of the methods of occurrence of each particular metal. It is emphatically not a description of the known metalliferous deposits of the world and the carrying out of the scheme has doubtless resulted in certain parts in an apparent want of balance, in that certain ores or groups of ores may appear to occupy too much or too little space in proportion to their actual economic value. But this simply means that if a commercially important product does not show much variation it is dealt with briefly, while a less important but more variable subject may occupy more space in bringing out these variations.

In the general section, occupying the first half of the book, an attempt has been made to deal with the genesis and characters, or what may conveniently be called the natural history of ore-deposits, in accordance with the latest theories and facts of chemistry and physics, along the same lines as have been followed so successfully by many recent writers on the geology of the igneous rocks. The literature of this subject, descriptive, experimental and theoretical, is now of vast proportions, so that it becomes almost invidious to mention any individual names. But the author cannot refrain from recording his deep indebtedness to the inspiring work of Brögger, Vogt and Harker in Europe and of that brilliant group of petrologists in America who have done so much to build up the theory of the origin of igneous rocks and of ore-deposits on a basis of field observation and laboratory experiment.

The second part of the book contains descriptions of actual occurrences of ores, usually, but not invariably, selected from those of high commercial importance, and arranged under the

headings of the different metals. The system of classification to
be adopted here was the subject of long and earnest consideration. Every possible basis, chemical, mineralogical, geological
and geographical, was subjected to criticism. Of all these
possible principles the geographical is easily the worst, and was
at once rejected. Geological and genetic classifications seem
logically sound, but also possess great drawbacks, which would
take too long to enumerate in detail. One very grave objection
however may be cited, namely, that in some instances great
doubt still exists as to how certain deposits were formed, and
an arrangement on these lines would entail dogmatism on an
unsound basis. The greatest objection to a classification on a
basis of metals is that many deposits produce more than one,
but this can be got over by a system of cross reference and
when the chief object is to describe only selected typical examples, very complex cases can be avoided. This classification
was therefore finally adopted. In every case what appeared to
be the best recent description was consulted; doubtless errors
have crept in, but in a work of this kind such are inevitable.
Statistics have been avoided as far as possible, except when they
served to illustrate some special point. Owing to war disturbances,
recent figures are often difficult to obtain, and in any case
statements of the outputs of so abnormal a period are of little
permanent value.

The number of illustrations has been kept low, in order to
avoid a book of undue size and cost. All borrowed figures are
duly acknowledged in the explanatory text; special thanks are
however due to the Director of the United States Geological
Survey, who kindly gave a general permission to copy any
figures in the official publications of that Survey. Grateful
acknowledgment is also made of the permission accorded by the
Director of H.M. Geological Survey and by the Controller of
H.M. Stationery Office for the reproduction of a large number
of figures from the *Special Reports on the Mineral Resources of
Great Britain* (Figs. 50, 51, 53, 54, 55, 56, 61, 62, 63, 64, 65, 66).
Kind permission to make use of published figures was also given
by the Editors of the *Mining Magazine* and of *Economic
Geology*, by Trafikaktiebolaget Grängesberg-Oxelösund, Stock-

holm, by the Institution of Mining and Metallurgy, by Dr Hatch and others.

It is regrettable that a most important work by Mr J. E. Spurr, entitled *The Ore Magmas* (New York, 1923), dealing with the theoretical side of the subject, appeared only after the manuscript of this book was in the printers' hands. Hence it was impossible to include any reference to Mr Spurr's conclusions as to the genesis of ores, which are closely similar to the ideas here set forth.

Finally, the author is much indebted to Mr W. H. Wilcockson, M.A., F.G.S., Lecturer in Geology in the University of Sheffield, who kindly contributed the section on Vanadium, and to Mr A. G. Brighton, B.A., of Christ's College, Cambridge, who drew almost the whole of the original diagrams contained in the book.

R. H. RASTALL.

CAMBRIDGE,
September, 1923.

CONTENTS

PART I. *GENERAL PRINCIPLES*

CHAP.   PAGE

I. INTRODUCTION   1

II. THE IGNEOUS ROCKS   12

The relation of igneous rocks to ore-deposits. The consolidation of igneous rocks. The lowering of the freezing-point and the eutectic ratio. The reaction-principle in crystallization. Differentiation of igneous magmas. Limited miscibility of solutions. Differentiation by absorption: assimilation. The separation of differentiated partial magmas. The mechanism of separation. Crystallization in intrusive masses. Pegmatites and aplites. Pegmatitic and aplitic textures. Mineralized veins. Pneumatolysis. Solfataric action. The petrology of the pegmatites.

III. THE SEDIMENTARY AND METAMORPHIC ROCKS   48

The sedimentary rocks. Metasomatic changes in sedimentary rocks. Metamorphism. Thermal metamorphism. Dynamic metamorphism.

IV. THE RELATIONS OF WATER TO ORE-FORMATION   57

The origin of natural waters. Ground-water. The relation of water to different types of rock. Special conditions in arid regions. The reascensionist theory of water circulation. Deposition of ores from solution. Secondary changes in ore-bodies due to water.

V. THE FORMS OF ORE-DEPOSITS.   70

General considerations. Veins. The lie of veins. Form and extent of veins. Nomenclature of veins. Saddle-reefs. Stockworks. Irregular and massive ore-bodies. Disseminations. Bedded ore-deposits. The influence of the country rock on ore-deposits. Ore-shoots.

VI. THE COMPOSITION AND CHARACTERS OF ORE-DEPOSITS   92

Mineral composition. The deposition of ores. Primary ore minerals. Gangue minerals. Vein fillings and veinstones. Stockworks and disseminations. Contact deposits and replacements. Other types of ore-deposit.

VII. THE CLASSIFICATION OF ORE-DEPOSITS . .   106

CHAP.   PAGE

VIII. THE RELATION OF ORE-DEPOSITS TO EXTERNAL
INFLUENCES 113

Oxidation and secondary enrichment. The zone of oxidation.
Gossan or iron cap. The relation of ore-zones to denudation.
Changes in a copper lode. The influence of climate on secondary
changes. Secondary zones in silver deposits. Zoning in mixed
lodes. The chemistry of secondary enrichment of copper ores.
Metamorphism and ore-deposits. Two classes of metamorphism.
The causes of thermal metamorphism. The effects of thermal
metamorphism. Dedolomitization. The zinc ores of New Jersey.
Weathering processes in rocks. Weathering of igneous rocks.
The weathering of sediments. Weathering and ore-deposits.
Rock-changes accompanying mineralization. Silicification.
Sericitization. Propylitization. Pyritization. Formation of
alunite. Carbonates in altered rocks. Dolomitization and
carbonatization.

IX. METALLOGENESIS. 144

General introduction. Iron in igneous rocks. Iron and sulphur.
Magmatic sulphides. Primary low-temperature sulphides.
Arsenides, selenides and tellurides. Non-ferrous oxide segregations. Native metals. Paragenesis of metals. Isomorphism in
the sulphides and arsenides. Sulphides of lead, zinc and silver.
Compounds of gold. Isomorphism in the secondary sulphides.
Metallogenetic provinces. Tin and its associates. The distribution of silver. Metallogenetic epochs. Metallogenetic epochs in
the United States. The ore-deposits of Montana. Generalizations.

X. METALLOGENETIC ZONES 181

The general problem. Variation in depth. An ideal case.
Variations in lead-zinc lodes. The ore zones of Cornwall. Zones
in the lead-zinc deposits of the British Isles. The generalized
succession of ore zones. American examples. China. Zones of
rock-alteration. Causes of primary zoning.

XI. MINERAL FORMATION 202

The physical chemistry of mineral formation. The phase rule.
Solidification of a single component. Crystallization from solutions. Solid solutions. Mixed crystals. Dimorphism and polymorphism. Formation of secondary minerals. The stability of
minerals. The separation of ore-minerals from silicate magmas.
Colloids and mineral formation. Formation of minerals from
gases. Minerals of complex composition.

PART II. *DESCRIPTIVE*

CHAP.   PAGE

XII. COPPER 227

Ores of copper. Types of copper deposit. Secondary alterations
in copper veins. The British Isles. Moonta and Wallaroo, South
Australia. Copper deposits of the United States. The Lake
Superior copper region. The copper deposits of Butte, Montana.
"Porphyry copper" deposits. Limestone replacement deposits.
Cupriferous pyrite ore-bodies. Rio Tinto district, Spain.
Copper-bearing sulphides in Norway. The copper deposits of
Namaqualand. Copper in sedimentary rocks. Copper in the
Trias of England. The copper deposits of Katanga. Copper in
red sandstones. The copper shale of Mansfeld, Saxony. Origin
of copper in sediments.

XIII. TIN 260

Tin minerals. Cornwall and Devon. Tin-bearing granites of
Devon and Cornwall. Pneumatolytic changes in the granites.
The tin ores of Cornwall. Tin in Bolivia. Tin in the Malay Pen-
insula. Tin in the Commonwealth of Australia. Northern
Nigeria.

XIV. LEAD AND ZINC 281

Lead and zinc minerals. General occurrence of lead and zinc
ores. Lead and zinc in Great Britain. Central Wales. Shropshire. North Wales. The Lake District. The Leadhills district,
south Scotland. Lead and zinc ores in the Carboniferous rocks
of North Wales. The Pennine region. Origin of the lead-zinc
ores of the British Isles. The zinc ores of Franklin Furnace,
New Jersey. The zinc ores of Missouri and neighbouring
States. Broken Hill, New South Wales. The Bawdwin mines,
Burma.

XV. IRON. 313

Sources of iron. Magmatic segregations. The magnetite ores of
northern Sweden. Gellivare. Grängesberg. The banded ores of
Sydvaranger. The Lake Superior iron region. Banded ironstones in general. Metasomatic iron ores. Haematite of Cumberland and Lancashire. Forest of Dean. Bessemer ores of
northern Spain. Stratified iron ores. The Coal-measure ironstones of Great Britain. The Jurassic ironstones of the British
Isles. The Frodingham stone of north Lincolnshire. The Middle
Lias. The Middle Lias ironstones of the Midlands. The Cleveland
ironstone. The Inferior Oolite. The iron ores of Lorraine. The
Clinton ores of the United States. The oolitic iron ores of
Wabana, Newfoundland. Microscopic structure of oolitic ironstones. The origin of oolitic ironstones. Bog-iron ores and lake
ores.

CHAP.   PAGE
XVI. NICKEL, COBALT, MANGANESE, CHROMIUM .   .   362

Nickel minerals. Nickeliferous sulphide ores. Nickel-bearing
sulphide deposits in Norway. The nickel ores of Sudbury,
Ontario. The nickel ores of Insizwa. Nickel in New Caledonia.
Arsenical nickel ores. Cobalt minerals. The cobalt-silver lodes
of Saxony. Cobalt in Canada. Manganese. Manganese ores.
Manganese production. The British Isles. India. Russia.
Brazil. Chromium.

XVII. MERCURY, ANTIMONY, ARSENIC, BISMUTH .   .   396

Mercury. The mercury mines of Almadén. The mercury deposits
of Idria. The mercury deposits of Tuscany. The mercury deposits
of the Pacific Slope. Mercurial tetrahedrite ores. Antimony.
The antimony deposits of China. France. Bolivia. Arsenic.
Bismuth.

XVIII. THE MINOR METALS   411

Classification. Tungsten. Tungsten in Burma. Wolfram lodes
without tin. Scheelite deposits. Origin of tungsten ores.
Secondary tungsten deposits. Molybdenum. The molybdenum
minerals. Types of molybdenum ore-deposits. Molybdenite in
Canada. Molybdenite in Australia. Molybdenite in Norway.
Vanadium. Titanium. Thorium. Tantalum. Zirconium. Uran-
ium and radium. Uranium in Cornwall. Uranium lodes in the
Erzgebirge.

XIX. ALUMINIUM   446

Occurrence and properties. Bauxite. Laterite. Bauxite in
Ireland. Cryolite.

XX. THE PRECIOUS METALS   454

Gold. Ores of gold. Types of occurrence. Quartz veins with
free gold. The Kolar goldfield, Mysore. The gold lodes of Cali-
fornia. The Porcupine district, Ontario. The saddle-reefs of
Bendigo. The Morro Velho mine, Brazil. The Passagem mine,
Brazil. Gold-telluride ores. Cripple Creek, Colorado. The
Transylvanian Erzgebirge. The auriferous conglomerates of the
Transvaal. Secondary gold deposits. Placer gold in California.
Gold placers in Australia. The Yukon and Klondike goldfields.
The beach placers of Nome, Alaska. Silver. Silver minerals.
Silver-lead ores. The Kongsberg silver deposits. The silver
veins of the Cobalt district, Ontario. Tonopah, Nevada. The
Comstock lode. Silver in Bolivia and Peru. Silver mines in
Germany. The platinum group. The platinum deposits of the
Urals. Platinum in Colombia. Platinum at Sudbury, Ontario.

INDEX   498

PART I

GENERAL PRINCIPLES

CHAPTER I

INTRODUCTION

In the nature of things we do not and cannot possess any definite information as to the primary source of the material of the ore-deposits. This is very much a matter of cosmical and astronomical speculation. As a matter of fact we do not know how the solid earth as a whole came into being. We know that there are in the universe masses of glowing gas, the nebulae, and other heated masses of gaseous, liquid or solid material, the sun and the fixed stars. There are other heavenly bodies, the planets and the moon, which are not hot enough to be self-luminous. All these show a kind of evolutionary series, and it is a fair inference that they represent different stages of development. This is the essence of the Nebular Hypothesis of the earth's origin. According to it, the earth is in a somewhat advanced stage of existence, having lost a good deal of its original heat by radiation. At one time this theory seemed quite satisfactory, but the discovery of radio-active disintegration of elements in recent times has thrown doubt on the general applicability of the progressive cooling theory; it is possible that the earth is actually growing hotter instead of colder, although the facts of geological history render this unlikely. Nevertheless it is clear that the cooling, if occurring at all, is extraordinarily slow. Geology also shows that the sedimentary rocks of the outer crust appear to be underlain everywhere by rocks which have obviously been formed by freezing from a molten state. Hence whatever may have been the events of the remotest past, we have evidence of a time when the earth was everywhere covered by a layer of crystalline material, formed by consolidation from a state of fusion. From this primitive crust the later sedimentary deposits were formed by the normal geological agencies.

Of late years another theory has received considerable support, especially in America, under the name of the Planetesimal Hypothesis. According to this theory the earth has been formed by the chance aggregation of vast numbers of small bodies, called planetesimals, analogous to meteorites, and possibly on the average comparable in size with grains of sand, careering through space in swarms and falling together by gravity to form larger masses, the suns and planets[1]. However in the opinion of competent cosmical physicists this theory is dynamically impossible, and in any case a globe so formed would have by its own evolution to pass through a gaseous phase so that the later stages, with which alone we are here concerned, would be the same as in the case of the pure nebular hypothesis[2].

When the earth is considered from the cosmical point of view one of its most striking features is its high density. The density of the earth as a whole is about 5·5, while the average specific gravity of the rocks of the accessible crust is about 2·75 or almost exactly one half. Experimental evidence has shown that this high density cannot be due to simple compression in depth, and it is inferred that the interior of the earth is composed of actually heavier material than the crust. This heavy material probably consists of metals, and possibly sulphides. This is confirmed by the composition of meteorites, which may be regarded as miniature planets, and often consist of nickel-iron, or of iron sulphides.

The outer and accessible crust of the earth, on the other hand, contains an overwhelmingly large proportion of silica, even down to the greatest depths laid bare by denudation and uplift. Some of the rocks now visible at the surface have been brought up from depths of several miles, and yet they are composed of silicates, similar to those in the shallow zones: many lavas must have come up from great depths and they also are siliceous. Hence it must be concluded that the silicate crust is thick, though perhaps small in comparison with the total radius of the earth. Observations on the propagation of earthquake waves also indicate that at certain depths there are abrupt changes in

---

[1] For a discussion of this theory see Chamberlin and Salisbury, *Geology*, vol. I, New York, 1903, or Hobbs, *Earth Evolution*, New York, 1921.

[2] See Eddington, *Nature*, Jan. 6, 1923.

the constitution of the earth, and especially changes of density, thus confirming the ideas above expressed.

If we assume that the primaeval earth thus consisted of a mass of fused metals, sulphides and silicates, the relations just sketched out are what might be expected on theoretical grounds. As will be explained later, we know that metals or sulphides on the one hand, and silicates on the other, when in the liquid state, are miscible in very limited proportions. This fact is the foundation of many metallurgical processes. In these we have a clear separation between metal or matte below and slag above: but, and this is an important point, each dissolves a little of the other. Thus on the metallurgical theory of the origin of the crust, we should have a metallic core below, with a little, probably very little, dissolved silicate, and a siliceous slag above with a little dissolved metal or sulphide. This slag forms the primary rock-crust, or earliest igneous rocks, and in it are disseminated, probably more or less uniformly, the metallic elements, which are later, in various ways, concentrated to form the ore-deposits as we see them to-day. This is only the beginning of the story, which up to this point is frankly speculative, but from this point onwards we are on surer ground. This primary crust would inevitably undergo many and far-reaching further changes, which can be reduced under known physical and chemical laws, leading to processes of concentration and segregation that have given rise to the workable ore-deposits of the present time. What these processes are will appear in detail in later sections.

We are thus led to the conception of an early stage in the history of the earth when the metals as we know them were more or less evenly distributed throughout the slaggy crust, forming disseminations of native metals, sulphides and oxides such as we now find over limited areas in certain types of igneous rocks; with an ever-increasing tendency, as time went on, to differentiation and segregation along with particular rock-types, according to their solubility in the more or less siliceous and more or less alkaline fractions into which the silicate crust when in a molten state tended to differentiate. This process, when once started, was doubtless cumulative, and eventually led to

the endless diversities of type now existing both in igneous rocks
and in ore-deposits.

A calculation of the composition of the whole accessible crust
of the earth, founded on a summation of the results of many
hundreds of analyses of rocks of every kind, has been made by
F. W. Clarke of the United States Geological Survey[1]. These
figures, omitting the details of the minor constituents, are given
in the following table:

	per cent.
Silica, SiO_2	59·77
Alumina, Al_2O_3	14·89
Ferric oxide, Fe_2O_3	2·69
Ferrous oxide, FeO	3·39
Magnesia, MgO	3·74
Lime, CaO	4·86
Soda, Na_2O	3·25
Potash, K_2O	2·98
Water, H_2O	2·02
Titanium dioxide, TiO_2	·77
Carbon dioxide, CO_2	·70
Phosphorus pentoxide, P_2O_5 ...	·28
Sulphur, S	·10
	99·44

From these figures two striking conclusions can be drawn. In
the first place only two of the metals of common technical
application, iron and aluminium, are really important constituents of the earth's crust, amounting to 4·18 and 7·30 per cent.
respectively. Secondly all the other valuable and useful metals
taken together must amount to less than one half of one per
cent. of the crust: that is to say that each actually forms a mere
trace from the analytical point of view, and it is only when
somehow exceptionally concentrated that they give rise to workable ore-deposits. It is the study of these exceptional concentrations that is the chief object of mining geology.

If however we are content to assume that the earth was once
a liquid globe it is not difficult to picture what must have been
the condition of the earliest crust, at least in general terms. We
have seen that the outermost layer so far as accessible to
observation is mainly siliceous and of low average density. We

---

[1] Clarke, *The Data of Geochemistry*, 4th edition; Bulletin 695, *U.S. Geol.
Survey*, 1920, p. 33 and Washington, "The Chemistry of the Earth's Crust,"
Journ. Franklin Inst., vol. cxc, pp. 757–815.

know also that the interior must be composed of much heavier material and may very probably be mainly metallic. It is therefore permissible to picture to ourselves at one stage the highly heated earth consisting of a superficial shell of fused slag with a metallic core. As the temperature fell this shell of slag would solidify to a crust of silicate rocks, probably undergoing a certain amount of differentiation in itself, so that on the whole in the silicate shell the acid or granite material tended to float to the top, while the heavier basaltic material sought the lower layers of the crust, thus giving rise to the two primitive earth-magmas, acid and basic, as pictured by Daly[1]. It must be admitted that we do not as yet possess any theoretical or practical data to explain the tendency of fused silicates on a large scale to separate into acid and basic fractions, but the facts of petrography show clearly enough that such a tendency does exist. It has been proposed to extend the idea of immiscible liquids to this case also, but hitherto no experimental evidence has been found in support of it. So far as we know all fused silicates are miscible in all proportions. The explanation is probably to be found in partial crystallization leading to sinking of the heavier, earlier formed minerals, which may melt again on reaching a region of higher temperature[2]. Silicate melts are usually very viscous and equilibrium is only restored by diffusion with extreme slowness.

Whether or not this process has occurred on a scale applicable to the whole original liquid slag-crust, it is certain that something similar has happened in fused silicate masses on a smaller scale, which have become isolated during later stages of rock-formation and in many instances far more than two differentiated fractions have been formed, as can be seen by reference to any text-book of general petrology[3]. This subject will be dealt with in detail in a later section, when we come to consider the relation of metalliferous deposits to different rock-types.

From the general average composition of the silicate crust, as

[1] *Igneous Rocks and their Origins*, New York, 1914, p. 162.
[2] Bowen, *Journ. Geol.* vol. XXIII, 1915, Supplement.
[3] See especially Harker, *The Natural History of the Igneous Rocks*, London, 1909.

set forth in the table already given, it is evident that certain metals have passed up into the slag in considerable proportions, namely aluminium, iron, magnesium, calcium, potassium and sodium. The presence of so large a proportion of aluminium may perhaps be explained by its strong tendency to form alumino-silicates, as shown by the abundance of felspars, micas and many other similar minerals both of primary and secondary origin. The elements aluminium and silicon evidently have a strong affinity for each other, at any rate in the presence of oxygen. From the abundance of silicates of magnesium, calcium and iron in nature it may also be concluded that there is a marked tendency for these metals to combine with silica, while potassium, sodium and calcium commonly function as bases in the alumino-silicate minerals. With the other metals however the relations are evidently very different. They occur in much smaller quantity in the slag crust, and when present, appear to be mostly combined as minerals of the sulphide group. The common practice of matte smelting and special researches directed towards this end have shown that fused sulphides and silicates form a system of two liquid phases, as with silicates and metals, and this fact is undoubtedly of the utmost importance in the genesis of sulphide ore-deposits[1]. The origin of such deposits is evidently a special process of wide application and many examples will be brought forward in the succeeding chapters of this book. From the frequency of sulphide ores it is clear that the majority of the important metals possess a strong affinity for this element under the conditions that prevail in the chemistry of the silicate crust, mainly however as a result of secondary and subsidiary processes, which are subordinate to the broader phenomena of crust-formation. Under certain conditions this function of sulphur as a metal-carrier can also be performed by arsenic, selenium and tellurium, though less commonly.

It must of course be understood that we are here speaking only of what are commonly known as primary ores. All those great and important deposits, perhaps the most valuable of all, belonging to the zones of oxidation and secondary enrichment

---

[1] Vogt, *Die Sulfid-Silikatschmelzlösungen*, Kristiania, 1919.

are excluded, since they originate from the primary ores when these are exposed to oxidation and other simple chemical processes in the superficial layers of the earth's crust under the influence of known and well understood causes. We are here dealing solely with the more recondite phenomena belonging to the chemistry and physics of the deeper layers, where direct observation and experiment are almost or quite impossible and we have to rely on speculation and inference and extrapolation from known to unknown processes and laws.

The general composition of the somewhat hypothetical successive earth-zones has been considered by Suess[1] and other geologists and systematised mainly on theoretical grounds. As a matter of convenience names have been invented for each of these zones, founded mainly on the initial letters or chemical symbols of the chief constituents. Thus the outermost zone or visible crust largely characterized by silica and alumina, is called *Sal*. The lower crust zone in which the silicates of magnesia are the most abundant constituents is called *Sima*. Below this come *Crofesima*, with chromium, iron, silica and magnesia, and *Nifesima*, with nickel, iron, silica and magnesia. Last of all the innermost core, supposed to be mainly a nickel-iron alloy, is called *Nife*. This rather characteristically Teutonic system of nomenclature certainly possesses some conveniences, but must not be applied too strictly. Nevertheless it does express what appears to be an undoubted fact, that the amount of silica, alumina and other light elements decreases in depth, while magnesia and iron are more abundant in the rocks with a lower silica percentage and higher density. It takes no account however of the distribution of calcium, sodium and potassium, which are also of much importance in rock-classification. The last two at any rate are certainly most abundant along with silica and alumina in the lighter and upper zone, while calcium is also common in the heavier Sima zone. Naturally the boundaries of these successive zones, and especially of the outer zones, cannot be sharp, owing to many disturbing factors, and it is a matter of everyday geological observation that portions of the Sima zone have broken through into the upper salic crust

[1] Suess, *Das Antlitz der Erde*, vol. III, Part II, 1909.

forming what are collectively known as the basic igneous rocks, both extrusive and intrusive. There is every reason to believe that the lower zones have behaved in a similar way and portions derived from the heavier zones with chromium, nickel, etc., have in all probability formed certain ore-deposits now existing near the surface, such as some primary sulphide masses.

The condition and state of combination of the metals in the first-formed slag-crust is of necessity wholly unknown to us. We have no reason to believe that any part of this crust is still in existence, or at any rate accessible to observation. Its outer layers have doubtless been worked over again and again by processes of geological destruction and reconstruction and in the course of these processes the metals may have been alternately concentrated and redistributed over and over again. At any rate in the parts of the crust now visible to us or reached by mining operations they are now found here and there, in fairly considerable quantity, and it is these masses of native metals or metals in simple states of combination that are called the primary ore-deposits. The second class, that of the simple compounds, is by far the most abundant: the primary native metals, except gold and platinum, are of little importance compared with the compounds.

For many years there have been great controversies concerning the origin of ore-deposits in general. These controversies have mainly centred around the part played by atmospheric water, circulating through the superficial layers of the crust. This agency is undoubtedly of first-class importance in secondary changes, but of late the general trend of opinion has been against the conception of it as the chief factor in the origin of primary ore-deposits. In a great number of cases, in fact it might almost be said in most cases, of primary ore-deposition the connection with igneous activity is obvious. Careful study of the age and distribution of the more important ore-bodies has demonstrated a close correlation between crust-movements, vulcanicity and ore-deposit and this may now be regarded as one of the fundamental principles of the science. Furthermore it has been shown that special types of ore-deposit are associated with special types

of igneous rock, and that the mechanism of the processes of
concentration and deposition varies according to the chemical
composition and physical characteristics of the associated
igneous rock-types. These conceptions clearly offer a large field
for systematic study and research.

Thus as a geological groundwork for the study of ore-deposits
we possess certain simple generalizations. Firstly, rocks can be
divided into two principal groups, igneous and sedimentary,
which are essentially, in their origins, primary and secondary
respectively. The igneous rocks have solidified directly from a
state of fusion, while the sedimentary rocks have been formed
of material derived ultimately from igneous rocks, either from
the primary crust or from later eruptions. To these must be
added the class of metamorphic rocks, which are either igneous
or sedimentary rocks altered by external agencies, usually heat
or pressure, or both, with concomitant physical and chemical
effects. Secondly, ore-deposits may be divided into simple groups:
(a) primary ores, those that were formed in the condition and
position in which we now see them; (b) secondary ore-deposits,
those that have undergone physical and chemical changes since
their formation. It is this latter class that gives most trouble in
definition, since here we have two distinct ideas to deal with:
some secondary ores are formed by chemical and physical
changes in primary ores in or very near the position in which
they were originally formed: such are the ores of the oxidation
zones of veins and lodes and the rich deposits due to secondary
enrichment of lodes by descending solutions; another important
group of secondary ores includes those which have been transported from their original home and redeposited by geological
agencies in a different situation: such are the placers of gold and
platinum and many tin-gravels. Two very different conceptions
are evidently here involved, nevertheless they have this in
common, that the metallic minerals have undergone some
change since their original formation. Still from the practical
point of view the results are very different. This illustrates well
the difficulties that beset any attempt at a scientific treatment
of the subject of ore-deposits.

Since ore-deposits form part, though but a small part, of the visible crust of the earth, and are in most cases in genetic association with the constituents of that crust, that is, the rocks, it is obviously essential for a proper understanding of the problems of the ore-deposits to be first acquainted with rocks in general, with their origins and distinguishing characteristics. This at once raises the fundamental question, what is a rock? This question is not by any means easy to answer. However there are two definitions, both of which are fairly satisfactory in practice. The first definition states: *a rock is any solid substance forming part of the earth's crust*. This is at any rate comprehensive and couched in plain everyday language. Furthermore it has the advantage of including certain masses, which are not hard and stony in the popular sense of these words, such as beds of loose sand and soft clay. To the geologist these are rocks, for want of a better word. The second definition is: *a rock is an aggregate of mineral particles*. This definition is more scientific, but it has the disadvantage of introducing another unexplained term, the word *mineral*, which in its turn needs definition. This again is not easy, but we may say that a mineral is a homogeneous substance, usually possessing a characteristic crystalline form and having a composition which is either constant or varies between certain fixed limits. The latter clause is necessary because a great number of minerals are "mixed crystals," consisting of two or more compounds in varying proportion. This is equivalent to saying that many minerals are isomorphous mixtures or solid solutions.

In this book we shall adopt the second of the definitions of a rock as above given, for the reason that ore-deposits are usually also minerals or mixtures of minerals, and it is with these as units that we commonly have to deal.

The scientific study of rocks is called petrology and the geology of ore-deposits is in the main concerned with petrology and mineralogy. Petrology is essentially the study of the origin and characters of the rocks or mineral aggregates composing the earth's crust, while mineralogy deals more in detail with the composition, properties and mutual relations of the mineral units which are the proximate constituents of the rocks. Besides these, a knowledge of the principles of structural geology is also essential, together with in some cases an acquaintance with historical geology. Petrology and mineralogy are in ultimate analysis special applications of certain branches of chemistry and physics towards a definite end, namely the elucidation of the laws of matter, energy, space and time so far as they have led to the formation of deposits of rocks and minerals. Obviously in such a connection philosophical or quasi-philosophical speculations into ultimate reality and so forth are out of place: we have to take things as they appear to us to be, and to describe them we must use the ordinary conventional language of descriptive science, in its utilitarian and materialistic applications. In a subject of this kind the higher flights of modern physics and chemistry are as yet out of place.

From the foregoing considerations it is manifest that the study of the primary ore-deposits is closely connected with the general problems of petrology, and that the first requisite for their successful study is a knowledge of the general principles governing the origin of rocks. Furthermore, the origin and characters of all the numerous types of secondary and derived ore-deposits depend also on petrological principles and the general laws of geology. It is evidently most logical to begin with a study of the primary types and the laws controlling their genesis, so far as these can be ascertained: the rest will follow from an application of these principles to the particular cases that may arise. Hence the earlier sections of this book will be concerned mainly with generalities: special cases as they arise will be treated in later chapters. Finally we shall deal in a systematic manner with the distribution and characters of the deposits of the different metals, mainly from a geological point of view, but paying due attention, when necessary, to industrial and economic factors.

CHAPTER II

THE IGNEOUS ROCKS

The Relation of Igneous Rocks to Ore-deposits. As a result of innumerable investigations carried on during recent years no possible doubt can remain that in a great number of cases the presence of ore-deposits is directly associated with igneous rocks; many lines of argument lead concurrently to this result. It would perhaps be too much to say in the present state of our knowledge that all primary ore-deposits are of igneous origin, but this is a view which may very likely be established as a result of future work. The arguments in favour of such an association are largely founded on considerations of space and time: mineralization on a large scale is for the most part confined to regions where the effects of igneous activity are manifest, and it is easily demonstrated in many instances that the mineralization occurred after, but not long after, the igneous phenomena. In fact, putting the statement in its most general form it may be said that mineralization is characteristic of the later phases of vulcanicity, using the latter word in its most extended sense, to include underground as well as superficial phenomena, or in petrological language, intrusion as well as extrusion.

Ore-deposits in many forms, as lodes, stockworks and disseminations, and masses of very varying shape and size are found within the igneous rocks themselves or in the country rock in their immediate neighbourhood, while the contact deposits necessarily occur at and near the junctions of the intrusion and the surrounding rocks. Some ore masses are also due to metamorphism produced by igneous intrusions and these are in many cases not to be sharply distinguished from contact deposits. In such cases it is generally quite certain that the metallic minerals were derived directly from the igneous rock, and it is necessary next to consider how their separation and concentration can have been effected. To do this it is essential to take into account the physical and chemical laws governing the cooling and consolidation of fused rock-material.

The Consolidation of Igneous Rocks. It is unnecessary to demonstrate at length the patent fact that the material of which the igneous rocks are composed was once in a liquid condition owing to its high temperature. This molten rock-material can be seen in the craters of many volcanoes or issuing from cracks in the ground in their immediate neighbourhood. The external form and internal structure of most intrusive masses also make it certain that when they reached their present position they were in a liquid state, and have since consolidated. Now fused material of this kind is from the physical point of view a solution, and in its cooling and solidification it must obey the established laws of solution founded on the general principles of thermodynamics. In such a case the ordinary conception of solute and solvent, arising from our acquaintance with water solutions, is inapplicable, and a molten rock-mass must be regarded as a mutual solution of all its constituents. To such a rock-solution the convenient term *magma* is commonly applied. A magma, then, having the composition of a granite, for example is to be regarded as a solution of certain constituents, of which the most abundant are silica, alumina, magnesia, lime, iron oxides, potash, soda and water, together with smaller amounts of a large number of other constituents, which though forming but a small percentage of the whole are of the greatest importance, since they give rise under certain conditions by their concentration, to ores and other substances of economic value. It is to these *accessory* constituents of the magma that our attention must be largely directed in the following pages.

The course of consolidation in such a magma is naturally controlled by the physical conditions prevailing at the time, of which the most important are rate of cooling and pressure. A silicate solution, when cooled, may according to circumstances consolidate as a homogeneous mass, known as a *glass*, or it may form an aggregate of crystals of one or more minerals. In order to form a glass, the cooling must be very rapid and glassy rocks are only known among the superficial volcanic products, as flows of obsidian and pumice, or as certain dykes of pitchstone, intruded very near the surface and thus rapidly chilled. These need not be considered further. But when we come to deal with the deep-seated intrusions, very different conditions are found to prevail: the rate of cooling is so slow that the rocks become wholly crystalline, being composed of aggregates of crystals of varying size: it may so happen that the composition of the original magma and the physical conditions are such that the whole consolidates as crystals of one mineral only, but such cases are very rare. Magmas are highly complex solutions containing many constituents, and the mineral composition of the resulting rock is usually also complex: but of still greater importance is the fact that with slow cooling there is a possibility of processes taking place within the mass leading to a variation of composition and a consequent concentration of some of the constituents in different parts, so that the magma becomes heterogeneous. The processes that give rise to this kind of concentration are known as the *differentiation* of igneous magmas and their study is of the utmost importance in connection with the genesis of ore-deposits.

As already pointed out, rock-magmas are highly complex solutions, containing a large number of constituents, and this has an important bearing on their consolidation. The freezing points of most of the common rock-forming minerals have been determined with a considerable degree of accuracy, but it must be remembered that the freezing point of a pure mineral determined at atmospheric pressure is by no means the same as the temperature at which the same mineral will begin to crystallize from a complex solution under the pressure of many thousands of feet of overlying rock. The effect of pressure on the freezing point depends on whether the substance expands or contracts on freezing. Most minerals are denser in the crystalline form than when fused and therefore their freezing point is raised by pressure. But experiment has shown that in the case of minerals this effect is usually small and another factor is of much greater importance. It is one of the fundamental laws of solutions that the addition of a second component to a fused mass of any pure substance capable of crystallizing lowers the freezing point of that substance, in a degree more or less proportional to the amount added, down to a certain limiting value, as will be explained later, and the more components are present in a solution the greater will be the lowering of the freezing point of each. Now rock-magmas contain a very large number of components, hence it follows on theoretical grounds that they should remain liquid down to comparatively low temperatures: observation has shown with a high degree of probability that certain granites have consolidated at temperatures as low as 350° C., or a very low red heat, and some of their differentiated products doubtless remained liquid at still lower temperatures. A further important point in the case of deep-seated intrusions is the fact that the thick cover of rock prevented the escape of water and other volatile constituents which have a very marked effect in lowering the freezing point. There is no evidence available as to the temperature at which such intrusions are injected, but it is unlikely to be lower than that of lavas, say 1300° C., and it is clear that the cooling of such an intrusion from 1300° to 350° or lower, must be a long-continued process, giving plenty of time for differentiation to take place and to produce important effects.

The Lowering of the Freezing point and the Eutectic Ratio. As already mentioned the depression of the freezing point of a pure substance by the addition of some other substance is a matter of high importance in the study of the solidification of magmas and must be considered more in detail. It can be dealt with very simply in the case of a system of two components, but the general principles involved in more complex cases are precisely similar. For the sake of simplicity the effects of pressure, which are not of great moment, are not taken into account, but pressure is supposed to be constant.

Let A and B be two substances having definite freezing points when pure and unable to form mixed crystals but completely miscible in the fused state[1]. In the diagram, Fig. 1, AB represents the percentage composition of any mixture of A and B, while a and b represent the freezing points of the pure substances. The addition of successive quantities of B to A lowers the freezing point of A along the curve ae; similarly the addition of A to B lowers the freezing point of B along the curve be; these

[1] The case of limited miscibility in the liquid state, which is of much importance in the study of the genesis of ores, will be considered later.

curves are usually not quite straight lines, but slightly convex upwards. It is obvious that these two curves must eventually intersect at some point *e*. This is known as the *eutectic* point and represents the composition of the mixture of *A* and *B* having the lowest possible freezing point; this composition in percentages is given by the perpendicular let fall from *e* to the line *AB*, while the horizontal line through *e* parallel to *AB* gives the freezing point of the eutectic mixture.

Let us now consider the course of crystallization of a cooling solution represented in composition and temperature by the point *h*, which lies well above the freezing point curve: the

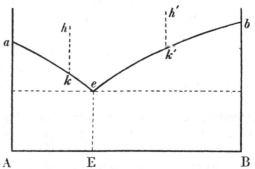

Fig. 1. Freezing point curves for a system of two components, with eutectic.

uniform fall of temperature is represented by the line *hk*, which eventually cuts the curve *ae* in *k*. During the interval of time represented by this line no separation of solid occurs, but on reaching *k* the pure substance *A* begins to separate. This separation of *A* alters the composition of the solution, while the temperature continues to fall, and the system may then be considered as moving along the line *ke*. When it reaches *e* crystallization of *B* begins concurrently with *A* and the final result is a rock consisting of crystals of pure *A* embedded in a matrix of *A* and *B*, this matrix having a constant composition, namely that of the eutectic mixture. Starting from the point *h′* precisely similar reasoning shows that the final result will be crystals of *B*, embedded in the same matrix *AB*. Hence it is seen that the order of separation of the minerals from a magma of two components is independent of the freezing points of the pure minerals and depends only on the proportions in which they are present in the solution. This accounts for variations in the order of crystallization of minerals which puzzled the earlier petrologists. When three or more components are present the case becomes more complex, but the general principle remains the same, namely, that the course of crystallization is towards the formation of a final residue having the composition of the eutectic mixture of the components. Many complications are introduced by the occurrence of isomorphism, solid solutions, intermolecular compounds and other features well known to metallographers, who have studied the crystallization processes of the metals in the most minute detail[1], while the silicates possess almost equally complex relations, as shown by the investigations conducted at the Carnegie Institution at Washington and elsewhere. The fact of prime importance from the point of view of ore-deposits is that the metallic constituents of the igneous magmas tend to concentrate in and are commonly associated with the extreme products of differentiation.

The Reaction Principle in Crystallization. The eutectic principle, as just briefly outlined, has been of immense service in the development of petrogenic theory as applied to igneous rocks and ore-deposits, but it has its limitations. The theory of eutectic crystallization in this simple form is only strictly true when applied to substances (minerals) possessing a perfectly definite and unchangeable composition. It is doubtless true, for example, of such a pair of minerals as quartz and orthoclase, or quartz and cassiterite, but it is not true for the plagioclase felspars, most of the ferromagnesian minerals, and many sulphides, for the reason that these are mixed crystals or isomorphous mixtures of variable composition. The theory of the crystallization of such mixed compounds was long ago worked out on thermodynamical grounds and confirmed by synthesis, and in a recent memoir of much importance Bowen has pointed out its importance in the study of differentiation[2]. The crystalliza-

---

[1] Gulliver, *Metallic Alloys*, London, 1913. Desch, *Metallography*, 3rd edition, London, 1922.

[2] Bowen, *Journ. Geol.* vol. xxx, 1922, pp. 177–198.

tion of a mixture of two isomorphous compounds, for example soda and lime felspars, or magnesia and iron olivines or minerals of the spinel group (magnetite and chromite) can be expressed as follows: in the diagram, Fig. 2, the abscissae represent percentages of the two substances, the ordinates being temperatures. The upper or liquidus curve represents the composition of the liquid that is in equilibrium at a given temperature with the solid at the same temperature whose composition is given by a point on the lower solidus curve on the same horizontal line. But as the temperature continues to fall the particle of solid

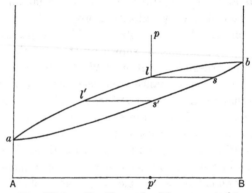

Fig. 2. Thermal equilibrium (liquidus and solidus curves) of two completely isomorphous substances, illustrating the principle of continuous reaction in crystallization.

that has separated is no longer in equilibrium with the liquid and begins to react with it, part of the component now in excess being leached out of the solid and going again into solution. Thus in the diagram if p represents the composition and temperature of the original liquid, the temperature steadily falling, l gives the composition and temperature of the liquid at the moment of separation of the first solid particle s. The separation of this particle has changed the composition of the remaining liquid, and as the temperature continues to fall, the liquidus point moves along la, while owing to the reaction before mentioned the solidus point moves along sa, provided always that the cooling is sufficiently slow to allow of a restoration of equilibrium. When the solidus has reached s' and the liquidus l' all the liquid will have been used up and the final product is a mass of mixed crystals of the composition and temperature given by s'.

As a rule however in natural crystallization the maintenance of equilibrium is imperfect: not enough time is allowed for the reaction to be quite complete and the result is often the production of zoned crystals, the centre being richer in B and the outer layers poorer in B than the average. There is also always the possibility that during cooling the system may pass beyond the field of stability of the first mineral formed, so that the outer layers may consist of a shell of some other mineral having different properties. In complex systems this seems to be often the case, resulting in the formation of a core of one primary mineral inside a shell of another. Bowen has laid much stress on the importance of this principle in differentiation and has worked out a definite succession of rock-forming minerals formed in accordance with it. The detailed application of the theory to metallic minerals yet remains to be investigated. This may eventually be found to have a most important bearing on the distribution of ores in zones, as well as on the well-known variations in the silver-content of galena, for example, and the gold-content of auriferous pyrites and other sulphides.

From the considerations very briefly outlined above it is clear that the course of crystallization in a magma would be very simple and definite provided no other factors had to be taken into account; an intrusion on cooling would then crystallize as a rock-mass or aggregate of crystals having a uniform composition, the minerals present depending wholly upon the composition of the original magma. Some small masses do appear to approximate to these conditions, but it is very rare to find a homogeneous rock-mass of any size: nearly always it is quite apparent that disturbing agencies have come into play, leading to variations of composition in different parts of the same mass. Furthermore it is often obvious that a magma has been disturbed while only partially solidified, structures already formed being broken up and parts or the whole of the remaining liquid portion being removed elsewhere. Again there may have been partial or complete re-fusion of an already solidified rock: all of these causes may lead to important results.

Differentiation of Igneous Magmas. Let us consider first the possible causes for the development of heterogeneity in a cooling magma under constant physical conditions. Differentiation of this kind has undoubtedly taken place and a large number of physical principles have been invoked by different writers in explanation. Some of these are obviously incompetent to produce the observed effects and may be disregarded. According to the most recent ideas the five following causes are applicable[1]:

(1) differentiation by diffusion during crystallization;

(2) differentiation by gravity during crystallization;

(3) differentiation by limited miscibility in the liquid state;

(4) differentiation by squeezing out of liquid from a partly solid magma;

(5) the absorption of masses of previously existing rock from the walls.

Of these explanations the first is supported by British petrologists, while the second and fifth are largely in favour in America. The theory of limited miscibility, as shown by Vogt, is certainly applicable to the separation of sulphides from silicate magmas.

(1) The theory of differentiation by diffusion is based on the principle of thermal equilibrium. Let us take the case of a homogeneous mass of magma injected into a basin surrounded by cool rocks: loss of heat is mainly by conduction from the margin, so that the outer part of the solution is cooled; it will then become supersaturated for that constituent having the highest freezing point, which will begin to crystallize: as soon as crystals are formed the equilibrium of the solution is disturbed and there will be a migration of molecules of that constituent towards the margin, followed by further crystallization as the temperature falls; eventually a point will be reached at which a second mineral will begin to crystallize, when similar considerations apply. Thus there is produced a marginal concentration of the minerals having the higher freezing points

[1] For a complete discussion of theories of differentiation, see Harker, *The Natural History of Igneous Rocks*, London, 1909. Bowen, "The Later Stages of the Evolution of the Igneous Rocks," *Journal of Geology*, Supplement, vol. xxiii, 1915. Daly, *Igneous Rocks and their Origin*, New York, 1914.

under the given conditions, with a concomitant central impoverishment of these minerals and concentration of those with
the lower freezing points. These considerations are applied
specially by Harker to the case of the gabbro of Carrock Fell,
where the margin contains 25 per cent. of ilmenite and much
augite with only 32·5 per cent. of silica, while the central portions
have about 59 per cent. SiO_2 and consist mainly of felspar and
quartz[1]. This is an example of differentiation in place, or laccolithic differentiation, as it is sometimes called, where there
has been no subsequent removal or other disturbance of the

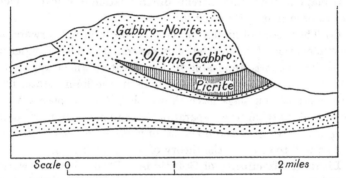

Fig. 3. Diagrammatic section of Tabankulu, Griqualand East, South Africa,
    a laccolithic intrusion with stratified arrangement of rocks of different
    density, apparently due to differentiation by gravity. (After du Toit,
    Trans. Geol. Soc. S. Africa, vol. XXIII, 1920, p. 19.)

magma. Similar basic marginal facies are found in almost all
intrusions of any considerable size and even in some quite thin
sills and dykes, an important feature being that the marginal
concentration is often found on both the cooling surfaces of
such thin sheets, even when standing vertically.

(2) The second theory of differentiation above mentioned is
based on the fact that as a rule the crystals separating from a
magma when cooling are denser or lighter than the remaining
solution and therefore tend to sink or rise under the influence
of gravity, thus producing a stratification in the resulting rock.
This principle is undoubtedly applicable within limits and
especially to horizontal masses such as sills and flat laccoliths,

---

[1] Harker, Quart. Journ. Geol. Soc. vol. L, 1894, p. 311.

but its advocates do not seem to have taken sufficiently into account the well-known fact of the extremely high viscosity of silicate solutions even at temperatures well above their general solidification point. This would soon limit the rising or sinking of crystals in a cooling magma when it approached the freezing point of the silicate constituents, such as felspar and nepheline. It would appear therefore that though this theory will explain the concentration in a magma basin of those minerals having a freezing point much higher than that of the silicates, possibly including some of the oxidic and sulphidic metallic compounds, it is inapplicable to the extreme differentiation of silicates, such as has been frequently observed.

(3) The theory of limited miscibility has been put forward as an explanation of differentiation in general. There is no evidence to show that it is applicable to the silicates, which appear to be mutually soluble in all proportions in the liquid state, but Vogt's studies on slags have shown that it must play a very important part in the separation of sulphides from silicate magmas[1]. Owing to the practical importance of this subject it will be well to consider the theory of it in some detail.

Limited Miscibility of Solutions. When ether and water are shaken up together at the ordinary temperature and then allowed to stand the mass separates into two layers, the upper one consisting of ether with a little water and the lower of water with a little ether. When the operation is repeated at a higher temperature each constituent dissolves more of the other till eventually a temperature is reached above which the two are miscible in all proportions, but again separate on cooling. Precisely similar reasoning is applicable to silicates and sulphides, except that here we have to take into account also the separation of solid phases. These phenomena are illustrated in the ordinary process of copper smelting, where the furnace charge separates into the matte below and the slag above. Analyses show that

---

[1] Vogt, "Studier over Slagger," *Bihang til k. Svenska Vet.-Akad. Handl.* IX, 1884. *Die Silikatschmelzlösungen,* Part I, pp. 96–101, Kristiania, 1903. "Die Sulfid-Silikatschmelzlösungen," *Vidensk. Skrifter, Mat. Nat. Klasse,* 1918, No. 1, Kristiania, 1919; also abstract with same title in *Norsk Geologisk Tidskrift,* vol. IV, 1917.

the metal always contains a small proportion of silicate, while the slag contains a little sulphide. Vogt has shown that at furnace temperatures the mutual solubilities are small, but the solubility of sulphide in slag is higher than that of silicate in metal. The diagram below, Fig. 4, represents the temperature-concentration relations of a system of this kind, the conditions

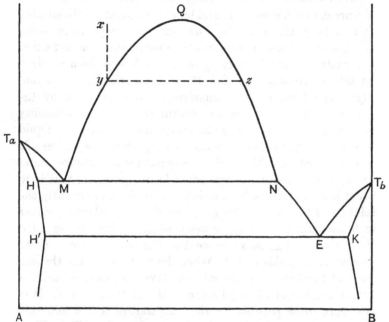

Fig. 4. Thermal equilibrium curves for two substances with partial miscibility in the liquid state.

being that the substances in question possess partial or limited miscibility in the liquid state, while in the solid state also they are able to form solid solutions within certain limits. The abscissae AB represent percentages of A and B in the mixture, while the ordinates are temperatures. T_a is the freezing point of the pure substance A, and T_b that of the pure substance B. The complex curve $T_a MNET_b$ represents the liquidus, that is, the composition of the liquid that is in equilibrium with solid at the temperatures indicated. The solidus however, or curve

showing the composition of the solid in equilibrium with liquid
at the same temperature is a matter of greater complexity, and
its form can best be illustrated by considering the order of events
in a particular case. The first thing to be considered is that
every system represented in composition and temperature by
any point in the diagram above the curve T_aMQNET_b is a single
homogeneous solution of A and B: that is to say, at all temperatures above this curve there is complete miscibility. Let
x represent such a solution, and let the temperature fall steadily
from x to y, till it reaches the point of temperature below
which a homogeneous solution of this composition cannot exist;
the limits of mutual solubility of A and B have been reached.
At this temperature therefore the solution splits into two conjugate solutions, whose compositions are represented by the
points y and z. As the cooling continues the mutual solubility
of A and B decreases, and the compositions of the two liquid
phases are represented by points moving along the curves yM
and zN; at the point M solidification begins, the solid that separates
having the composition represented by the point H; that is to
say it is a saturated solid solution of B in A. Consequently the
solution M becomes poorer in A and therefore relatively richer
in B, its composition being represented by a point moving along
the horizontal line MN; hence also the temperature remains
constant during this period. When the point reaches N the last
trace of the left-hand solution will have disappeared and the
system consists of the solid solution H and the liquid N. It is
now once more possible for the temperature to fall and this
decrease of temperature is accompanied by separation of the
solid solution H, with in some cases a concomitant change in the
composition of the solid solution as shown by the line HH'.
Eventually the composition of the remaining liquid reaches the
point E which is a eutectic point, at which H' and K, the corresponding solid solution of A in B, possess the same solubility
and therefore crystallize together. At this point also there is a
temperature-arrest, until the whole has become solid; it is to be
specially noted that the eutectic consists of two solid solutions,
of A in B, and B in A, and not of a mixture of the two pure
compounds. Furthermore, on further cooling after consolida-

tion, the relative proportions of A and B in the two solid solutions may undergo further variations, as shown by the divergence of the lines below H' and K. Such further changes are however extraordinarily slow in most solids, unless they are kept for some time at a temperature just below that of solidification and thereafter cooled very slowly. Such changes may, and probably do take place in igneous rocks and magmatic ore-deposits, the general effect being a segregation of minute particles of one substance within larger crystals of the other.

When the composition of the original system is represented by a point lying to the left of the vertical through H, or to the right of the vertical through K, there will be no separation into two conjugate liquids, but the whole will separate as an unsaturated solid solution of B in A or A in B respectively: in the space between the verticals through N and E, however, *two* solid solutions will be formed, H being deposited first and finally the eutectic of H and K. The curved area MQN represents the conditions of concentration and temperature within which separation into two layers can occur, and consequently this represents also the conditions under which natural differentiation or artificial smelting are possible. From general considerations of the investigation of smelting processes and from the study of alloys, as well as from the petrological data available as to ore-deposits it appears probable that in actual cases this area occupies most of the diagram, the limits of solid solution and the eutectic points being very near the boundaries. From Vogt's researches it is known for example that the mutual solubilities of fused silicates and sulphides are very small, in some instances almost vanishingly so at furnace temperatures, and the amount of sulphide normally present in igneous rocks is only small: according to the table given by Clarke[1] the average proportion of sulphur in all igneous rocks taken together is 0·10 per cent., while Vogt states that the amount of silicate dissolved in copper matte amounts to only a small fraction of 1 per cent. Hence it appears that in nature also the separation of metal and slag must be very complete, varying no doubt somewhat with the composition of the silicate fraction, as in artificial smelting.

[1] "The Data of Geochemistry," Bull. 695, *U.S. Geol. Survey*, 1920, p. 28.

Differentiation by Absorption. Assimilation. It is a
matter of common observation that large igneous masses frequently enclose recognizable fragments of other rocks, sometimes
nearly or quite unaltered, sometimes highly metamorphosed:
such are called *xenoliths* by Harker. Moreover in many instances
it is possible to trace every degree of alteration in such xenoliths,
even up to complete absorption, their former presence being
only disclosed by some slight local peculiarity in the igneous
rock. There is thus every stage from an obvious lump of rock
through all grades of visible patches down to an enclosure only
recognizable as such by refined methods. This process obviously
leads to local variations in the composition of the rock, and it
has been suggested by petrologists, especially in America, that
this process of absorption or *assimilation* has been a potent
factor in the production of heterogeneity in igneous rocks.
Such absorbed rock-masses must be torn from the walls or roof
of the magma-basin and Daly[1] has suggested for the process the
name of *stoping* by analogy with the mining term. It is still a
matter of discussion to what extent this idea is applicable on a
large scale as an agent of differentiation. Daly, who is a strong
advocate of it, considers it to be of great significance and suggests
that important effects must necessarily follow on the absorption
of large quantities of sedimentary rock, and especially of limestone, by molten magmas. He thinks that the fluxing action of
the lime would be very efficient in producing partial magmas of
abnormal composition, especially some of the syenitic types
very rich in alkalies, which as he points out, are rare in comparison with granite and basalt.

There seems to be little doubt that the occasional presence in
granite, for example, of a class of minerals rich in alumina and
usually found only in metamorphic rocks indicates the absorption of highly aluminous sediments: such minerals are andalusite,
cordierite, corundum, spinel, etc. Morosewicz showed over
twenty-five years ago[2] that when excess of alumina is added to
a melt having the composition of a normal rock such minerals
actually crystallize out, and recent elaborate investigations at

[1] Daly, *Igneous Rocks and Their Origins*, New York, 1914, p. 194.
[2] Morosewicz, *Tschermaks Mittheilungen*, 1898.

the Carnegie Institute at Washington have enlarged and amplified his conclusions. As natural examples we may quote the andalusite found in some Cornish and other granites, the pinite pseudomorphs after cordierite in the granites of the Land's End and of Table Mountain, South Africa, and the well-known sapphires in a dyke in Montana. From isolated single crystals of the foregoing minerals we can trace every gradation through obvious xenoliths consisting of aggregates of crystals of many kinds up to sediments riddled by innumerable granite veins, and gneissose rocks as to which it is difficult to decide with certainty whether they are fundamentally of igneous or sedimentary origin. In the instances here considered it is clear that absorption has produced a local variation of composition, at any rate on a small scale, and it is natural to suppose that a similar principle may have worked on a larger scale also, leading to essential differences in the composition of large bodies of rock, whether followed or not by actual physical separation of the differing fractions. It is also worth while to remember that assimilation of already existing ore-deposits by an intrusion may help to account for the presence of various metallic constituents either occurring in localized patches, disseminated through the mass of the rock, or concentrated in magmatic residues to form the material of new veins or other segregated deposits.

The Separation of Differentiated Partial Magmas. In the preceding sections it has been shown how various physical causes may lead to the production of heterogeneity in crystallizing magmas, so that different constituents become concentrated in different parts of the basin. In the first case the constituents of highest freezing point will tend to concentrate in the marginal portions without reference to the orientation of the mass; in the second and third cases the heavier portions will be below with the lighter above. Both these conditions are found in nature in cases where the magma has remained undisturbed during the whole course of crystallization. But more commonly other factors come into play leading to internal disturbances of the state of equilibrium. In a cooling magma of any considerable size convection-currents must play an important part, and these mix up the separated layers and draw them out into streaks, forming the banded structures so common in the marginal parts of large intrusions[1]. But more important still is the fact that during crystallization magmas are frequently removed either wholly or in part to some other position. If the whole body is intruded into a new situation it is obvious that its already separated layers will tend to become mixed and drawn out during flow: this is the cause of a large proportion of the parallel structures with bands of varying mineral composition that are so common in the gneissose rocks, giving rise for example to many of the important deposits of iron ore in Scandinavia. Some of these are described in detail elsewhere (see Chapter XV). Then again it may happen that only one portion of the differentiated magma is drawn off, thus making the separation still more effective: or there may be successive drafts of magma of varying composition from one original source, leading to a series of intrusions of varying composition, but usually having some characters in common to show their affinity. The most famous example of this kind is afforded by the Devonian igneous rocks of the Kristiania district, described by Brögger[2]. Furthermore it is obvious that each of these partial magmas may again undergo differentiation after it reaches its new position.

It is thus clear that it is necessary to draw a distinction between differentiation before and after intrusion. Although essentially the same in principle their effects are different. The former gives rise to a series of intrusions, often showing a gradation in character when taken in series, but forming separate masses: this is a case of discontinuous variation. The latter causes a continuous variation of composition within the limits of a single mass. Examples of both types are of importance in connection with ore-deposits.

The Mechanism of Separation. So far we have considered mainly the results of the separation of differentiated magmas, without taking into account the means by which such separation

---

[1] Grout, *Journ. Geol.* vol. xxvi, 1918, pp. 439–58 and 481–99.

[2] Brögger, *Die Eruptivgesteine des Kristianiagebietes*, 4 vols, Kristiania; also "Die Syenitpegmatitgänge des Südwestlichen Norwegens," *Zeits. für Krist.* vol. xvi, 1890.

may be brought about. On this subject much uncertainty prevails. It is at any rate clear that if a part of a magma is removed bodily to a new position some mechanical cause must operate and the most obvious cause is the disturbances of the earth's crust which we know from other evidence to be closely associated with igneous activity of all kinds. Magmas during cooling are often subjected to crustal stresses, which must produce marked effects on them in their semi-solidified state. At a temperature within its range of crystallization, which is probably pretty wide, a magma must be partly liquid and partly solid. Dr Harker states that its condition must be pictured as that of an open fabric or sponge of crystalline matter with the interstices occupied by liquid. Powerful mechanical forces due to earth movement applied to such a semi-solid reservoir will tend to squeeze out the liquid material and to transfer it to places of less pressure: the action is much like that of squeezing water out of a sponge[1]. Here then we have an efficient agent of differentiation as well as of transfer. The part squeezed out would necessarily consist of the constituents with the lowest freezing points, which we have reason to believe include many compounds of metals. This action is invoked by petrologists to explain the vast networks of pegmatite dykes found penetrating the crystalline schists of many areas, and it can often be demonstrated that the injection of the pegmatites was closely associated with the dynamic metamorphism that impressed their special characteristics on the rocks. The still molten material would necessarily be carried in definite directions, controlled by the crustal stresses[2]. This process may account also for certain disseminations of ores throughout large volumes of rock, the material being carried by solutions and deposited in the solid form on reaching a region of lower temperature and diminished pressure.

The simplest case, however, may be pictured as follows: a

---

[1] For a convenient summary, see Harker, *Compte-Rendu XIIe Session du Congrès géol. internat.* 1913.

[2] Barrow, "On an Intrusion of Muscovite-biotite Gneiss in the South-eastern Highlands of Scotland," *Quart. Journ. Geol. Soc.* vol. XLIX, 1893, pp. 330–58.

deep-seated magma basin, either wholly liquid or partially solidified and containing suspended crystals of various minerals will be subjected to high pressures partly due to the weight of overlying rock and partly to its own steam and other volatile constituents. Owing to external crust disturbances this pressure becomes partially and locally relieved by the opening of a crack in the surrounding rocks, usually upwards towards the surface. Along this crack the molten magma passes in obedience to the laws of hydrodynamics and is either extruded at the surface as a lava flow or intruded into some weaker portion of the crust as a laccolith, sill or dyke. Many of the vast bathyliths found in highly disturbed regions appear to be due to magma welling into the cavities formed under the cores of anticlines forming mountain ranges. This will go on till stopped either by renewed pressure or by the choking of the ducts due to cooling of the magma. Further solidification of the magma then proceeds, possibly under physical conditions very different from those prevailing before the transfer; crystals already formed may be more or less melted down owing to relief of pressure and the magma generally may solidify as an aggregate of minerals differing from those forming in the original basin, though the bulk composition remains the same. From a study of a rock formed in this way it is often easy to demonstrate that the solidification took place in two stages, under very different conditions as to pressure and rate of cooling. It is quite obvious that this process will in most cases tend to emphasise the effects of any differentiation that may already have taken place, since it is unlikely that the portion of the magma drawn off will be an average sample of the whole. The removal of a part of the magma will also disturb the equilibrium of the remainder, so that thereafter the processes of differentiation may proceed along new lines.

Another cause of removal of a portion of a differentiated magma is connected with re-fusion of already consolidated rock-masses. Geological study of the sequence of rock-types and the general characters of intrusions in certain regions suggests that the facts can be best explained on the hypothesis of the complete solidification of a deep-seated magma-basin in which some form of differentiation, either diffusional or gravitative, has taken place, followed by successive re-melting of portions of the same, usually from below upwards, with contemporaneous removal of the re-melted portions under the influence of crustal stresses and fractures. If in such a basin the original differentiation led to the accumulation of basic material at the bottom and acid material at the top, the successive tappings during refusion would lead to a series of intrusions of increasing acidity, as in the Isle of Skye, where Harker has shown there was a sequence of plutonic intrusions ranging from peridotite through gabbro to granite. The peridotitic partial magma showed a marked concentration of chromium, as chromite and picotite. This was followed by a sequence of minor intrusions, dykes and sills, in reverse order from acid to ultrabasic. These were probably drawn off during the second solidification of the deep-seated basin which would progress from above downwards[1].

It is quite clear that this or some similar process has taken place in a large number of instances, since the igneous rocks of any one region, when examined in detail, commonly show a regular sequence of rock-types of graduated character, but having sufficient common peculiarities, chemical and mineralogical, to indicate a genetic connection between them. This principle is obviously of great importance in the study of the origin of primary ore-deposits and needs further consideration. But it is necessary first to discuss more in detail the course of crystallization in a single large intrusive mass.

In all discussions of the solidification of magmas it is necessary to form a clear idea of the conditions that are likely to prevail. It appears to be a common conception that a magma basin consists of a mass of intensely hot liquid enclosed by walls of cool rock, or in other words that there is a strong contrast of temperature between the magma and its surroundings, but Harker has pointed out that this idea is probably erroneous. Geological evidence shows that a parent reservoir may persist as a source of intruded and extruded magmas for a very long time: such a degree of permanence indicates something like thermal equilibrium between the contents of the basin and the

<hr />

[1] Harker, "The Tertiary Igneous Rocks of Skye," *Mem. Geol. Survey*, 1904.

crust around it, otherwise the magma would freeze or the walls would be melted. Hence the general temperature of both must be somewhere near the freezing point of the magma at the time; and it would appear that the contrast of temperature cannot be very great, except in special cases, where the whole cools rapidly, as when a magma is intruded quite near the surface. Such an intrusion should show little differentiation and a low degree of metamorphism in the surrounding rocks, owing to the short time available for chemical reactions to take place.

Crystallization in intrusive masses. As the simplest possible case let us consider the course of crystallization in a mass of magma intruded as a laccolith into sedimentary rocks, and finally solidified in its present position without further disturbance due to re-fusion or any other cause. Such an intrusion will crystallize according to the principles already explained. It may and commonly does show more or less banding due to flow of heterogeneous and partly differentiated material, such bands being generally arranged roughly parallel to the boundary walls, and being less conspicuous in the centre. This is the result of processes that have taken place before and during the intrusion. But we are at present concerned more with the phenomena that result from processes occurring after the intrusion. It is obvious that the mass may undergo differentiation in place, giving rise to stratification and to marginal variations from the mean composition. As a matter of fact the latter are generally more conspicuous, for the simple reason that it is but rarely that it is possible to see a complete vertical section of a large intrusion, while the margins are usually quite accessible. An excellent example of a vertical section showing variation is however provided by the Palisade Traps of New York and New Jersey, which form a sill about 800 feet thick, ranging from olivine dolerite near the bottom to a granitic rock at the top[1]. This appears to be due to gravity differentiation. Nearly every large rock-mass shows variations at the margin, partly structural due to rapid cooling ("chilled edges") and partly mineralogical owing to concentration of basic minerals of high freezing point due to

[1] Lewis, "Petrography of the Newark Igneous Rocks of New Jersey," *Ann. Rep. State Geologist N.J.*, 1907.

diffusion-differentiation. In many instances marginal concentrations of sulphides form important ore-deposits: it is however in many cases doubtful whether these are directly due to differentiation of the type here dealt with or should rather be referred to contact deposits. This subject is discussed elsewhere. In such a laccolithic mass, then, the course of crystallization would be somewhat as follows: in the early stages more or less differentiation takes place producing marginal facies, which as a rule are not very extensive, probably owing to general increase of viscosity preventing ready diffusion. After a certain point of viscosity is reached we may suppose that the composition remains fairly uniform throughout, at any rate so far as the silicate minerals are concerned. These then tend to crystallize in the order determined by their concentration in the solution according to the rules already set forth, the composition of the liquid residue always tending more and more towards the complex eutectic. Some minerals such as zircon, apatite and sphene always seem to crystallize at a very early stage from mixtures of any composition, since they appear to be extremely insoluble in silicate solutions. But the silicate minerals, such as felspars, felspathoids, micas, amphiboles, pyroxenes and olivine, crystallize generally in the "order of decreasing basicity" of Rosenbusch. That is to say each successive mineral usually contains more silica than the preceding one, while quartz, being pure silica, is usually the last of all, or at least contemporaneous with the felspar. This rule is however purely empirical, being founded on observation, and was formulated at a time when little was known as to the behaviour of silicate solutions under varying conditions. The rule works in practice mainly because the range of chemical composition in magmas is not very wide, and it fails in just those exceptional cases where failure might be predicted on chemical grounds. In this way are formed the common and widely spread rock-types, such as granite, syenite, monzonite, diorite, gabbro, norite, peridotite in all their varieties, to so many of which systematic petrographers have given confusing and unnecessary names, often of the most limited local application, thus concealing their true affinities. The greater part of these rocks then consist of a comparatively small number of standard minerals in varying proportions and on these proportions their petrographical classification is mainly founded. This principle of classification is perfectly logical, since the mineral composition is an expression of the chemical constitution of the original magma. These standard minerals consist of a comparatively small number of elements, often combined in varying proportion in different mineral species. But most rock-magmas contain in addition small and varying quantities of many other elements, which are nevertheless of the utmost importance from many points of view: it is with these that we are here more specially concerned, since it is they that give rise to a very large class of ore-deposits. These constituents tend to accumulate in the last liquid residue of the magma, for two reasons; some because they are present in such small quantity that they remain dissolved in the last remaining liquid fractions, others because they possess such low freezing points that they themselves form the last fractions, dissolving the other group. Thus it is seen that there is a concentration of the less common constituents in the mother-liquor from which the common minerals have crystallized. Other things being equal they would naturally tend to concentrate towards the centre of the mass, or near its base, where the high temperature would be longest maintained, but here other factors come into play. This concentration of highly volatile substances, and especially steam, gives rise to considerable pressure. As the whole intrusion cools it contracts and cracks are formed, through which the still heated material from the interior escapes, cooling and crystallizing in its passage, partly in the cracks, where it forms veins and dykes of peculiar character and partly passing beyond the margin of the intrusion where it also finds its way into cracks and fissures, forming dykes, veins and lodes in the country rock, and during its escape setting up those marked chemical and mineralogical changes in the rocks which are summed up under the general name of *pneumatolysis* or gas action. The dykes formed in this way possess special structural and mineralogical characters, often containing rare and valuable minerals: they are called pegmatites, aplites and so on, and graduate by imperceptible transitions into ordinary *mineral veins*.

Pegmatites and Aplites. Even the most cursory examination of any fairly large exposure of intrusive rock is usually sufficient to show that it is not homogeneous throughout. The main mass of a granite, for example, is traversed by veins, sheets and tabular masses of coarser or finer texture and often containing somewhat different mineral assemblages. To these variants special names are applied, the coarser ones being called *pegmatites* and the finer ones *aplites*. These veins or dykes are by no means confined to the intrusion itself, but frequently pass outwards into the surrounding sedimentary or other rocks. As will be seen later, both pegmatites and aplites are often associated directly with ore-deposits, hence their study possesses great practical importance.

(a) *Pegmatites.* The detailed field and petrographical examination of a great number of occurrences of pegmatites has revealed certain facts throwing a great deal of light on their real nature and origin. In the first place, as just stated, they often pass beyond the boundaries of the intrusion, continuing into the schistose or sedimentary rocks beyond and frequently showing a gradual change of character when followed laterally for some distance. For instance, pegmatite dykes in granite can often be traced into what are for convenience called mineral veins, or even beyond this into barren quartz veins: secondly, their texture shows features and mineral-relations which we have reason to believe from analogy are characteristic of the crystallization of eutectics: and thirdly, they carry minerals which are known to contain an unusually high proportion of the rarer and more volatile constituents of the magma. All these facts taken together suggest that the pegmatites are in point of fact the solidified form of the magmatic mother-liquor previously spoken of. At one time there was a good deal of controversy as to the origin of the pegmatites, some writers considering them to be of igneous origin, while others held that they were formed by deposition from watery solution. However, with the recognition of the important part played by water in the economy of igneous magmas, this cause of difference of opinion has disappeared, since it is evident that both groups of antagonists were right. The lateral variation in the composition of the dykes and veins is due to the gradual cooling down of the magmatic solution as it travelled further and further from its source, depositing its dissolved contents in the order of their solubility; the last to remain is in most cases silica, which is deposited as quartz or one of the many other forms of crystalline and amorphous silica. One of the most conspicuous features of many pegmatites is the occurrence in them of the so-called graphic structure, an intimate intergrowth due to the simultaneous crystallization of two or more minerals. On analysis of several samples it is found that this graphic mixture possesses a constant composition implying a definite ratio between the amounts of the minerals and it is now generally believed that this indicates eutectic proportions. But from our present point of view the most important feature in pegmatites is the presence of certain minerals absent from the main mass, or in other cases a concentration of such minerals as are sparingly disseminated throughout. This group includes a considerable number of valuable ores, such as tinstone, wolfram, chalcopyrite, mispickel, molybdenite, compounds of silver, antimony, bismuth, tantalum, niobium and uranium, and many others, as well as minerals containing the rare earths in large quantity. Besides these are found tourmaline, topaz, beryl, fluorspar, apatite and other compounds containing the highly volatile constituents of the magma, such as fluorine, chlorine, boron, phosphorus, lithium and beryllium. Descriptions of many ore-bearing pegmatites will be found in other sections of this book. In the present theoretical discussion the important point is that the metallic elements are concentrated in the last residues of the solidifying magma; here they combine with non-metals to form compounds volatile under the prevailing conditions, and are carried out in these forms to become part of the peculiar coarse-textured dykes and veins called pegmatites. The characteristic elements of the pegmatites vary in different classes of rocks: in the acid rocks are found especially minerals rich in fluorine, boron, lithium, beryllium, with tin, tungsten, arsenic, molybdenum and copper. The intermediate types, such as syenites and diorites, carry gold, silver, copper and the rare earths, while the basic pegmatites are rich in chlorine, phosphorus and titanium.

(b) *Aplites.* The veins and dykes of finer texture than the main mass, called aplites, are of less importance from the economic point of view than the pegmatites, since they do not often contain ore-deposits of much value. As a general rule the aplites are somewhat more acid than the main mass of the rock, with a lower proportion of minerals rich in iron and magnesium. Aplites are best developed in and around granites and not much is known as to their occurrence in connection with basic rocks. It is quite clear however that, like the pegmatites, the aplites are due to concentration of certain constituents in the later liquid residue of the cooling and crystallizing magma, with subsequent injection into cracks and fissures in the mass of the earlier solidified rock.

Pegmatitic and Aplitic Textures. There has been at various times much discussion as to the reason for the remarkable difference of texture observed in the coarse pegmatites and the finely crystalline aplites. The true pegmatites are specially characterized by coarse and irregular textures, the minerals often showing graphic relations and cavities being common. The aplites on the other hand are notable for the fineness and evenness of their grain and the usual absence of open spaces of any kind. These differences are probably to be correlated with the presence or absence of fluxes in each case. Besides lowering the freezing point, a flux of any kind also tends to increase the mobility of the solution; this favours diffusion of molecules and consequently assists the formation of large crystals. On the other hand a silicate solution without a flux is always very viscous, so that diffusion is hindered and numerous centres of crystallization are set up, resulting in a fine and even grain. The presence of highly volatile fluxes also naturally helps to originate the formation of gas-cavities in cooling rocks. Furthermore fluxes tend to keep metallic compounds in solution and to concentrate them in the last fractions, whereas in a "dry melt" the metals would remain evenly disseminated. The marked difference of texture between the pegmatites and aplites is therefore to be correlated with the presence or absence of fluxes, or volatile substances, and especially water, in the particular solutions from which they are respectively derived.

Mineralized Veins. Numerous examples are to be found in all districts where ore-deposits occur in connection with igneous rocks of what are commonly described as veins, carrying more or less metallic minerals. Many of these are essentially pegmatites, while others are better described as quartz veins. There is however no real distinction between pegmatites and veins, since a complete transition is found between all the different types. It was formerly supposed that some fundamental difference existed between dykes and mineral veins, the former being considered as of true igneous origin, while the latter were universally attributed to deposition by aqueous solutions. It was generally implied, if not expressly stated, that this water was not of volcanic origin, but rather meteoric or ground-water. However in more recent times it has been recognized that very many quartz veins are actually igneous phenomena, comparable to pegmatites, and in fact forming merely a somewhat later stage in the solidification of the magma. In some instances a body of this nature can be followed along its strike, showing a gradual change of character from an undoubted pegmatite carrying minerals characteristic of plutonic rocks, through every stage to a vein consisting only of quartz, with or without ore-minerals. Instances of this will be described later. The main point to be established now is that many metalliferous quartz veins are of undoubted magmatic origin, and are products of an extreme type of differentiation by crystallization.

Pneumatolysis. Besides the effects just briefly outlined in the production of peculiar minerals and structures of primary origin, the last highly volatile constituents of the magma often give rise to a variety of secondary changes in the earlier-formed minerals and in the country rock of the intrusion. These changes are summed up in the general term *pneumatolysis* or gas-action. By some writers this term has been used to include also the formation of primary pegmatites, aplites and quartz-veins and the genesis of peculiar minerals in the igneous rocks as a direct product of crystallization, but it is better to restrict its use to secondary changes[1]. It must be admitted, however, that in some instances it is difficult to be certain whether a given

[1] On this point see Harker, *The Natural History of Igneous Rocks*, p. 299.

mineral is primary or secondary, at any rate in the igneous rocks, but the general principle is clear. Pneumatolysis in the sense here employed is a very common and characteristic phenomenon in connection with many types of mineralization, as will appear in later sections: it is not proposed to give here any detailed account of this phenomenon, only a few general principles may be mentioned.

Pneumatolysis consists essentially of mineralogical changes brought about by vapours and solutions derived from igneous intrusions and remaining till the latest stages of crystallization. The effects are mainly chemical and a change is usually brought about in the total bulk-composition of the rocks acted on. The most common effect is the formation of minerals containing some of the elements that are concentrated in the magmatic residues, which naturally tend to be those substances that are either volatile in the free state or can form volatile compounds. These include especially boron, fluorine, chlorine, sulphur and phosphorus among the non-metals, with beryllium, lithium and other metals of low atomic weight, while perhaps the most important of all is water, which at high temperatures and pressures is a chemical agent of great power. The part played by silica in these processes is still somewhat of a mystery. Some heavy metals also take part in pneumatolytic processes by means of volatile compounds; among these are tin, tungsten, molybdenum, uranium, and apparently copper, lead, zinc and even iron. Ferric chloride is well known to be volatile at a high temperature.

The type of change brought about in any given case naturally depends on two principal factors: the composition of the fluids that bring about the changes and the composition of the rock acted on. Both of these factors vary widely, but it has been found by observation in many cases that a particular group of agents of pneumatolysis is characteristic of a particular kind of igneous rock: in fact the type of pneumatolysis is to a large extent correlated with the silica percentage of the rock. Thus fluorine and boron are especially associated with granites, chlorine and phosphorus with basic rocks, and so on. Furthermore in each main group there are sub-groups, according to the presence or absence of particular elements. As examples we may mention the formation of tourmaline and topaz rocks with granites, scapolite rocks with gabbros and so on. With each type certain metals are associated, hence the study of pneumatolytic phenomena and their characteristic products is a valuable guide in prospecting. **Solfataric Action.** Hitherto reference has been made only to processes of a pneumatolytic nature occurring in connection with deep-seated intrusions of igneous rock, but there exists also a class of phenomena of great interest and importance from the point of view of ore-formation in connection with volcanic rocks of the eruptive type, or true lava-flows. The molten material of lava-flows, before eruption, is exactly the same as that of intrusions and has the same origin. It therefore contains the same gases. These gases, however, when the lava is erupted at the surface, escape into the atmosphere, and hence have less opportunity of bringing about changes in the rocks. Nevertheless after the eruption some gaseous material remains in cavities in the lava, and, what is perhaps of more importance, for some time after, and often for a long time after the eruption, there is an escape of gases and steam from the volcanic vent. These vapours often bring about far-reaching changes in the minerals of the ejected lavas and of the surrounding rocks, and these changes are frequently accompanied by ore-deposition. The classical example of a volcanic gas jet is the Solfatara, near Naples, and from this instance the name has been extended to the whole group of similar phenomena. In general terms, the effect of these processes is to bring about a decomposition of the rocks, often of a very far-reaching kind, as will be explained in detail in a later section. Among rocks rich in alkali-felspar, a frequent result is the formation of products of the china-clay type, mainly hydrated silicates of alumina, the alkalis being removed, while in the lavas of more basic composition the effects are often summed up in the convenient but somewhat vague term propylitization. Rocks showing this type of alteration very often carry ores and a common effect of such late-volcanic activity is that the surrounding rocks become impregnated with large quantities of sulphides, sometimes merely pyrite, but commonly containing much copper, gold, and other valuable metals.

II] THE IGNEOUS ROCKS 41

The Petrology of the Pegmatites. In an earlier section we have discussed in a general way the nature of the processes that have taken place in large intrusions during the later stages of cooling and crystallization, leading to the concentration in the final residue of certain constituents of a special nature, often including important proportions of the useful and valuable metals. It was shown also that these residual concentrations tend to form bodies often of a dyke-like and vein-like form by filling cavities and fissures produced within the intrusion itself by shrinkage on cooling and by other stresses, these dykes and veins also often extending into the surrounding rocks. Furthermore emphasis was laid on the important fact that no hard and fast line can be drawn between pegmatites, mineral veins and quartz veins of the current nomenclature. It now becomes necessary, on account of the economic importance of these pegmatitic bodies, to consider somewhat more closely their characters and composition, and especially their relation to ore-deposits.

The ordinary conception of a pegmatite is that of a tabular or sheet-like body composed of crystals of unusually large size and often possessing a graphic structure. Sometimes, however, pegmatites occur as quite irregular masses, showing little or no relation to any visible geological structures. Nevertheless they are invariably connected with intrusions of igneous rock of some kind. The most typical pegmatites are undoubtedly those of composition allied to the granitic rocks: syenite-pegmatites are also well known and of much interest, but as to the occurrence of truly basic pegmatites little is known. This is in accordance with the principles already set forth, since the pegmatites consist essentially of the last residue of the crystallizing magma and therefore tend to contain more of the last products to consolidate which are more siliceous than the average, the basic constituents consolidating first and thus leading to one type of differentiation.

In accordance with these facts, then, pegmatites may be regarded as being characterized by certain special features, of which the chief are: late consolidation, coarse texture, and the prevalence of minerals of low freezing point and exceptional chemical composition. They are, in point of fact, natural concentrations of the rarer constituents of the magma, dissolved before consolidation in an aqueo-siliceous solvent. On cooling these crystallize out with the structures and characters appropriate to such a solution under the given conditions.

The typical pegmatites therefore are those of acid (granitic) composition: in these the most characteristic minerals are quartz, felspar and mica, with or without accessories. Among the simpler types we find the mica-bearing pegmatites, which yield the commercial output of mica. Although of great importance, such do not strictly come within the compass of this book. The world's supplies of felspar for the ceramic and other refractory industries are also mainly derived from granitic and syenitic pegmatites, and attempts have been made to obtain potash on a commercial scale from them, especially during the war, when supplies of German potash salts were cut off. The processes however proved too expensive for general use.

Of more immediate interest are the types of pegmatite containing ores of the metals: these may be divided for our present purpose into three principal groups, as follows:

    (1) the tourmaline-tin-tungsten type;

    (2) the rare-earth type;

    (3) the apatite dykes.

Between these there are necessarily intermediate and transitional forms, and doubtless it would be easy to find numerous examples which do not fit readily into either group: nevertheless this generalization does seem to correspond closely to the facts of nature.

Furthermore we may recognize another group:

    (4) the quartz veins with precious metals.

The veins of this type are not commonly recognized as pegmatite at all, being generally attributed to water action, although at a high temperature. But this is a point that must be strongly emphasized in a scientific treatment of the subject, namely that there is no real distinction between the pegmatitic and the so-called hydrothermal phases of ore-deposition: the difference is one of degree and not of kind. To this category belong many gold-quartz veins, for example, which contain a certain, though often small amount of felspar, mica and other distinctively igneous minerals.

The pegmatites of the first three classes are characterized in most cases by their excessively coarse texture, which is conditioned, as before explained, by the presence in large amount of various substances acting as fluxes; among these are specially to be mentioned fluorine, boron and water, together with a variety of elements of low atomic weight, such as lithium and beryllium. With these are associated certain metals that are capable of forming volatile compounds with the non-metallic elements just mentioned. In the second group also we find a remarkable association of elements of the "rare earth" group, such as zirconium, cerium, lanthanum, didymium, together with uranium and thorium and their numerous radio-active disintegration products. The physical chemistry of this facies is more obscure than that of the first group. Uranium compounds are also found along with the first group.

Among the basic rock-masses pegmatitic structures are less definitely marked, nevertheless there does exist a type of magmatic concentration of accessory constituents, which shows somewhat similar characteristics; among the elements that may be concentrated in this way are chlorine, phosphorus and titanium, but the effects tend to show themselves rather in the form of contact metasomatism, generally known as pneumatolysis. Special minerals and especially ore-minerals, are not found to any great extent in basic pegmatites. This is probably correlated with the limited miscibility of metals in basic silicates, as discussed elsewhere, leading to an early separation from magmas low in silica.

It is clear then that the distinctive characteristics of the true pegmatites are two-fold; coarse texture and uncommon minerals. The cause of the coarseness of texture has already been discussed and attributed to the presence of fluxes, keeping the solution very mobile during crystallization. This coarseness sometimes reaches an extraordinary degree of development, individual crystals being measurable by feet, or even yards. This is closely analogous to the occurrence of fine mineral specimens in the cavities of veins of true aqueous origin, where the solutions depositing minerals become concentrated, giving rise to formation of various chemical compounds by a process of slow precipitation and molecular growth in very perfectly developed forms.

It appears then that in the commonest type of pegmatites the characteristic minerals are quartz, felspar and mica, with, among the non-metallic minerals tourmaline, topaz, beryl and sometimes apatite. The felspar is usually of the most alkaline type, orthoclase, microcline, albite, and perthite, while the micas are often of peculiar varieties, lithium-bearing varieties, such as lepidolite, being quite common. Among the metallic minerals of this group the most characteristic are cassiterite, wolfram, molybdenite; varieties with columbite and monazite form a transition to the rare-earth group. Very exceptionally cryolite is found in quantity, as at Ivigtut in Greenland. This is described on p. 452. One of the most remarkable examples of pegmatite in the world is found at Etta Knob in the Black Hills of South Dakota. This is a roughly circular mass about 200 feet by 150 feet in superficial area: it consists mainly of large crystals of quartz, felspar and mica, and nearly forty mineral species have been identified: one of the most notable elements is lithium and individual crystals of spodumene measuring 40 ft × 6 ft × 3 ft have been measured. Cassiterite, wolframite, columbite, molybdenite, arsenopyrite, galena, bismuth, zircon, rutile, spinel, tourmaline, beryl, sphene and garnet have also been recorded, together with others that obviously belong to the surrounding schists[1]. In this region many other pegmatite dykes and quartz veins contain cassiterite, wolfram and columbite: a mass of the last-named mineral has been found weighing a ton. This pegmatite is an excellent example, though on a rather exaggerated scale, of a very common type of chemical and mineral paragenesis, characterized specially by lithium, tin, tungsten, tantalum and molybdenum: a similar assemblage is found in numerous localities in many parts of the world.

Another very striking example of a pegmatite is to be seen at Baringer Hill, in Llano County, about 100 miles N.W. of Austin, Texas[2]. In this neighbourhood is an isolated inlier or

---

[1] Hess, Bull. 380, *U.S. Geol. Survey*, 1909, p. 158.

[2] Hess, Bull. 340, *U.S. Geol. Survey*, 1908, p. 286. See also Hidden and Warren, *Amer. Journ. Sci.* vol. XXII, 1906, p. 515.

"island" of Precambrian rocks, surrounded by a wide extent of Palaeozoic strata and invaded by porphyritic granite and pegmatite. Some of the dykes of pegmatite are as much as 60 feet wide; some are nearly pure quartz, being representative of the most acid products of differentiation, while others are almost entirely felspar. Baringer Hill itself is a mass of pegmatite of most remarkable type, more or less circular or oval in form, and forming a low mound rising about 40 feet above the flood-plain of the river. At the edge of the mass is a zone of very well-developed graphic intergrowth of quartz and felspar, from 1 to 6 feet thick: inside this comes a mass of enormous crystals, chiefly an intergrowth of microcline and albite, single crystal edges being sometimes a yard long: there is also fluorspar in cubes with sides up to 1 foot in length, varying in colour from very pale violet to nearly black; plates of biotite 3 feet across and an inch thick; and ilmenite in bunches of blades up to 10 inches long. But the most important feature is the occurrence of very large crystals and lumps of gadolinite and many other rare-earth minerals, including allanite, cyrtolite, gummite, thorogummite and other uranium minerals. Molybdenite has been observed in hexagonal platy masses up to 10 lb. in weight, but wolfram, tourmaline, monazite and garnet are entirely absent, while much doubt exists as to the reliability of one record of cassiterite. This remarkable deposit is exploited by the Nernst Lamp Co., of Philadelphia, for gadolinite as a source of yttrium. This may be taken as a somewhat exaggerated example of the type of pegmatite specially characterized by rare-earths and radio-active minerals.

Closely analogous to the foregoing are the syenite-pegmatite dykes of southern Norway, exhaustively studied by Brögger many years ago[1], and exploited to some extent as a source of rare earths, especially thoria and zirconia, in the early days of the incandescent gas-lighting industry, before the large-scale utilization of monazite sands. These pegmatites contain an immense variety of minerals, mostly of very complex composition and of limited occurrence. A special point of interest is the

---

[1] Brögger, *Zeits. für Kryst.* vol. XVI, 1890, where full descriptions of all the minerals will be found.

association of this type of pegmatitic concentration with a highly alkaline series of rocks, many of which are specially distinguished by a notable excess of soda over potash, giving rise to nepheline-syenites and other cognate types. Pegmatites of this sort may be accompanied by deposits of sulphides, such as molybdenite, blende and galena, but rarely on an important scale. In this instance Brögger describes the deposition of rare minerals as having taken place in three stages: the first includes minerals rich in fluorine, boron, chlorine and water, with sulphur and arsenic, such as fluorspar, eucolite, lâvenite, rosenbuschite, zircon, apatite, katapleiite, astrophyllite and others with much zirconium: then comes deposition of melinophane, leucophane and the sulphides, followed by extensive formation of zeolites in druses and other cavities.

The granite-pegmatites of southern Norway also contain many rare minerals, and have been exploited industrially both for felspar and for rare earths. These granites belong to the intrusions of the later Pre-Cambrian period and occupy large areas among the ancient rocks on both sides of the Kristiania Fjord and its approaches. On the eastern side the Fredrikshald granite is accompanied by many pegmatite dykes often of very coarse texture, workable for felspar and mica: these contain also in abundance topaz, zircon, beryl, monazite, fergusonite, cleveite, fluorspar, samarskite and columbite, with bismuth and molybdenum sulphides in fair quantity. On the western side the Telemarken district contains great masses of granite of varying ages, but all apparently Pre-Cambrian. Around the margins of this area is a great development of pegmatites of granitic composition, often rich in columbite, samarskite, monazite, gadolinite, orthite, thorite, xenotime and tourmaline. It is clear that in both these areas there has been a segregation in a granitic magma of large amounts of tantalum, zirconium, thorium and the rare earths generally, along with fluorine and the other usual concomitants of the pegmatitic type of differentiation. From these dykes also there has been a considerable output of felspar for technical purposes[1].

Another type of considerable interest is the apatite dykes

[1] Brögger, *Vidensk. Selsk. Skrifter, Math. Nat. Klasse*, 1906, No. 6.

found in southern Norway and Canada in association with gabbros. Here the most characteristic minerals, besides apatite, are scapolite and various titanium compounds. As a rule the gabbro has undergone considerable alteration in the immediate neighbourhood of the dykes, the augite being converted to hornblende and the felspar to scapolite. This phenomenon is in fact a kind of metasomatic alteration or pneumatolysis, of a type apparently confined to basic magmas, the chief elements concentrated being chlorine, phosphorus and titanium. In the strip of country about 70 miles long and 15 miles broad between Bamle and Lillestrand on the western side of the Skagerrak are hundreds, perhaps thousands, of apatite dykes, often of very coarse texture and many rich enough in apatite to have been extensively worked at one time as a source of phosphate. The apatite in Norway is principally chlorapatite, with little or no fluorine, which is a great advantage in the manufacture of artificial manures. Other characteristic minerals are scapolite, as before mentioned, rutile, ilmenite, sphene and other titanium compounds, zircon, tourmaline, pyrrhotite, chalcopyrite, galena and pyrites, together with much magnesia-mica, enstatite, hornblende and felspar[1].

[1] Vogt, *Zeits. prakt. Geol.* 1895, p. 367.

CHAPTER III

THE SEDIMENTARY AND METAMORPHIC ROCKS

IT is not intended in this book to describe in detail the general geology of the great groups of sedimentary and metamorphic rocks which constitute so large a proportion of the outer and visible crust of the earth. This subject will be found treated in general terms in any text-book of geology, and in a more detailed manner in various special treatises[1]. We are here only concerned with the origin, characters and distribution of such rocks in so far as they are directly connected with ore-deposits, both as the country rock of veins and other deposits of analogous character, and as the basis of a large group of contact and replacement deposits, or even as themselves constituting important sources of ores of metals, such as ironstones, manganese ores or bauxite, as well as the valuable group of placers and other superficial and alluvial deposits that yield so much of our available supplies of gold, platinum and tin. It is thus clear that the relations of sediments and metamorphic rocks to ore-deposits embrace a wide range of variation, including as they do natural formations belonging to many and very different geological categories.

It may be well to state clearly at the outset that it is a difficult matter to draw up a satisfactory definition of a sedimentary or metamorphic rock, and still more difficult to demarcate these two groups clearly one from the other. Generally speaking the sedimentary rocks are understood to be those formed by superficial geological processes from the material of pre-existing rocks, including therein various deposits of chemical and organic origin formed under similar conditions, while the metamorphic group includes those rocks of either igneous or sedimentary origin which have undergone a notable degree of alteration since their original formation. The former group is made difficult to define by the fact that many of its members

---

[1] See especially Hatch and Rastall, *The Petrology of the Sedimentary Rocks*, 2nd edition, London, 1923.

contain material of direct igneous origin, but of fragmental character, that is to say, volcanic ash. It is often almost impossible to say whether a deposit laid down in a lake or in the sea near a volcano should be described as of sedimentary or volcanic origin: in point of fact it is both. The definition of the second group is made difficult by the impossibility of deciding what degree of alteration is necessary to remove the rock from its original category to the metamorphic division. Hence the systematic study of the rocks here dealt with is obviously a somewhat unsatisfactory subject, and we have to be content with a generalized treatment based on somewhat vague and ill-defined conceptions. The one clear character that they all possess in common, however, is that they are not, *in their present condition*, the direct results of the consolidation of igneous magma: in the case of the true sediments this statement requires no qualification; many of the metamorphic rocks were originally igneous but have now undergone structural and other changes of sufficient importance to justify their relegation to a special class.

The Sedimentary Rocks. The term sedimentary as applied to rocks has to be understood in a very broad general sense. The word in its usual application connotes solid material deposited in water, and the majority of them do come under this heading; but some rocks universally understood to be sediments are not formed in water; for example, wind-blown sands. Again some calcareous rocks and salt deposits are formed by direct deposition of coherent, solid, usually crystalline, material from solution, and this is hardly sediment in the ordinary sense of the term. Some rocks also, such as coral-reefs and coal are the direct results of the activities of animals and plants.

Nevertheless, in spite of these exceptions, the meaning of the word is generally understood by geologists. Another alternative term is *stratified*, but this again is open to quite as many objections and exceptions.

Although the materials forming the rocks of this group show a very wide, or almost indefinite range of variation, still the great majority of them belong to a comparatively small number of types. They may be most simply regarded as being formed from materials precisely similar in character to those now found on the surface of the land or on the floors of lakes or of the sea. The commonest kinds of material thus found are beds of boulders and pebbles, gravel, sand, mud, clay and (especially in the sea) different varieties of calcareous matter, such as shell-banks and coral-rock. In the course of ages these masses of incoherent material have gradually become buried to a greater and greater depth and have been dried, hardened, solidified and compressed to solid masses, and welded together by different cementing materials into solid rocks, giving rise to the older sedimentary, often stratified, rocks as we see them to-day.

On this basis we obtain a simple classification of the commoner types, as follows:

Boulder and pebble rocks: conglomerates.

Sand rocks: sandstone, grit, greywacke, arkose, quartzite.

Mud rocks: clay, shale, mudstone, slate.

Calcareous rocks: limestone, dolomite-rock, oolite, chalk.

Ferruginous rocks: bedded ironstones.

Out of rocks of these simple types the vast majority of the sedimentary or stratified formations of all ages are built up. The names applied to each particular type depend very largely on the degree of cementation and the amount of subsequent alteration undergone by each: thus to take a concrete case we may recognize a series of progressive alteration including sand, sandstone and quartzite. Sand is loose and incoherent, sandstone is more or less solidified but still soft and crumbly, while in a quartzite the mass has been cemented into a hard and solid mass by deposition of secondary silica as quartz. Some quartzites have been crystallized by heat and pressure and must therefore be regarded as metamorphic rocks. Similarly when mud is partly dried it becomes clay; when quite hard and solid it is called mudstone, and when cleaved by metamorphic pressure it forms slate. Limestones when intensely heated or compressed are changed into crystalline marble.

It does not require much consideration to show that these processes of alteration must have effects of the highest importance with regard to the permeability of such rocks to mineralizing solutions and are therefore the controlling factors in the formation of many mineral deposits. Furthermore the chemical composition of the rocks is also of dominating effect. Of all ordinary rocks limestones are most easily acted on by solutions and are therefore specially liable to replacement by ore-bodies.

It must also be remembered that in nature rocks are not divided up into clean-cut and well-defined groups: on the contrary innumerable transitional forms exist, so that almost every case has to be treated on its own merits. In fact in the sediments the range of variation is actually much greater than in the igneous rocks. The petrology of the sedimentary rocks is thus in reality an extremely complicated subject.

Another very important side of the case is the study of characteristic textures and structures, including not only the porosity and permeability of the rocks with regard to solutions and vapours, but also the existence or otherwise of planes of discontinuity either actual or potential, such as joints and other forms of fracture with or without actual displacement, and especially faulting and folding. It would require much more space than is here at our disposal to discuss these subjects even in general terms. It is, however, well known to all mining geologists that in many regions the lie of joints and faults is the controlling factor in the distribution and extent of ore-deposits. It is therefore essential for all such to be well acquainted with the general principles of structural geology: without this knowledge the interpretation of surface or underground features is impossible. In close correlation with this part of the subject also stands the study and interpretation of geological maps, which is also of first-class importance for the correct understanding of underground structures.

From the practical point of view therefore the chief characteristics of the sedimentary rocks that have to be taken into account are those determining the relation of the rocks to ore-deposit. In the first place rocks which carry large open joints facilitate ore-formation of all kinds: such are pre-eminently sandstones and limestones in their various degrees of alteration. Close-textured rocks on the other hand, especially clays and shales, are unfavourable, since it is difficult for open spaces to exist in them under any degree of pressure. Again the actual

porosity of the rock itself to solutions is of importance, especially
in regard to replacement deposits: this depends primarily on the
degree of cementation. Here again sandstones are most favourable. In the case of limestones the question is complicated by
the solubility of the calcium carbonate in solutions. Rocks of
the clay and shale group are very impervious to solutions, quite
apart from the non-existence of open fissures in them. As a rule
sediments, with some important exceptions, are more pervious
to solutions than igneous rocks, and more liable to chemical
replacement.

Metasomatic Changes in Sedimentary Rocks. Certain
types of sediments may undergo changes of chemical and
mineralogical composition at the ordinary temperature and
pressure that have an important influence on their relations to
ore-deposit. Such changes may be either favourable or otherwise. In one extreme case the alteration may make the rock
entirely insusceptible to either permeation or replacement by
mineralizing solutions: in the other extreme the change may
itself convert the rock into a valuable ore, as in the case of the
formation of certain metasomatic ironstones. The following are
examples of a few of such changes belonging to both categories.

Dolomitization of limestones. This consists in the replacement
of the whole or part of the calcium carbonate of a limestone by
the mineral dolomite, the double carbonate of lime and magnesia.
The exact process by which this change is brought about is still
rather obscure and need not be discussed here. However, a
common practical result is to render the rock decidedly cellular
and porous, and therefore easily permeable by ore-bearing solutions. Also the value of the rock as a flux for various metallurgical purposes is considerably modified.

Silicification of limestones. This type of metasomatic alteration
is exceedingly common, and in some cases has had an important
influence on the distribution of ores, by rendering the rock
impervious and resistant to chemical replacement. A silicified
limestone is commonly called *chert*, and the change appears to
take place by complete replacement, molecule by molecule, of
calcium carbonate by silica, which is commonly not quartz, but
some form of chalcedony or opal.

Metasomatic ironstones. The question of the origin of bedded ironstones from limestones is discussed in considerable detail in the chapter on iron ores and need not be dealt with here.

In the other principal rock-groups, owing to the greater stability of the original minerals, metasomatic changes are of very much less importance and do not require further discussion. Furthermore to consider here all possible changes of this class brought about even in limestones alone would involve a description of many processes of ore-formation that are dealt with in other sections of this book. Again the igneous rocks are by no means exempt from changes of this kind: in rocks of all classes an extensive introduction of sulphides at a date subsequent to their formation is a common phenomenon. This class of change at the ordinary temperature merges by imperceptible degrees into the high-temperature metasomatism that produces most of the so-called contact ore-deposits. Igneous rocks also are very often highly silicified at some period after their original formation and this process is sometimes accompanied by mineralization, *e.g.* deposit of gold.

Metamorphism. During the course of geological time rocks of whatever origin have often undergone changes of a far-reaching character, so that in extreme instances it may be difficult or impossible to determine their original nature. These changes are conveniently summed up in the word *metamorphism.* This term is however of necessity somewhat vague in its application, and opinions differ widely among geologists as to what degree of alteration is necessary to bring rocks within this category. The agencies concerned are high temperature and high pressure and the general effect may be summarized as the production of new minerals or new structures, or both, in pre-existing rocks.

High temperature and great pressure, however produced, may obviously act upon rocks of any kind whatever, and the resulting products naturally vary widely according to the original character of the material acted on. Furthermore, in certain circumstances there may be brought about at the same time an actual change in the total bulk composition of the rock. This type of alteration is now usually classed as *metasomatic metamorphism.* This is of great importance in ore-formation, but is here for the moment left out of account, attention being confined to those cases where the effect of the heat and pressure is merely a rearrangement of the rock-material, without addition from other sources. In general terms it may be said that the effect of heat alone, without notable increase of pressure, is to induce a higher degree of crystalline structure in the rock. As simple examples, a limestone is converted into a crystalline marble and a sandstone becomes a quartzite.

The effect of pressure on the other hand is to induce new and characteristic structures, usually with a conspicuous tendency to a parallel orientation of minerals, or of the rock-building elements in general, if these are of complex character. This usually brings about a tendency for the rock to split into sheets or slabs in a direction at right angles to the pressure. Thus are formed that great class of rocks known collectively as the slates and crystalline schists, and some forms of gneiss. In accordance with the general ideas here laid down, metamorphism may be subdivided into two headings, thermal and dynamic.

Thermal Metamorphism. The usual cause of thermal metamorphism is the intrusion into the older rocks of the crust of masses of heated igneous magma: the study of thermal metamorphism is therefore of special importance in connection with ore-deposits, since the same phenomenon is also the commonest cause of mineralization. The finding of metamorphic rocks of thermal type in prospecting is a good indicator of possibilities of ore-deposits in the immediate neighbourhood. Each intrusion of any size is surrounded by a sheet of altered rock, the thickness of the altered sheet naturally varying according to many different factors and the alteration progressively decreasing away from the margin of the intrusion, according to the temperature gradient. The cross section of such a sheet around a boss or laccolith, when exposed at the surface by denudation is known as a *metamorphic aureole*. In the case of a dyke or a sill the altered rocks will form a sheet on either side of the intrusion. In the case of irregular intrusions the altered regions may of course assume an endless variety of forms. Again the susceptibility of different rocks to thermal alteration varies within very wide limits, in accordance with chemical and physical principles. Generally speaking the sediments most liable to notable change are the limestones, and the argillaceous or clay rocks: the former are very easily converted to marbles, while the latter have a special tendency to development of a spotty texture, the spots being usually single crystals or aggregates of crystals of various aluminous silicates, such as andalusite, cordierite or mica. Garnets are also very common in highly altered rocks. On the other hand as the sandy rocks are usually of very simple composition, little change can take place in them: the alteration is often confined to a mild degree of crystallization giving rise to quartzite. Ironstones have a tendency to be converted to crystalline haematite or magnetite. Igneous rocks again are not very liable to alteration, unless already decomposed, since in their normal condition they already consist of minerals stable at high temperatures.

Dynamic Metamorphism. The changes produced by this agency are mainly structural, the formation of new minerals being somewhat subsidiary, and probably conditioned largely by the heat due to friction produced by pressure. Perhaps the simplest type of dynamic metamorphism is the shattering of rock along the lines of faults and zones of fracture in the rocks. This process is of immense importance in ore-formation, since it gives rise to the extremely important class of brecciated lodes. It is but rarely that a fault shows a clean-cut fissure: there is nearly always more or less shattered material, or fault-breccia; frequently an important line of fracture does not show any definite fissure at all, but consists of a belt or zone of shattered rock, with secondary matter, often ore-bearing, deposited in the interstices, and the shattered rock itself showing more or less chemical and mineralogical alteration. In such cases there are often developed on the rock walls or on the broken fragments different kinds of polished and striated surfaces (slickensides) often showing clearly the direction of movement. An extreme form of this process results in a rock composed of lenticular crushed masses with a roughly parallel arrangement and this passes by insensible gradations into normal schistosity, where the rock consists of flattened, more or less parallel mineral bands, often crumpled and contorted in a very complex manner. The most characteristic minerals of the true schists are mica and various members of the chlorite group, as well as actinolite and hornblende. Garnets are also common and give rise to a lumpy structure.

The effects of dynamic metamorphism on different kinds of rock naturally vary mainly according to their original texture and the hardness of their minerals or other components. Simple sandy rocks again generally crystallize as quartzites, and limestones being soft commonly also become crystalline marbles, often flowing in a very notable manner to conform to the pressure. Fine-textured muddy rocks undergo cleavage and form slates or phyllites, according to the degree of crystallinity. Igneous rocks, including volcanic ashes, are generally converted into some kind of schist or gneiss, according to the coarseness of their texture. The very common types of chlorite and hornblende schists are usually derived from basic igneous rocks, such as gabbro, dolerite or basalt, while ultrabasic rocks become talc-schists.

It must always be remembered that many of the names applied to altered rock types are very vague in their definition, and this remark applies with special force to the terms gneiss and schist, which grade into one another imperceptibly and their application in any given case is often a matter of individual taste. Furthermore some gneisses are not metamorphic rocks, as elsewhere explained.

CHAPTER IV

THE RELATIONS OF WATER TO ORE-FORMATION

WHILE it would be going too far to say that water plays a part in all instances of ore-deposition, there can be no doubt that its influence is of the highest importance in the majority of cases. There has been for many years past a great deal of discussion on this subject in all its bearings and on some points there is still a considerable division of opinion. The questions at issue are however concerned more with the comparatively minor point of the origin of the water present in particular instances, rather than with the fundamental fact of its participation in the processes of ore-deposit: on this there is on the whole very general agreement. Owing to the obvious importance of the subject it is necessary to enter into it pretty fully.

It is a matter of daily observation that in most parts of the world abundant water exists in rocks below the surface, and the data of water-engineering show that in the superficial layers of the crust water is in active circulation, and is, so to speak, in continuity with the surface waters forming rivers, lakes and the sea. Another simple and suggestive fact is that heated waters are common in depth, and sometimes come to the surface as hot springs. Furthermore natural underground and spring waters are sometimes conspicuously saline. Actually the difference between salt and fresh water is only one of degree. All natural waters contain dissolved salts in varying amount and it is only when this amount is large enough to impart a taste that the water is commonly called salt. The facts above mentioned, as well as some others to be referred to presently, suggest that there may be fundamental differences in the origin of natural waters.

The Origin of Natural Waters. It is known from innumerable analyses that all igneous rocks contain a considerable amount of water. Thus according to Clarke[1] the average of

[1] Clarke, "The Data of Geochemistry," Bull. 695, *U.S. Geol. Survey*, 1920, p. 33.

many hundreds of determinations made on rocks from all parts of the world is about 2 per cent. In the plutonic rocks as a rule it is somewhat less, while glassy lavas, like obsidian, may contain two or three times as much. Mention must also be made of the vast clouds of steam emitted by volcanoes, although it is not clear that this water is actually brought up from magmatic sources: there is always the possibility that it may have merely penetrated from above into the volcanic fissure. Much of the water of igneous rocks is actually enclosed in the minerals in the form of bubbles, as can often be seen in sections of quartz under a high power. In other instances the water is in some form of chemical combination in the minerals, perhaps as the OH group, as in micas[1]. In any case this water must be derived from the magma. Unfortunately we have no real information as to the proportion of water actually present in magmas, since there is always a possibility and often a certainty, that much of it has escaped before complete crystallization of the rock. At any rate it is beyond doubt that igneous magmas on the average contain at least 2 per cent. of water and probably much more. This at once suggests that magmatic water may be and probably is of much importance in underground processes.

This idea is confirmed by the peculiar character of many waters occurring in volcanic regions, which are often very rich in dissolved material; as examples of these we may mention the geysers and hot springs of Iceland, the Yellowstone Park and New Zealand, which deposit vast amounts of silica, and the hot springs of Italy, which form travertine, a mass of carbonate of lime; also the boron springs of Tuscany. A further point of much importance is that many hot springs are highly radio-active. It is probable that this property can be used as a criterion for the magmatic origin of waters, though it must be admitted that at present the proof of this proposition is rather a case of arguing in a circle.

On such grounds as these it was long ago suggested by Suess[2] that the hot and highly-saline springs of Karlsbad in Bohemia

---

[1] From some minerals the water of constitution can only be driven off at red heat.

[2] Suess, *Geographical Journal*, 1902, p. 517.

were actually supplied from magmatic sources, and for such waters he suggested the term *juvenile* (juveniles Wasser). The underlying idea here is that this is *new* water which has never before been at or near the surface, and may have been quite recently formed by direct combination of hydrogen and oxygen on fall of temperature and pressure.

At any rate it is clear that water of magmatic origin plays a great part in the formation of pegmatites and similar coarse-grained segregations from igneous rocks: these are often specially rich in those minerals that contain much water of constitution, and also rich in quartz. Since in many cases the temperature of formation of such masses must be above the critical temperature for water (375° C.), at which point the distinction between a liquid and a gas disappears, it is quite immaterial whether we speak of it as water or steam. The high-temperature deposits of which we are now speaking are as a rule highly siliceous, and much more information is desirable as to the behaviour of water and silica under such conditions: it is possible that most of the material really exists in the form of a sort of jelly of hydrated silica. Arrhenius has also shown that at very high temperatures and pressures water becomes a much more powerful chemical agent and may have the power of combining with bases to form hydrates, so that a magma at great depths may be regarded as consisting of hydrates and silica instead of silicates and water[1]. This however is rather a speculative proposition and it is better to regard the magma as a homogeneous system in which water is present along with all the other constituents without regard to their actual manner of combination or association.

With regard to the general direction of movement of magmatic water, this must be in the main upwards, towards the surface. The most important factor here is undoubtedly vapour pressure, or in other words the expansive power of steam. This often produces explosive effects, as in volcanic eruptions and geysers, and some hot springs are flowing upwards with considerable force. The highly heated steam below is constantly exerting

---

[1] Arrhenius, *Geol. Fören. Stockholm Förh.*, vol. XXII, 1900, p. 395. See also Morrow Campbell, *Bull. Inst. Min. Met.* No. 193, October, 1920, p. 11.

pressure and endeavouring to escape and when it reaches a certain temperature it is condensed to water, which continues to move upwards under pressure of the steam below. Moreover the fissures available for its passage commonly run upwards, increasing in size and number near the surface. Owing to the weight of overlying rocks, fissures with any approach to horizontality must be rare or non-existent at great depths, while the stresses tending to close steeply-inclined or vertical fissures must be much less powerful.

Hence the tendency of magmatic water in the general scheme of circulation must be upwards.

As the second great class we have surface waters, mainly derived from rainfall or snowfall as the case may be. It has been estimated that in a normal temperate climate, such as that of the British Isles, of the total precipitation one third is evaporated back into the air, one third runs off as streams, and one third sinks into the ground, much of it again issuing as springs after penetrating to a shallow depth only. It is with this last fraction of the meteoric waters that we are here concerned. It will be considered in detail in a subsequent section.

Of other possible sources of water concerned in ore-formation one only need be briefly mentioned. There is always a possibility that water might soak downwards through the floor of the sea and it has been suggested that this may be at any rate a contributary cause of volcanic eruptions[1]. In support of this idea attention has been drawn to the occurrence of chlorides as volcanic products. However among ore-deposits chlorides are not common and it is unlikely that sea-water has much or anything to do with ore-formation[2]. Besides, many ore-deposits exist in regions which were far above sea-level at the time of ore-deposition and have never been submerged since. These show no special distinguishing features.

In some of the older sedimentary formations there are con-

[1] Arrhenius, *loc. cit.*

[2] The occurrence of chlorides, bromides and iodides in the oxidation zone in some arid regions is discussed elsewhere.

siderable masses of water, deeply buried and apparently in a
stagnant condition. This water is believed to have existed
in them from the time of their formation and to have been
sealed up and preserved by deposition of later sediments.
It has been suggested that some at any rate of the artesian
water of Australia is of this character. Such water however is
necessarily immobile and is unlikely to play any part in ore-
formation.

Ground Water. Under this general term we include all the

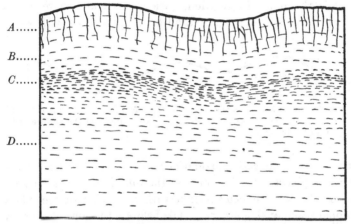

Fig. 5. Distribution of underground water. *A*, zone of active circulation.
B, zone of oscillating water level. *C*, top of zone of permanent satura-
tion, passing down continuously into the deep zone. *D*, of diminishing
ground-water.

water of the superficial layers of the earth, which is not obviously
of direct deep-seated origin[1]. It is essentially that fraction of
the rainfall that sinks into the ground, as above mentioned. It
is of course obvious that the proportion of the rainfall that
sinks in must be very variable, according to the general character
of the climate, the permeability of the rocks and the slope of
the surface. In some cases, such as the horizontal well-fissured
limestone plateaux of the north of England, almost the whole

---

[1] As a general name for this type Pošepný has proposed the name of
vadose water.

of the rainfall descends into the depths till it encounters an impervious stratum: in many light sandy areas also the proportion must be very large. With a heavy clay subsoil on the other hand very little actually sinks, as downward circulation is necessarily very slow in fine-grained material. Under average circumstances, however, as is well known to every miner, the rocks commonly contain much water down to considerable depths. This water-bearing region can usually be divided into two distinct zones. In the higher zone the rocks are incompletely saturated, the water is in active circulation and air is present. This is the zone of oxidation. Below this comes a region in which the rocks are saturated, all fissures are filled with water

Fig. 6.   Diagram showing general relation of water-table (broken line) to land-surface. In the deeper valley the water-table intersects the surface and a river or lake is formed.

and there is no active oxidation: indeed the conditions are commonly those of reduction. At the top of this stagnant zone comes the zone of cementation with formation of secondary sulphides and analogous ores, while below is the zone of primary ores extending down to an indefinite depth. In deep mines in rocks without large fissures it is often found that the amount of water falls off in depth and the lower levels are completely dry. Hence it appears that there is no definite downward limit to the ground-water zone. Its upper limit must necessarily be that of the lowest point of natural drainage of the district: in fact this zone is exactly analogous to a mine where pumping has stopped and the workings have filled with water up to adit level.

The upper surface of this zone, called *ground water level* or the *water table*, is not necessarily a horizontal surface, but usually follows roughly the contours of the ground, though in a less accentuated form. This irregularity is due to the friction that opposes the free passage of water through rocks, and the extent of the banking up thus produced depends on the porosity and degree of fissuring of the rocks. Its level will also vary from time to time according to the amount of rainfall. The intersection of the water table with the ground surface is the determining factor in the issue of springs and often fixes also "adit level" of a mine.

The Relation of Water to different Types of Rock. Both the circulation of water in the uppermost zone and the amount of water contained in the rocks of the lower zone as above defined, are naturally controlled to a large extent by the character of the rocks with regard to porosity and permeability. From this point of view rocks vary very widely indeed[1]. All natural rocks under normal conditions contain more or less water, but the physical condition may be such that the water is unable to circulate, remaining in a stagnant state. Again some rocks which are not themselves pervious to water, may be shattered and fissured so that they are full of open channels. As a third group we have rocks which are porous, the water being able to circulate through the interstitial spaces between the grains of which they are composed. These characters depend on the origin of the rock and the processes of change it has gone through.

The following types are recognized by H. B. Woodward[2]:

(1) Water-bearing rocks, or those which yield supplies of water:
    (a) Porous and permeable rocks.
    (b) Impervious but fissured rocks.

(2) Non-water-bearing rocks:
    (a) Absorbent and partially pervious.
    (b) Absorbent but impervious.

This classification is of course essentially that of a water-engineer, since the main distinction is founded on the possibility of obtaining supplies of water, by pumping or well-sinking, but it is also applicable to the present subject.

---

[1] In this discussion the term rock is used in the widest possible sense, to include various deposits which may not be actually consolidated, or only partially so.

[2] Woodward, H. B., *The Geology of Water Supply*, London, 1910, p. 17.

Woodward assigns the following rock-types to the different classes:

(1) *a*—porous and permeable:
Sand, soft sandstone, gravel, loose breccia, sandy limestone, chalk, oolite, marlstone, dolomite-rock, brown ironstone, decomposed parts of granite and greenstone; and veins in metamorphic rocks.

(1) *b*—holding water in fissures:
Quartzite, grit, conglomerate, hard limestone and marble, slate, granite, greenstone, gneiss, crystalline schists (the last four when shattered).

(2) *a*—partially pervious:
Very fine sand, brickearth.

(2) *a*—impervious:
Clay, shale, marl, flint, chert, slate, schists, granite, greenstone (the last four when not shattered), hard limestone and marble when not jointed, iron-pan.

The above classification can only be approximate, since the degree of jointing or shattering varies in different cases. Also the amount of cement, in sandstones and conglomerates for example, is very variable. In general terms, the finer the grain of the rock and the harder its texture, the less water it will hold, and the less free is the circulation, apart from fissures. In the above list lavas are not mentioned: they behave in much the same way as granite and greenstone, the water circulation depending mainly on the degree of decomposition and of fissuring. Frequently the intrusion of sills and dykes into permeable rocks opposes an obstacle to free circulation.

Since rocks vary so widely it is clearly impossible to lay down any general laws of universal applicability to the water circulation in the earth's crust, except the obvious statement that water can pass through pervious rocks and is stopped by impervious ones. Nevertheless it has to be recognized that there is a very general movement of water throughout the upper part of the crust in most parts of the world. This water always contains more or less dissolved material and often a good deal. The dissolved material may be abstracted from one place and deposited in another, as can be seen in the case of the travertine or calcareous sinter deposits found at the surface in many limestone districts. As the process of mineral deposition can in many cases be watched at the surface, it was a natural inference that similar processes going on underground might lead to the formation of vein deposits carrying ores, and many instances have been described where drainage waters did deposit metallic compounds in mines. These cases however have not much bearing on the origin of primary ores. They are mostly oxidized compounds, and nobody disputes the formation of secondary minerals of this kind in the natural zone of oxidation. It is the influence of surface waters of meteoric origin on the formation of the unoxidized ores of the depths that is the matter of dispute.

To sum up this part of the subject we may say that while it is clear that an active water-circulation goes on in the permeable rocks comprised in the superficial zones of the crust, the evidence available suggests that there is very little transfer of water and dissolved material going on in the deeper water-logged layers, where the absence of free oxygen is also a hindrance to chemical changes.

Special conditions in Arid Regions. In certain parts of the world, owing to abnormal meteorological conditions, the rainfall is very scanty. A notable example of this is found in parts of the Andes in South America, in Peru, Bolivia and northern Chile, as well as in much of the Great Basin region of the western United States, both of which happen to be important mining districts. Western Australia also is extremely dry. As a direct consequence of the dryness two special conditions arise: the zone of oxidation is unusually deep, and what water there is, is usually very rich in dissolved salts. These salts often play a conspicuous part in the formation of new oxidized minerals. For example this state of affairs, especially the saltness of the water, accounts for the abundance of horn silver (cerargyrite, $AgCl$) and the corresponding bromine and iodine minerals in the silver deposits of Bolivia. The silver salts set free by oxidation of primary sulphides are at once precipitated as silver chloride, etc. Similarly a basic copper chloride, atacamite, is also found in considerable quantity in northern Chile, whereas in regions of more normal rainfall the commonest oxidation products of copper ores are the basic carbonates, malachite and azurite. In some regions, however, especially in the western United States, the waters are alkaline rather than salt, carrying dissolved sodium and potassium carbonates instead of chlorides and sulphates. This naturally favours the formation of copper carbonates and has a bearing on the great abundance of malachite and azurite in some of the copper deposits of the south-western States.

The Reascensionist Theory of Water Circulation. Many geologists have felt a difficulty in subscribing to the idea of the magmatic origin of hot and highly-saline springs, and especially to the theory that magmatic waters are the source of many ore-deposits. In all the suggestions put forward to explain the formation of primary ores by meteoric waters there has always been a difficulty in accounting for the presence of metallic compounds and such substances as fluorine in sufficient quantity in the superficial rocks. To get over this, it has been suggested that meteoric waters may descend to great depths in the crust, thus becoming highly heated and dissolving material from the hot regions below. This water then ascends and deposits its mineral content as veins. This theory has something to recommend it, but there are certain obstacles to be overcome before it can be accepted as a sufficient explanation. In the first place it is not very clear why this circulation should occur at all. It is not easy to see any reason why the water should both descend and ascend, presumably in much the same region and as a continuous process. The force, whatever it may be, that causes its ascent would surely operate to prevent its descent. Again there is the important objection that the lower levels of so many deep mines are dry. It is of course possible that water may descend to considerable depths down fault-fissures, for example, but in all probability these do not remain open indefinitely in depth, owing to pressure and rock-flow. Even if they did, the water would soon be forced up again by steam-pressure. It seems very improbable that surface water is able to descend to sufficient depth to become very highly heated, except under exceptional circumstances. In regions of recent volcanic activity, where heated rock lies near the surface it doubtless does so, but the idea is difficult to apply to regions where there has been little or no evidence of exceptional temperatures due to this cause. Arrhenius, in the paper already quoted, regards the floor of the sea as a kind of semi-permeable membrane and considers that water penetrates downwards to regions of highly-heated rock by means of osmotic pressure. It seems difficult to suppose that this downward osmotic pressure, if it exists, could overcome the upward vapour pressure of the steam generated from the first portions of water to arrive in a highly-heated region, and the process would probably soon cease to operate, even if once started.

Deposition of Ores from Solution. Whatever view may be taken as to the actual source of the water concerned, there can be no doubt that a vast number of ore-deposits of the vein type are actually formed from aqueous solution. Furthermore in the case of replacement and metasomatic ore-bodies it is certain that the processes of chemical exchange are mostly affected by the medium of water. These facts are sufficiently apparent from the detailed structure of the deposits themselves, and the common banded, crustified and nodular appearance of both ore and gangue. Now most ore-minerals as well as gangue-minerals, except perhaps the carbonates, are distinctly insoluble substances, and the question arises how they were deposited from solution, or rather, in what form they actually existed in the ore-bearing solutions. In the case of deposits formed by magmatic emanations the difficulty is less acute, since these contain considerable quantities of various substances possessing powerful chemical affinities, capable of forming compounds with even the most refractory elements: many of these compounds, as for example those of fluorine, boron and chlorine, are markedly volatile and the magmatic extracts are doubtless highly concentrated and highly acid solutions, or in certain cases possibly alkaline: at any rate it is the concentration that is the important factor, together with the high temperatures and pressures that prevail. But it is when we come to consider deposit of very insoluble substances from meteoric waters at small or comparatively small depths that difficulties arise. Such is the case for example with the formation of lead and zinc ores in limestones, if this explanation is accepted, as it seems to be, very widely for, among others, the ores of Missouri. A very striking case is the common occurrence of barytes as a gangue mineral. It is well known that barium sulphate is one of the most insoluble compounds in existence, and it cannot have been present in quantity in water as such. However, other salts of barium, *e.g.* the chloride, are freely soluble and in this case there is a simple explanation in the meeting of two solutions, one containing a soluble barium salt and the other a sulphate: from these barium sulphate is precipitated on mixing, but the deposition must be very slow or the precipitate would not form good crystals. We are therefore driven to the conclusion that the solutions must be very dilute. Further, it is not difficult to conceive of conditions such that dissolved zinc sulphate is reduced to sulphide and precipitated as blende, but such a process must have gone on at some depth in order to ensure absence of free oxygen.

From the above considerations it is pretty clear that the conditions prevailing during ore-formation of almost any kind must be somewhat peculiar, and in order to explain the occurrence of large bodies of insoluble minerals we have to suppose a rather elaborate system of water circulation, with the meeting of different solutions carrying soluble compounds which are precipitated on mixing. This requires a rather nice adjustment of natural conditions. But after all, ore-deposits are local in their distribution and there are large areas without any, which is what might be expected on these grounds.

It is not easy to discuss the part played by water in the formation of primary deposits without at the same time discussing the whole subject of ore-formation, which is dealt with elsewhere in detail, so further treatment of the subject here would involve unnecessary repetition.

Secondary changes in Ore-bodies due to Water. There can be no possible doubt of the important part played by water in inducing changes in ore-minerals in the superficial layers of the crust. This is of the very essence of the processes that go on in the zones of oxidation and cementation. In the uppermost layers the water is in free circulation and contains abundance of dissolved oxygen, leading to a whole series of changes of both chemical and physical nature which can be conveniently summed up in the expressions solution, oxidation and hydration. The nature of these changes is discussed in general terms in the chapter on secondary changes in ores, and in sufficient detail in numerous instances in the descriptive sections of this book. It must suffice here to say that in these phenomena water of meteoric origin is of paramount importance, and in many cases the processes that take place are assisted by substances dissolved in the water and derived from the surrounding rocks. The nature of these varies greatly from place to place and is determined by local conditions. The most strongly marked chemical characteristic of most of the newly-formed compounds is the presence of oxygen, often with water of constitution or water of crystallization. Chemical changes are often very largely assisted by carbon dioxide dissolved in the water, and carbonates are extremely common as secondary minerals.

In general terms it may be said that the water of the zone of free circulation is acid and oxidizing, while the stagnant water of the lower levels is alkaline and reducing: thus there is a marked contrast between the chemical processes at work in the two regions. The acid and oxidizing conditions of the upper zone facilitate decomposition and solution of primary ores, which are in part reprecipitated at once as insoluble compounds, and in part carried down by descending water to the reducing zone, where the metallic compounds are again largely precipitated as compounds free from oxygen, either as native metals or as secondary sulphides, often of complex character. In the stagnant waters of the lower depths the simplest compounds seem to be most stable, the primary ores having as a rule the simplest constitution.

CHAPTER V

THE FORMS OF ORE-DEPOSITS

General considerations. To the mining engineer the form of an ore-deposit is one of the most important factors to be taken into account, if not the most important of all. The determination of the metal-content of a deposit and the chemical condition in which this metal-content occurs is the province of the assayer. The determination of its form and extent belongs to the geologist. From the evidence afforded by the investigations of both of them, taken together, the engineer is enabled to lay out his plans for winning the ore and for preparing it for market. Hence it is not surprising that the earliest classifications of ore-deposits were in the main founded on form; that is to say, they were morphological rather than chemical or genetic.

The diversity in the forms of ore-deposits is infinite; it is impossible to draw up a simple scheme which will comprise all variations. Nevertheless it is possible to recognize certain typical and dominant forms, to one or other of which most actual examples may be referred. The basis of such a classification is purely geological, being founded on the shape and position of the ore-body. By these terms we are to understand its relation to the vertical and horizontal planes, its extension in definite directions and its dependence on other known or determinable geological structures.

One of the earliest attempts at a classification by form is that of von Cotta[1]. This author divided the types known to him in the following simple manner:

> I. Regular deposits.
>   A. Beds.
>   B. Veins.
> II. Irregular deposits.
>   C. Stocks.
>   D. Disseminations.

[1] *Die Lehre von den Lagerstätten*, Freiberg, 1859.

The underlying ideas here are that the regular deposits have a tabular or sheet-like form, beds being originally horizontal, veins originally more or less vertical or at any rate steeply inclined. Beds may of course eventually come to lie at any angle with the horizontal and may be folded in various ways. Of the irregular masses, stocks are considered to be those with definite boundaries, while disseminations shade off more gradually into the surrounding rocks.

Broadly speaking, group A is intended to include all those ore-deposits which are interbedded with other stratified sedimentary rocks, or rest on the earth's surface in more or less horizontal sheets. It thus includes the placers and alluvial deposits, as well as sedimentary iron, manganese and other ores, besides a great variety of the so-called non-metallic deposits. The true bedded ore-deposits of this scheme are always sediments in the geological sense, and are newer than the rocks below them and older than those above, if any such exist.

The second class of regular deposits, the veins, on the other hand, are always newer than the rocks that enclose them on either side: they are in fact similar to the dykes and sills of the geologist: that is, they are intrusive in character; some of them are actually of direct igneous origin, while others are formed by deposition from water, but in any case they are fissure-fillings. The term has also been extended to include the fillings of cavities which are not strictly sheet-like in form, and even cylindrical pipes are commonly assigned to this class, as well as various irregular masses occurring at the intersections of joints and along the contacts between different rock-masses, the so-called contact-veins.

The third group, or stocks, is the least satisfactory of von Cotta's categories, since it is almost impossible to draw up a reasonable definition, owing to the great diversity of form possible in this class. On the one hand they grade into the pipes and contact deposits, while many of them are in actual fact metalliferous igneous intrusions of varying and most irregular form.

The term disseminations almost explains itself: it refers to masses of rock, either igneous or sedimentary, in which the mineral is scattered in sufficient quantity to be payable, the workable portion not possessing any definite outline. The ore may be of either primary or secondary origin: frequently it is an original constituent of an igneous rock.

Veins. Perhaps the most important and most widely distributed of all ore-bodies are those comprised under the general heading of *veins* or *lodes*. In its simplest form a vein is a tabular or sheet-like mass of minerals, usually thin in comparison with

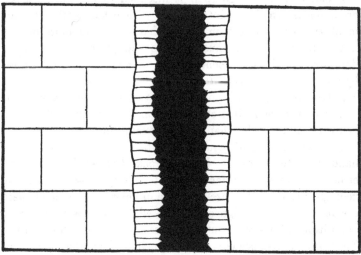

Fig. 7. A mineral vein of the simplest possible type: a vertical fissure in the country rock, lined with crystalline quartz and the centre occupied by ore.

its lateral extension, and differing in character from the rocks in which it is enclosed. The essential feature of veins is that they are the fillings or partial fillings of cracks or fissures in the earth's crust. They may lie in any position but most veins are either steeply inclined or actually vertical. An inclination of less than 45° with the horizontal is comparatively uncommon, although important instances of *flat veins* are known, especially in regions of highly disturbed rocks, such as the Great Flat Lode of Cornwall. Another important type is afforded by the *flats* of the Derbyshire lead district, which are simply fillings of horizontal, originally open joints formed by solution of the limestone.

It is of course also possible that an originally steep or vertical vein may be tilted by earth-movements after its formation, or an originally flat vein may thus become steeply inclined.

The direction and lie of the veins of any particular district is naturally dependent on the geological structure of the enclosing rocks, since they usually form in the larger faults, joints and other discontinuities. Now as a result of earth-movements faults and the larger open joints usually tend to form two sets at right angles to each other, parallel respectively to the dip and strike in inclined strata. It is common for large vertical veins to send

Fig. 8. Mineral vein sending off branches (flats) along bedding planes of sedimentary rocks.

off small offshoots along specially weak bedding planes or cleavage planes, as shown in Fig. 8. The rectangular arrangement in plan of two vein-systems is often well displayed in geological maps: perhaps the most celebrated instance is that of Devon and Cornwall, which is figured in every text-book. In other instances veins develop in parallel series in one direction only: Fig. 9 shows in plan the outcrops of a series of veins cutting across the general strike of the country rock, as occurs in central Wales.

Many of the most important veins known are formed in fault-fissures: these can often be identified by the occurrence among the vein-minerals of fragments of the country rock: the vein-minerals in fact often act as a cement for this fault-breccia.

Closely connected with these are the so-called brecciated lodes, where there is in point of fact no definite fault-fissure, no vein in the strict sense of the term, but the ore-body consists of a sheet of shattered rock, a crush breccia without definite boundaries, cemented and possibly more or less replaced by various ores and gangue-minerals. Such are some of the great tin-lodes of Cornwall. Occasionally it is possible to distinguish a narrow crack in the middle of the lode, the actual fault-fissure, called the *leader*, from which the mineralizing solutions spread out

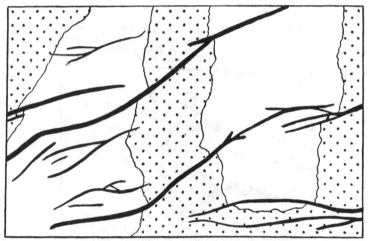

Fig. 9. Plan of vein-system of a district, with more or less parallel arrangement of veins, cutting across the general strike.

into the broken country rock around. Such lodes have arisen along what are called shatter belts, zones in which the rock has been shattered by a grinding movement, without much lateral displacement such as is found in a true fault. In cases where a large number of such brecciated lodes occur in close proximity the intervening rock may be so highly mineralized that the whole becomes payable and can be worked as a single ore-body. Such a mass of parallel or interlacing lodes is generally known as a *stockwork*, while the *carbonas* of the St Ives district of Cornwall are essentially similar. These might however almost equally well be classed with the disseminations.

It is evident that in a fault-vein with displacement the country

rock on the two sides of the vein may be different and this often
has an important bearing on the winning of the ore, since one
side may be better suited than the other for mining operations.
Or again one side may be more permeable by the solutions and
more easily mineralized, or more likely to be penetrated by
offshoots of the vein. When the displacement is such that the
two sides are in similar rock, the existence of faulting can often
be proved by striations and polished surfaces (slickensides) on
the walls of the fissure, due to grinding during movement and
these are often accompanied by layers of soft rotten clay known
as *flucan* or *gouge*. When there are two or more sets of fault-

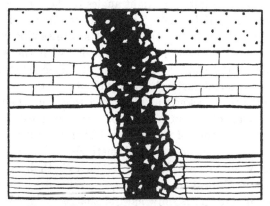

Fig. 10. Brecciated lode in horizontal strata.

veins crossing at an angle the earlier set must obviously be shifted
by the later: in such a case the determination of the direction
and amount of the displacement is of the utmost importance,
since this determines the direction in which the faulted continuation of the older vein must be searched for by driving or sinking.

The Lie of Veins. In order to estimate the tonnage of ore in
a vein it is necessary to know its volume and its position in
space; that is to say its horizontal and vertical extent, its thickness perpendicular to the two principal bounding surfaces, and
the relation of these quantities to some fixed directions. Just
as in the case of stratified rocks this is most easily effected by
means of the conceptions of *dip* and *strike*. Considering the vein
for a moment as a plane, the dip is the steepest line that can be

drawn in this plane; the strike is a horizontal line in the plane at right angles to the dip. The dip of rocks is measured in degrees from the horizontal: the same convention can be employed in the case of veins, and is really the simplest. Many miners however employ the term *hade*, which is the complement of the dip, or the number of degrees departure from the vertical. In this book the term dip will be employed. When a vein is inclined the rock-surface above is called the hanging wall, the one below the footwall. The *direction* of dip and of strike are always expressed by means of the points of the compass. Thus if we say that a vein dips N.W. at 70° its whole position is fixed, since the strike is by definition at right angles to the direction of dip or, in this case, N.E.–S.W. This however refers only to the underground extension of the vein. Its intersection with the ground-surface, or outcrop, only coincides with the strike on level ground: on a slope it will vary in a manner determined by several independent factors[1]. It is evident also that in order to calculate the true thickness of an inclined vein from the apparent thickness encountered in a shaft or level we must know the direction and amount of the true dip. Then the calculation can be made by simple geometrical methods. This is especially important where the workings are oblique to both dip and strike.

Form and Extent of Veins. Since veins are as a rule the fillings of fissures, it is only natural that they should possess certain irregularities of form. Commonly the thickness is not uniform throughout, though often fairly constant for long distances. More usually however there are expansions and contractions of quite irregular character. When a vein cuts through strata of different kinds it is often wider in some beds than in others: for instance a vein that is wide in limestone or sandstone often contracts greatly or even disappears on entering a shale or slate or an igneous dyke or sill: in these types of rock it may be represented by a narrow crack only, again expanding on the other side. This depends on the character of the country rock; some rocks, when fissured, can maintain open cracks

---

[1] For a full discussion of this subject see Elles, *The Study of Geological Maps*, Cambridge, 1921; or Harker, *Notes on Geological Map Reading*, Cambridge, 1921.

almost indefinitely, or even allow them to be widened by solution, while other rocks flow under pressure and the crack soon closes up. The effect on the later deposition of vein-fillings is obvious. Cracks in rocks formed by faults are rarely quite straight and uniform, but usually more or less irregular on a small scale, and when there is displacement it is clear that the projections of one side may be brought against other projections on the other side and similarly with concave curves, as shown

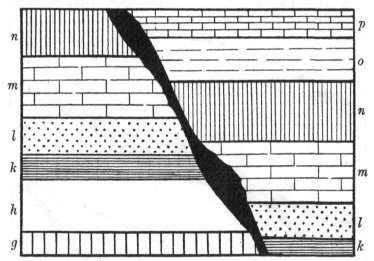

Fig. 11. Mineral vein occupying a fault-plane, and showing variations in thickness consequent on the unevenness of the fissure.

in Fig. 11, thus leading to a succession of ore-shoots and unproductive portions. It must also be remembered that the displacement of a fault may be either vertical, horizontal or oblique, so that these irregular ore-masses may lie in any direction and have any form. Hence fault-veins of this kind are often very variable in productiveness.

All veins must eventually be limited in horizontal extent, and in fact very many of them are distinctly lenticular in form, thick in the middle and tailing off to nothing at the ends. It is also often found in a mining field that lenticular veins are arranged *en échelon* in more or less parallel series, so that where one runs out another may begin a little way off. Precisely

similar considerations also extend to the persistence of veins
and lenticles in depth, but of course such veins are much more
difficult to find underground, and their discovery often entails
costly exploratory work of a rather speculative character. With
regard to the actual limits of downward extension of veins very
little reliable information exists. It is found as a result of much
experience that most veins fall off in productiveness at great
depths. The richly metalliferous portions are generally limited
to a range of a few thousand feet. The Morro Velho gold vein

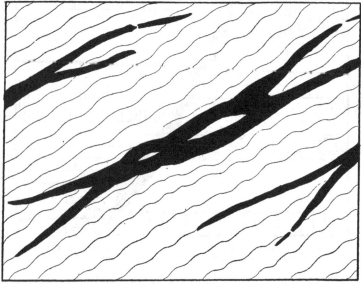

Fig. 12. Ideal plan of lenticular ore-bodies arranged *en échelon*.

in Brazil is still fully payable at a depth of 6700 feet vertical
from the surface, but most veins cease to be workable at much
less depth. This fact is correlated with the existence of mineralized
zones, as explained elsewhere (see Chap. X).

Besides the intersections of one vein by another, as previously
mentioned, veins are often found to branch; sometimes they split
into two or more portions for some distance and again coalesce.
When the portion of country rock between the two branches is comparatively large it is called a *horse*, and there is every grade between
these and the small rock-fragments enclosed in brecciated veins.

Where two sets of cracks or open joints cross there is often a concentration of ore, and if one set is later than the other, it may be found that the earlier veins have been enriched for some distance around by the solutions that filled the later ones. When

Fig. 13. Pipes in a granite boss, with step structure due to arrangement of major joints.

two sets of intersecting joints exist in soluble rocks, such as limestone, a vertical shaft of more or less circular form is often found at the intersections; when these shafts have remained empty they are called swallow-holes or pot-holes: when filled

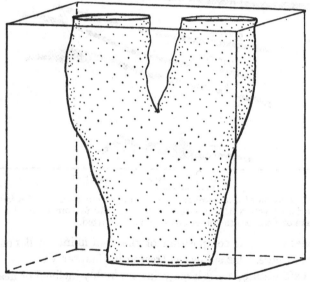

Fig. 14. Stereogram of a lenticular ore-body, splitting upwards.

with vein-minerals they form pipes; if the joints are inclined, so may the pipes be. Structures of somewhat similar appearance are also found in igneous rocks; here they are due to the replacing and mineralizing effect of vapours or solutions escaping from the cooling magma. These pipes are naturally often very irregular in form and change their inclination rapidly when

followed downwards, since they are formed along cracks and
fissures running in any direction in the igneous rock, due to
contraction on cooling. Pipes of this kind were never open
cylinders, as some of the others were, but mainly consist of
alteration products of the igneous rock itself, the chemical
changes working outwards from very small fissures. Since we
are dealing in this chapter only with the forms of ore-bodies,

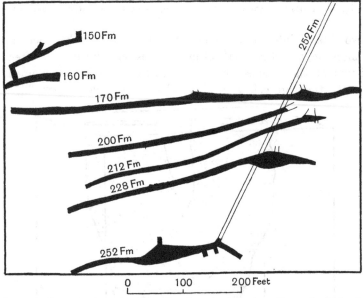

Fig. 15. Portion of a mine-plan, showing stoped-out portions of a lode at
various levels, with cross-cut to shaft on 252 fathom level. For the
sake of clearness the other cross-cuts are omitted.

the character and composition of the vein-fillings will not be
discussed here. This subject is treated in Chapter VI.

As before stated most large veins are highly inclined or nearly
vertical: an inclination of about 70° is, for some obscure reason,
very common. But the inclinations of veins are not at all constant.
The dip of the same vein when followed down sometimes alter-
nates a few degrees on either side of the vertical. Again in any
given district the veins may vary in direction of dip. The great
tin lodes of the Camborne-Redruth district in Cornwall, for
example, may be divided into two sets, those with north and

south underlie respectively. A few less important lodes are vertical. Of these, the lodes dipping N. are as a rule the most productive. In many American publications interesting figures, the so-called stereograms, can be found, showing the forms of important ore-bodies. These stereograms are in fact drawings of models of the ore-bodies, and they show some very peculiar forms; very instructive models of mines can also be constructed by drawing plans or sections of levels and stopes in coloured

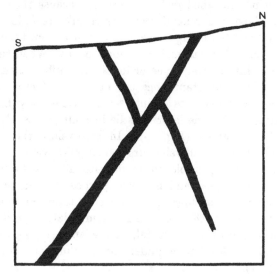

Fig. 16. Vertical section of a north-dipping lode heaved by a later lode dipping south.

ink or paint on sheets of glass, which are then fitted into frames and viewed as transparencies. The usual mine plan of a vein, showing shafts, levels and stopes in the plane of the vein does not convey much information of a useful kind except to the expert. The graphic representation by a model, or even by a drawing of a model is very much more effective.

Nomenclature of Veins. In different parts of the world a large number of terms are used to describe the varieties of the vein-like deposits: many of these are very vague and ill-defined. The terms vein and lode are often used almost interchangeably, but it would be better to confine the term vein to those structures with definite boundaries and a sharp distinction between vein-stuff and country rock. The term lode should be used for those ore-bodies consisting mainly of mineralized or brecciated material, where the workable ore is actually altered country rock, without definite boundaries and shading off gradually into unpayable stuff. In such lodes the workable width at any given time generally depends on conditions of cost of production and market price. In many gold-mining districts the gold-bearing quartz veins are called reefs, apparently because they tend to stick up above the general rock-surface, like reefs in the sea, and they are often conspicuous at a distance from their white-ness. In some districts such are called ledges. When there are two sets of intersecting veins or lodes with different strike the later and less important ones are often called cross-courses. Sometimes also two sets of intersecting veins may have the same strike, and if the later ones lie in fault-planes, they may heave the earlier set (Fig. 16). In Derbyshire the more or less vertical principal veins are called rake veins or rakes. It is however impossible to enumerate all the local terms that have been used and no good purpose would be served by so doing. A considerable number of terms are also em-ployed to express the distribution and relative positions of valuable and worthless mineral, or ore and gangue, in veins. These will be dealt with under the heading of vein-fillings (Chapter VI).

Saddle–reefs. A very special and important type of ore-deposit is that known as a saddle-reef. These are found only in folded regions, where stratified rocks have been bent into sharp anticlines and synclines. Such folding leaves weak regions, or even empty spaces along the crowns of the arches and to a less extent in the troughs of the synclines. Such places are eventually filled up by masses of quartz or other mineral, often ore-bearing. A cross-section will appear as in the figure, and in plan the reefs form narrow masses of indefinite length, lying parallel to the strike of the folds. The best example is afforded by the gold-bearing saddle-reefs of Bendigo, Victoria, where many of these structures are found lying one above the other to a depth of some thousands of feet. Saddle reefs are in all respects analogous to the type of igneous intrusion called by Harker phacoliths[1].

Stockworks. This term, originally of German origin (*Stockwerk*), is very largely used to describe a type of ore-deposit of rather indefinite character. A stockwork in the strict sense may be best described as a mass of rock intersected by large numbers of small veins or stringers, sometimes more or less parallel, sometimes crossing and interlacing in all directions. It will be obvious that there is no real sharp distinction between this type of deposit and disseminations as before defined. In the true

Fig. 17. Ideal section of the Saddle Reefs of Bendigo, Australia, in highly folded Ordovician strata.

stockworks however the ore is later than the rest of the rock, whereas in most disseminations it is contemporaneous. The classical example of a stockwork is found in the tin-deposit of Altenberg in the Zinnwald, Erzgebirge, where a mass of granite is penetrated by innumerable tin-bearing veins and converted into greisen, and similar structures occur in the tin-wolfram districts of Devon and Cornwall. The Hemerdon stockwork, near Plymouth, contains wolfram only, with little or no tin. In many instances the whole of the altered country rock between the veins is so highly mineralized that the whole mass forms payable ore. The Altenberg mass was quarried as a large open-work, and the same system has been employed at Hemerdon.

[1] Harker, *The Natural History of Igneous Rocks*, London, 1909, p. 77 and fig. 14.

Irregular and Massive Ore-bodies. Besides the more or less definite types just described there are very many ore-bodies which possess no defined shape, their form depending mainly on the geological structure of the rocks enclosing them. These include ore-bodies of different origins, such as the magmatic segregations, contact-deposits, pegmatitic and other intrusions and many replacement deposits. Magmatic segregations are as a rule entirely shapeless: if the rock from which they have crystallized out possesses flow-banding the ore-masses also are often more or less elongated and streaky in character, sometimes much contorted. Such are many of the iron-ore segregations of Norway and Sweden. The sulphide deposits of Sudbury form a highly irregular layer on the footwall side of a great sill or sheet of igneous rock, which has been on the large scale folded into a synclinal basin, but in detail the ore-layer is highly irregular, since it sends out many projections into the rocks enclosing it. Very similar in general character are many of the so-called contact or replacement deposits; sometimes these follow more or less closely the boundary between the intrusion and the older rock: in other instances where they are actually due to chemical replacement of one particular rock, such as limestone, their shape is determined by the area over which the intrusion happened to come in contact with the limestone, or near it. The larger masses of pyrite and cupriferous pyrite often take the form of great lenticles arranged *en échelon* parallel to the bedding or foliation of the country rock.

Again some ore-bodies are pegmatites or other igneous intrusions, and these do not necessarily take on the regular form of dykes or sheets, but may have any shape whatever, depending on the viscosity of the material and the position of lines of weakness in the rocks into which it injected. Many of the great ore-bodies of the Cordilleran region in South America appear to be masses essentially of volcanic origin, but intruded at a late stage in various forms into the lavas and ashes of the eruptive phase. These are sometimes sheet-like in form, but often quite irregular.

It will be easily apparent that every individual case has to be treated on its own merits and that no regular scheme of nomenclature can be drawn up.

Disseminations. Many of the ore-bodies assigned to this class are masses of igneous rock containing enough of some ore-mineral to be workable at a profit. In such cases the ore-mineral is as a rule an original constituent of the rock, formed by crystallization direct from the magma. Such are, for example, certain tin-bearing granites and the chromiferous serpentines. In some instances the ore-minerals, though present, are not sufficiently abundant in the mass of the rock to be workable unless they are concentrated by some natural or artificial means. Rocks containing very small proportions of gold, platinum, tinstone or monazite may give rise to highly important alluvial or other superficial deposits, by denudation and deposition: rocks containing magnetite can be concentrated by magnetic separators to a high grade iron ore, and so on. In point of fact, nearly all ore-deposits, whether disseminations, veins, or any-thing else, require some form of artificial concentration: it is only a matter of degree. But many natural disseminations could not be profitably concentrated by artificial means: it is only when they have been worked over by the slow but sure opera-tions of nature that they become sufficiently rich for man to try his hand on them with artificial processes. There is thus a double concentration.

Bedded Ore-deposits. Under this category come all those workable ores which are in fact sediments in the geological sense, that is stratified rocks and other unconsolidated bedded deposits. We can recognize here two main types: one of them includes all the alluvial deposits and other superficial accumulations of recent date which are worked for gold, platinum, tinstone, monazite and other valuable minerals. These are as a rule sands or gravels, formed by rivers or on the shores of lakes or the sea. We must include here also all those workable deposits of a residual character, such as the weathered material lying on the outcrop or *back* of a lode, or on the slopes immediately below, and for example certain tin-bearing granites in tropical climates which are so deeply weathered and decomposed that they can be worked by the methods suitable to alluvial propositions. These are of course not bedded, but they are most conveniently classed here. The second great subdivision includes the truly stratified rocks of ancient date, which can be worked bodily as ores. Of these the ironstones are by far the most important. These are often horizontal or very gently inclined and are mined by methods very similar to those employed in coal-mining. Some of the most important manganese deposits also belong here and such things as the copper-bearing shales of Mansfeld. Somewhat intermediate in character are the deposits of iron-ore, commonly haematite, formed by the replacement of irregular portions of stratified limestones. These may be of any shape, as also may the natural concentrations of iron-ore in the ancient sediments and schistose rocks of Lake Superior and elsewhere.

From the general statements in the foregoing sections it is clear that the forms of ore-bodies show endless variations: concrete examples of the types here briefly described and of many other related forms will be found in the descriptive chapters of this book. If it is permissible to generalize at all, it may perhaps be said that the commonest and most important forms are, on the one hand, the sheet-like or tabular, vertical or highly-inclined veins or lodes, which yield the greater part of our supply of the precious and base metals, except iron, and on the other hand the more or less horizontal bedded ironstones, which are of such enormous importance as sources of iron ore, being second only to coal in value in the mineral production of the world.

The Influence of the Country Rock on Ore-deposit. The actual formation of primary ore-deposits may be controlled in a variety of ways by the character of the country rock, both chemical and physical factors coming into play. In the first place an unstable or easily soluble rock-mass will be more readily attacked by the mineralizing solutions than stable and resistant rocks: a simple case is afforded by the actual replacement of limestones by ores, such as the formation of masses of malachite and azurite in the oxidation zone of copper lodes in limestones. It is hardly necessary to multiply instances of this kind of thing, since the effect is obvious and inevitable, but mention may also be made of some of the magnetite deposits of Sweden (*e.g.* Dannemora). But even among the highly siliceous and aluminous rocks there are variations in the precipitating powers of the different types, and it is well known that in many areas profitable ores are more or less strictly confined to a particular kind of country rock.

Of equal or perhaps even greater importance are the variations in the behaviour of different rocks towards stresses: some rocks readily undergo jointing and fracture, while others tend to flow under pressure, without formation of permanent open fissures. Rocks in which open channels can exist evidently lend themselves more readily to vein-formation than those in which the fractures are at once healed up by flow, and as a general statement, though not without exceptions, it may be said that hard and coarse-textured rocks are more likely to carry veins and lodes than those that are soft and of fine texture. Thus igneous rocks and coarse sediments such as quartzites and grits are usually favourable, while shales and clays are the worst. Limestones again are specially favourable, since joints and fissures once produced tend to become enlarged by solution.

An admirable example of the principles here set forth is afforded by the mining district of western Shropshire, south of Minsterley. Here there are famous veins worked for lead since the time of the Romans and carrying also much zinc and barytes. The workable deposits are entirely confined to one particular stratum, called the Mytton Grits, belonging to the Arenig division of the Ordovician. There are no ores in the soft Cambrian shales below, and when the veins come up against the overlying Hope Shales they suddenly pinch out, doubtless because open fissures could not exist in these soft beds. The workable ground is therefore confined to a definite horizon, about 2000 feet in thickness, and is apparently controlled entirely by the lithological character of the country rock, which lent itself particularly well to the formation of open fault-fissures easily penetrated by solutions (see also p. 289 and Fig. 52).

Again, hard rocks are most favourable to the formation of fault-breccias which can afterwards be cemented by ore and gangue minerals, or penetrated and replaced by actively mineralizing solutions, whereas fissures in softer rocks are often choked up and rendered impervious by the formation of a clayey mass, flucan or gouge, which may prevent mineralization and may even have important effects on the present underground circulation of water, as in some of the cross-courses in Cornwall. Some of these form effective barriers between the drainage areas of adjoining mines, and such structures must obviously also have a controlling influence on the horizontal circulation of mineralizing solutions.

To sum up this part of the subject then, the determining factors to be considered are the chemical and mineralogical composition of the rocks, and their textures; that is to say permeability to solutions and hardness in so far as this influences their behaviour towards stresses of whatever kind tending to produce open fissures. Solubility is also of importance.

Since earth pressures increase downwards it is evident that in depth every rock, however hard, must eventually reach the limiting value of pressure beyond which no opening can exist, but this is probably only attained at great depths. Adams has shown on experimental grounds, supported by the mathematical investigations of King, that open spaces may exist in granite to a depth of probably 11 miles, but experience in deep mines shows that nearly all known rocks are deformed at depths greater than about 5000 feet. Hence we conclude that openings of any size do not in practice exist much beyond this depth.

Ore-shoots. As a rule the distribution of ore in a vein or other deposit is not uniform, but nearly always the metallic minerals occur in patches of varying size and concentration. Such rich portions of a deposit are called *shoots,* or *chutes* as it is sometimes spelt. This term is generally restricted to masses of considerable size, while small locally rich patches are called *pockets* or *bunches.* In America a specially-rich portion of an ore-body is often called a *bonanza.* Ore-shoots may be divided into different categories with regard to character and to origin. In the first place it is necessary to distinguish between those of primary and of secondary origin. The latter group is discussed fully in the section on secondary enrichment of ore-deposits (Chapter VIII) and need not further concern us here. There is however an important morphological difference between the two groups, in that primary ore-shoots are commonly steeply inclined, with their maximum extension downwards, while secondary ore-shoots usually run more or less horizontally, or roughly parallel to the surface of the ground. The reason for the last-named fact is obvious from a consideration of their mode of origin as detailed in the chapter just mentioned. In this section attention will be confined to primary ore-shoots.

The second morphological characteristic of ore-shoots is of genetic significance: we have to distinguish between two fundamentally different conceptions. An ore-shoot may consist either

0　100　200　300　400　500 Feet

Fig. 18. Ore-shoots in Nevada City and Grass Valley Mines, California (after Lindgren, 17*th Ann. Rep. U.S. Geol. Survey*, Part II, plate xviii *b*.)

of an unusually wide or an unusually rich portion of a vein. In the first category it may be simply a local swelling or expansion in a vein of varying width, while on the other hand if the vein remains of constant width the shoot may consist of a region where the ratio of ore to gangue is unusually high, or in other words the metal content of the vein-filling shows local concentration and poverty. In many primary ore-deposits the shoots are the only workable portions.

It is of course to be understood that local enrichment and impoverishment are to be found in nearly all ore-deposits of whatever kind, and it is not·possible to enumerate all known varieties. One very simple case, though by no means the most common, is to be seen in magmatic segregations, where the ore-mineral has sunk to the bottom of the intrusion, or diffused to the margin, or, if subsequent disturbance has occurred, the ore may lie in streaks and bands within the intrusion, owing to flow or convection currents. Such structures are often seen in masses of magnetite, ilmenite, or sulphides of magmatic origin. Many such are described elsewhere.

Local expansions of veins may be caused in a variety of ways. One such has already been discussed, where an irregular fault fissure brings together badly fitting portions, leaving gaps to be filled by vein-stuff. Again, where a fissure cuts across rocks with open joints, expansions and flats are formed. Rich bunches of ore are frequently found at the intersections of two fissures and if the fissures are of considerable vertical extent the result may be a definite pipe. The chemical character of the country rock is of very great importance; when a vein crosses a bed of limestone, for example, owing to its solubility chemical reactions involving replacement cause expansions of the vein. This merges into a much larger

Fig. 19. Stereogram of a double pitching ore-shoot.

question of the restriction of payable veins to certain rocks only, and will not be pursued further at this point. One special feature of the crossing of two fissures must be mentioned as a possible source of rich ore-deposits, namely that each fissure may give passage to a solution of different composition, which by mingling cause precipitation of insoluble compounds.

Taking the case of a simple vertical or highly-inclined vein, a typical ore-shoot may be regarded as an enriched portion lying in the plane of the vein and usually of considerable downward extension in proportion to its width. The greatest length may coincide with the dip of the vein, or as is often the case, it may lie obliquely to the dip. This is described as *pitch*. The boundaries of large ore-shoots are generally indefinite, merging gradually into the poorer material; only rarely are the margins sharply marked off. The forms of ore-shoots are very irregular and variable. Numerous examples can readily be found by examining plans and vertical sections of long-established mines, where the stoped-out portions are usually the shoots. Many very important mines as a matter of fact consist only of workings along one particular ore-shoot. A very striking example of this is the Morro Velho mine of the St John del Rey Co., in Brazil, described on p. 463. This consists of one gigantic ore-shoot pitching steeply in the plane of a vertical vein. The well-known ore-shoots of the Kolar Goldfield in Mysore pitch similarly in the plane of an inclined vein. This is perhaps the commonest type. In both the cases just mentioned the bottom of the shoot has not yet been reached at depths of several thousand feet, but this is very exceptional: in nearly all cases there is a gradual impoverishment in depth, connected of course with the phenomena of primary ore zones.

An interesting example of ore-shoots of a somewhat different type is to be seen in the auriferous conglomerates of the Witwatersrand (see p. 470). Here the beds, the so-called reefs, dip due south, but the richer payable portions occur in belts running S.E., that is obliquely to the dip. This is considered to be due to original deposition of the gold-bearing gravels in deltas or on beaches by streams running from the N.W. It is in fact an example of fossil placers of extremely ancient pre-Cambrian date.

It is well known that in modern alluvials of all kinds, as well as in deep leads, the distribution of the ore is not uniform. The ore-minerals are all heavy and tend to sink to the bottom of the deposit, or to settle down in hollows and natural riffles on the bed rock, or to accumulate above impervious strata. All such accumulations are to be regarded in a sense as ore-shoots, but the details of their formation do not need discussion here.

CHAPTER VI

THE COMPOSITION AND CHARACTERS OF ORE-DEPOSITS

Mineral Composition. The substances capable of being worked as ores of metals are very numerous and of very varied character. As a rule they are also accompanied by many other substances not in themselves valuable, but which can only be separated from the ore by various special processes after the ore has been won. Thus we obtain at once a broad distinction into ore-mineral, the valuable part, and gangue, the commonly worthless part. Some gangue minerals have, however, a value of their own for some purpose other than as sources of metals. It must always be remembered, likewise, that in the miners' sense a mineral is only an ore, when it is worth working, or payable, and this payability depends on local economic conditions. To take a concrete example, that of a vein of copper pyrites and chalybite. In a remote locality with high working costs and inefficient transport, it may only be possible to work the vein for copper, the chalybite being thrown on the dump, while in a better situation the chalybite also may be a valuable ore of iron and pay for transport. Thus the same mineral under different conditions may be either ore or gangue. In many cases a gangue-mineral, such as barytes or fluorspar, may form a valuable secondary product of the mine, and enable a somewhat low-grade ore to be worked at a profit, which would not be possible if the gangue was worthless and merely entailed expense in dumping or filling.

As we have said, an ore must be payable. Now the proportion of any given metal which renders an ore payable depends on several independent factors; besides the selling price of the metal, which is the most obvious, there is also cost of production, which is made up of labour, transport, taxation, royalties, and so on. Hence the lowest payable proportion is an economic question. But this evidently varies largely according to the intrinsic value or market price of the metal, and within wide limits. Thus a mineral or rock with 5 per cent. of iron would be utterly valueless: 5 per cent. of copper would be a very rich ore, while 5 per cent. of gold would be beyond the dreams of avarice and does not exist on a large scale. One ounce of gold to the ton is good payable stuff and a considerably lower proportion can be worked at a profit. From these considerations it follows that some ores are nearly all metalliferous mineral, others nearly all waste stuff, with a minute proportion of metal. Hence the necessity for careful assaying and for refined methods of concentration in the case of the rare and precious metals.

Another important consideration is this: ores are frequently complex, yielding more than one metal, and the presence of a small proportion of a precious metal may have great influence on the success of a base metal mining enterprise. Hence it is necessary to take into account carefully the rare constituents of the deposit as well as the common ones. The rarer metals are often invisible to the naked eye, or even under the microscope, and their presence is only revealed by chemical or metallurgical methods. In some cases, where too many metals are present, an ore may be of small value owing to the trouble and expense of separation and metallurgical treatment.

An ore therefore is not usually a simple mineral: it is in general a mixture of minerals, or in the geological sense of the word, a rock. Some of the minerals have a perfectly definite and invariable chemical composition, but more commonly they are more or less mixtures: a chemically pure mineral is a rarity in nature. However, they often approximate closely to purity. A very large number of minerals belong to the class of what are now commonly called mixed crystals; that is to say they are mixtures in varying proportions of a certain limited number of chemical molecules usually possessing some properties in common. Minerals of both these classes possess definite physical properties, expressed in their crystalline structure, density, hardness and so forth. There is also a large class of substances which are not crystalline, but amorphous, and whose physical properties are much less constant: such for example are many forms of silica and of iron oxide. It is important to note, however, that crystallinity consists not merely in a regular outward form; it is a fundamental property of molecular structure. When a substance can exist in both the crystalline and non-crystalline states, the physical properties of the two forms, density, hardness, etc., are always different and their chemical behaviour often differs also. Some substances also possess two or more different crystalline states with different forms and different properties, such as calcite and aragonite, which are both $CaCO_3$, but differ in optical properties, density and chemical behaviour.

In a book like the present it is impossible to give a full account of the composition and properties of the minerals that constitute ore-deposits: such information must be sought in text-books of mineralogy, of which there are many, of various sizes[1]. Here it will be assumed that the student is familiar with at any rate all the common ore and gangue minerals. The rarer minerals will only be referred to incidentally, as occasion arises.

At this point it may perhaps be well to repeat once more that some ores are *minerals*, while others are *rocks*. The distinction is important in this way, that as a rule rocks vary much more widely in composition than minerals do. A rock never has a definite composition which can be expressed by a chemical formula. It is true that some rocks consist entirely or almost entirely of one mineral, but they are always aggregates and may contain anything, in any proportion. Igneous rocks containing only one mineral are very rare; sediments of such nature are more common, but even the purest limestones and quartzites have at least 1 per cent. of some foreign matter. This may be of much importance, for example in fluxes and glass-sands, while in some instances, such as the gold in quartz conglomerates, it is the small percentage of impurity that is of value. But a similar consideration also applies to many minerals. Some sulphides, otherwise of little or no value, are rendered worth working by the small amount of gold or silver that they contain, while a minute percentage of platinum may yield a large profit as a by-product. Hence the miner and the metallurgist have

[1] Two excellent small books for this purpose are Rutley's *Elements of Mineralogy*, 20th edition, London, 1919, and Hatch's *Mineralogy*, 4th edition, London, 1912.

to learn to take care of the side-shows as well as the main object or apparent object of their enterprise.

The Deposition of Ores. The heading of this section opens up some very large questions, involving theoretical considerations of the highest importance. These matters however are discussed elsewhere in this book from this point of view. Here we are not concerned with the exact mechanism or the physics and chemistry of the processes involved. We start from the undeniable fact that the ores got there somehow: we have to consider here mainly what they are and what they look like. From this point of view it is of no importance whether the metal was carried to its present position in the fused state, as a volatile gas or an aqueous solution, as a silicate or a sulphide or a carbonate. Here we deal mainly with the characters of the deposits as they exist to-day, regardless of their past history.

Primary Ore Minerals. The primary ores of the metals, as distinguished from those produced by their subsequent oxidation and other changes are usually comparatively simple in their composition. The most important group is undoubtedly the sulphides and the closely-related arsenides[1]. These include the important primary ores of silver, copper, lead and zinc, as well as the less important metals antimony, arsenic, bismuth, mercury and others. Among the oxides the leading place is taken by iron, aluminium and tin. Most of the carbonates are secondary, but iron carbonate at least is often a primary vein mineral. Tungstates, tellurides and (rarely) silicates also constitute important primary ores. Finally, the native metals include gold, silver, platinum and copper, though native copper is usually secondary.

Gangue Minerals. The number of minerals usually classed as gangue occurring in veins, is very large, but the following short list includes the most important of them. In the veins with clean-cut sides and definite fillings the commonest minerals of this class are quartz and other forms of silica, calcite, dolomite, chalybite, barytes and fluorspar. More characteristic of meta-

[1] Most of the more complicated sulpharsenides, sulphantimonides and other related minerals are clearly secondary; in a few instances a primary origin has been suggested for them, but the question is still under discussion.

somatic lodes are quartz, felspar, mica and chlorite, with tourmaline in special cases, while the pegmatites and other igneous veins contain quartz, felspar, mica, apatite, tourmaline, beryl and a great variety of rare minerals, often containing the less common elements. Taking all veins and lodes together, the most abundant mineral is undoubtedly quartz, which is almost universal in its occurrence, and calcite probably comes next in amount.

Vein-fillings and Veinstones. In the case of true metalliferous veins it is always perfectly obvious that the material of the vein is newer than the country rock (except in so far as fragments of the country rock may be included). Furthermore in many instances it is clear that the filling of the fissure has taken place in two or more stages; owing to subsequent earth-movements a fissure already filled may be re-opened and again filled and so on indefinitely[1]. But besides this there is commonly a regular order of deposition in a single filling. This is not usually prominent in veins which are essentially dykes of igneous rock; such are commonly filled by a single injection in a very short interval of time, and any appearance of bands or layers is due to flow or to cooling and crystallization from the highly-heated material. But the filling of veins from solutions is apparently often a slow process and the character of the material deposited may and commonly does undergo changes. Hence such veins usually show layers of different minerals, and sometimes the filling is incomplete, hollows or *vugs* being left, often lined by beautifully perfect crystals. The best mineral specimens seen in museums come from these hollow spaces.

The next main type of veinstone is found in those lodes where the workable material consists of altered and mineralized country rock. The formation of such veinstones is essentially some process of metasomatism, involving a radical change in the character of the rock itself with addition of metals and other substances from some outside source. The mineralogy of these changes is often highly complicated, and the study of them is still a somewhat neglected branch of petrology.

---

[1] In this discussion attention is confined for the present to primary vein and lode material. The changes subsequently brought about by oxidation and other processes are dealt with hereafter.

As an example of this type of veinstone we may take some of the tin-bearing lodes of Cornwall. These often consist of sheets of rock differing widely in appearance from the surrounding granite or slate, but shading off quite gradually into them at the edges. Confining our attention for the moment to those in the granite, we find that the first sign of mineralization is the appearance of narrow cracks filled with tourmaline or chlorite:

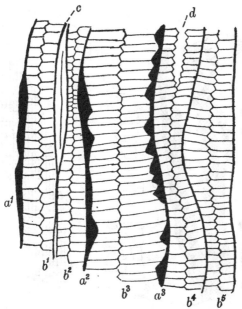

Fig. 20. Complex lode at Wheal Julia, Binner Downs, Cornwall (after De la Beche). a^1–a^3, blende and copper pyrites; b^1–b^5, quartz; c, hardened slate; d, open space (vug) lined with quartz crystals.

these become more and more abundant and the dark minerals spread more and more widely from them into the white granite, which is also often stained red or brown by iron oxide or green by chlorite. In the richer and more central parts of the lode near the main fissure the whole of the felspar and original mica of the granite have commonly disappeared, being replaced by an aggregate of quartz, secondary mica, tourmaline, chlorite and ore-minerals, often with fluorspar. This makes a rock of a dark blue or dark green colour with various spots and patches.

These peculiar veinstones are usually called by miners *capel* or
peach, and unfortunately some confusion exists as to the application of these two names. Thus the peach of one may be the
capel of another authority. The practice is now steadily gaining
ground of calling the blue varieties rich in tourmaline capel,
while the name peach is reserved for the green varieties rich in
chlorite. Many exceptions will be found to this rule however,

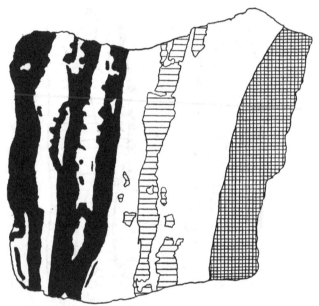

Fig. 21. Complex vein-filling, Kaiser Wilhelm mine, Clausthal, Harz.
Cross-hatched, blende in quartz, single lines, chalcopyrite, black,
blende in calcite, white, calcite.

even in modern writings. In some instances material of this
sort occupies the whole width of the lode; in other cases it
forms layers on either wall, the middle portion consisting of
coarsely crystalline quartz and ore-minerals. It is obvious that
the varieties of lode-material of this general type are endless,
depending on the original composition of the altered rock,
the character of the ore-bearing solutions, their temperature
and the degree of chemical alteration that they have brought
about.

Sometimes the vein-material is in direct contact with the country rock on one or both sides, or to use a miner's expression, the ore is "frozen to the country." In other instances one side of the vein may be formed of a clayey material called *flucan*, or in America, *gouge*. This is as a rule nothing but comminuted and decomposed country rock, and it is specially common in fault-veins[1]. Sometimes the ore and gangue minerals are mixed up indiscriminately in the vein: in other cases there is a more or less regular arrangement, in layers. The order of deposition is

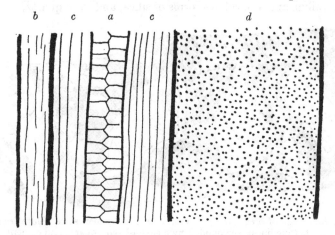

Fig. 22. Lode at Godolphin Bridge, Cornwall (after De la Beche). *a*, quartz in good crystals; *b*, massive quartz; *cc*, chalcedonic silica; *d*, copper pyrites with some quartz.

quite variable: the ore may be on the outside or the inside and not uncommonly there are alternations, especially in the case of repeated fillings. Again, it is not at all rare for the successive layers of ore to contain different metals: this is correlated with the definite succession of metal-deposit that·has been worked out in many regions. For example, when copper, lead and zinc occur in the same vein, the copper is generally the oldest, while the lead and zinc are later and more or less contemporaneous. The subject of the succession in time and arrangement in space of the ores of the different metals is of great importance, and is

[1] In broad brecciated veins of the shatter-belt type there may be many alternations of vein-stuff and gouge,

treated elsewhere (see Chap. IX). A similar succession is often apparent in the gangue minerals: much variation occurs and it is difficult to generalize, but on the whole the carbonates are usually later than quartz, occupying the middle portions of the veins, though the reverse sometimes occurs. However it is often very difficult to determine which of the gangue minerals are truly primary and which belong to the secondary, or even to very late stages of deposition. Barytes and fluorspar are probably nearly always primary, but the carbonates, the crypto-crystalline and amorphous forms of silica, and even quartz, may

Fig. 23. Block of brecciated ore, Van mine, Montgomeryshire. Slate fragments (fine lines), surrounded by a shell of pure quartz and cemented by zinc-blende (black) and quartz. Two-thirds natural size.

be deposited at any time by meteoric waters percolating downwards from the weathered surface rocks.

The simplest case of all is afforded by some of the so-called brecciated veins, which are simply fault-fissures more or less filled by fragments of rock broken from the sides of the fissure and cemented by ore-minerals. In such veins there is sometimes no gangue at all in the strict sense of the word, but the rock-fragments are worthless material and have to be separated by dressing in the same way as gangue-minerals. In other cases there are in addition gangue-minerals, and sometimes the various minerals are deposited in successive layers around the rock-fragments, the last to be deposited filling up the interstices. This process leads to the formation of various types of *ring-ore*. The layers of minerals are of course in reality shells enclosing the fragments, but in section they look like rings (Fig. 23).

When a vein has been opened and filled more than once it often happens that the earlier-deposited minerals are broken up and brecciated and then cemented afresh by the later filling. Such a vein may obviously be torn open in the middle, or along one wall, or quite irregularly, so that the sheets of breccia may have any relation to the newer non-brecciated minerals, and the structures of veins may thus become very complicated.

Fig. 24. Block of ore from brecciated lode, Mt Elburus, Caucasus; white, rhyolitic country rock; black, zinc-blende; half natural size.

When minerals are deposited on the walls of an open fissure by slow deposition from solutions, crystals usually tend to show a certain regularity of arrangement. Minerals possessing a prismatic habit, such as quartz, usually grow with their long axes at right angles to the walls of the fissure: when, as is usually the case, they are closely crowded together they will only show well-developed faces at the free ends. Such arrangements of large numbers of parallel prisms when seen in section often have a rather comb-like appearance, and the whole layer is sometimes called a comb of quartz (see Fig. 20, p. 97). Other minerals such as pyrites, galena, blende and fluorspar, belonging to the cubic system with three equal axes, do not show this feature so readily, while the carbonates and barytes often show no external crystal form, but occur in massive layers with well-developed cleavages. The amorphous and cryptocrystalline forms of silica (known as chalcedony, etc.) form massive layers often with a peculiar nodular or botryoidal surface.

Stockworks and Disseminations. The term stockwork is commonly used in the older literature of mining geology to describe a type of ore-body of indefinite shape which really consists of a great number of veins situated close together either with a general parallelism of arrangement or often interlacing and forming a network. So far as their actual fillings are concerned the individual veins of a stockwork are generally quite similar to simple veins, though often small in size and best described as *stringers*. In many instances they are undoubtedly of direct igneous origin and formed at a high temperature by pneumatolytic action. This is shown by the fact that the whole mass of the rock between the stringers is often converted into greisen or topaz-rock or some similar alteration-product, besides being impregnated by disseminated ore-minerals, so that the whole mass is often workable ore, yielding say tin or tungsten. A good example of this is afforded by the famous stockworks of the Erzgebirge. In this country perhaps the best example of this kind of thing is seen in the *carbonas* of the St Ives district in Cornwall. The St Ives Consols mine lies in granite near its junction with slate and greenstone. Its chief peculiarity is in the enormous irregular masses or pockets of tin ore which branch off from some of the lodes and swell out, so that in their working caverns 50 feet high and wide, and two or three hundred feet long were formed. These masses consist of altered and highly-tourmalinized granite with much cassiterite. It is not clear why the ore-bearing solutions spread outwards into the rock from the lodes in this irregular manner[1]. Somewhat similar masses were also found at Wheal Basset and at Balmynheer mine in Wendron[2]. An interesting example of a deposit of tin and wolfram in the general form of a stockwork has recently been discovered and developed at Hemerdon in Devonshire. The mass of ore-bearing veins lies in granite close to its contact with

[1] "The Geology of the Land's End District," *Mem. Geol. Surv.* 1907, p. 87.
[2] "The Geology of Falmouth and Truro," *Mem. Geol. Surv.* 1906, p. 164.

Devonian slates, and extends a short distance into the latter; the veins in the slate contain mainly tin, the wolfram being apparently confined to the granite. Over a large area the veins vary in width from less than an inch to 4 feet and are so closely associated that even 1 foot of intervening granite is rare. They intersect and ramify in a remarkable fashion showing that the rock must have been greatly shattered and fissured after its consolidation and then penetrated by ore-bearing solutions.

In all the cases just described there is no doubt that the ore is of primary origin, having been deposited during the later pneumatolytic stages of the igneous intrusions; the shattering of the rock is probably due to contraction on cooling, but it is not clear why the pneumatolytic effects should be so strongly localized.

Contact Deposits and Replacements. The ore-bodies coming within this category are somewhat difficult to deal with satisfactorily, since they show endless variety and in many instances their origin is not well understood. We are not dealing here with the replacements that occur in the walls of more or less well-defined tabular veins and lodes, but with those ore-masses, often of perfectly irregular form, that occur for example at the intrusive contacts of igneous rocks with sediments. This type of deposit is naturally common at the contact of intrusions with limestones and other fairly soluble rock-types. Here the character of the ore and gangue will obviously depend on the special conditions of the case. A very common type is deposits of pyrites or other sulphides, often containing copper or nickel. Similar masses often form lenticles in the country rock, as in the Huelva district and in Norway. With many of these there is still much discussion as to whether they are to be regarded as true intrusions of a sulphide magma, or as replacements of country rock by sulphide solutions. Some of these are in actual contact with intrusions, while others lie at some little distance from them. On the exact mode of origin of these bodies it is as yet too early to dogmatize.

It is unnecessary at this stage to describe in detail any deposits of this kind: certain examples will be found in the descriptive portions of this book. Furthermore it is almost impossible to draw up any effective generalities on this rather obscure subject. All that can be usefully said now is that certain ore-bodies, including many of great technical importance, appear to have been formed by actual replacement of country rock by ore-minerals derived from solutions. It is as to the character and origin of these solutions that the greatest doubt exists in many cases. Some of them are certainly of direct magmatic origin, and the phenomena in connection with them are often well described by the term pneumatolysis. Others again may be conveniently classed as metasomatic, without restriction as to

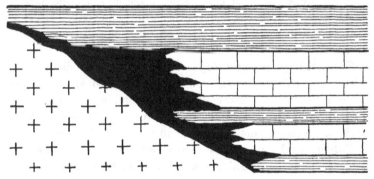

Fig. 25. Ideal section of a contact-metamorphic ore-deposit. Mineralizing solutions from a granite intrusion (left) have attacked bands of lime-stone and formed an ore-body by replacement, leaving shales un-affected.

the physical conditions under which metasomatism can take place, especially with regard to temperature. The detailed petrology of such deposits is a subject which still requires much further investigation and opens up some very interesting lines of research.

With regard to the forms of contact and replacement deposits, there is evidently scope for great variation. When they are formed along the contact of an intrusion with a uniform type of sediment, for example, they often show an outward resemblance to some vein-deposits, being more or less tabular or sheet-like in form, but more commonly the country rock is variable in character, possessing different degrees of susceptibility to change and the ore-body will then follow the variations, extending further in certain favourable directions and becoming narrower, or dying out altogether when in contact with resistant rocks. Hence contact deposits are as a rule of markedly irregular form. True replacements often conform pretty closely to the bedding of sediments, and the extreme form of this is found in some of the metasomatic iron ores, which are simply replaced beds of limestone. In some of the sulphide ore-bodies, *e.g.* Rio Tinto, the lenticular form is very marked and is strongly suggestive of the laccolithic type of intrusion. In some instances replacements are very closely allied to the category of disseminations; these do not possess any definite boundaries, but gradually shade off into the unaltered rock.

From the mineralogical and petrological point of view the contact and replacement deposits do not differ in any notable respect from the vein deposit type. The actual ores present are of much the same character, including among the primary minerals the usual sulphides and oxides. The gangue minerals however are perhaps usually of a somewhat different type: quartz and calcite are equally common, but such minerals as barytes and fluorspar are perhaps less usual. A very notable feature in limestone replacements is the occurrence in large quantities of such metamorphic minerals as garnet, wollastonite, diopside and epidote, derived from limestones and dolomites.

Other types of Ore-deposit. No good purpose would be served by attempting to deal here in a general way with the composition and characters of the various kinds of ore-body grouped together under the heading of secondary deposits. They are formed by simple geological processes and are for the most part in reality sediments, therefore obeying the laws of sedimentary geology. The special characteristics of their contents and their manner of formation and distribution are described in sufficient detail in the appropriate sections of the descriptive chapters in Part II. For superficial and alluvial deposits special reference may be made to the chapters on gold, platinum and tin, and for consolidated ore-bearing sediments to the description of the bedded iron ores in Chapter XV. Here full details will be found in each case, with in most instances a discussion of the general principles involved.

CHAPTER VII

THE CLASSIFICATION OF ORE-DEPOSITS

In considering what features should be adopted as a basis for the classification of ore-deposits it is essential to take into account the purpose for which the classification is required and the point of view from which the subject is to be regarded. In the first place the mining geologist proper and the prospector look at an ore-deposit from an angle quite different from that of the mining engineer. The former are chiefly concerned with locating the ore-body from its outcrop, its mineral contents and its possible extension, while the engineer is more concerned with its form and its relation to the surroundings. Both of course wish to gain some idea of size and value: but to put the matter in more scientific language the interest of the prospecting geologist is largely chemical and mineralogical and therefore genetic, while the engineer's point of view is largely morphological. A genetic basis of classification is undoubtedly more scientific, but for many purposes a morphological arrangement is more practical. It is therefore not surprising that most of the early attempts at this direction were founded on the basis of form rather than of origin or composition, being based on such simple conceptions as veins, beds and the like, taking into account also regularity or irregularity of shape. Next arose a school of geologists who began to classify ore-deposits mainly by their relation to the surrounding rocks, and especially according to whether they were contemporaneous with, or newer than the country rock. This led to a primary twofold division into *syngenetic* or contemporaneous and *epigenetic*, or subsequent, deposits. In the writer's opinion this idea was exceedingly harmful, and did more than anything else to retard a truly scientific treatment of the subject, since it laid down arbitrary rules and brought together into one category some deposits of the most varied origin and geological relations, as will appear later.

Another method and one which at first sight appears very

simple, is a classification founded on metallic contents. To a wide extent this is applicable and has been largely employed of recent years. In some ways this is the most practical of all methods and the chief difficulty arises in assigning to their proper group many instances where two or more metals occur together and the actual output consists of several products. In such cases there is bound to be more or less overlap and repetition in descriptive writings, but the practical application is as a rule fairly easy. This is as a matter of fact the method adopted in this book, after long and careful deliberation. It is also employed by several of the most recent American writers, for example, by Ries and by Emmons.

On scientific grounds undoubtedly the most logical classification is one founded on origin: this has however at any rate two great drawbacks: in very many instances the mode of origin of ore-deposits of the highest importance is still quite uncertain, and in any book of reference the occurrences of the several metals become inconveniently scattered, and various rather incongruous things are brought together.

All systems yet devised have obvious drawbacks, and in the present state of our knowledge this cannot be remedied. Nevertheless there can be no doubt that a study of the genetic relationships of ore-deposits must be of great value to mining geologists, since it helps us to form an opinion as to what is likely to be found in any given district, and also, what is almost equally important, it allows conclusions to be drawn as to what ores are not likely to occur and need not be looked for. Negative evidence is proverbially dangerous, but if founded on adequate knowledge it has undoubted value. It may now be said that the scientific study of the ore-deposits has advanced far enough to permit of the application of the negative method in many instances.

In devising a genetic classification the most obvious starting-point is the physical conditions that prevailed during the formation of the deposit, mainly the temperature and pressure. These two fundamental conceptions, which are the basis of all modern physico-chemical and thermodynamical investigations, are conditioned in the case of ore-deposits by numerous and complicated factors, some of which are still very much a matter of speculation rather than of actual knowledge. It may however be accepted as established that both temperature and pressure tend to increase downwards. With regard to temperature this is borne out by a multitude of facts of observation, familiar to every mining man, and in the case of pressure it is merely a matter of common sense. The greater the thickness of rock-cover the greater must be the pressure to which the rocks and minerals below are subjected. This is a rule of universal application. But in the case of temperature local complications are introduced by the phenomena of vulcanicity. It is unnecessary to prove at length that some parts of the crust are hotter than others at equal depths from the surface. Hot mines are only too well known to everybody, while some districts are specially favoured by a low temperature-gradient. We have every reason to infer that at greater and as yet inaccessible depths similar differences exist. The most obvious cause for these differences is the existence below the crust of masses of heated rock, perhaps recent intrusions which have not yet had time to cool down.

Now in most of the highly-mineralized regions of the world we find evidence that in the past the crust has been in a heated condition, since the association of ore-deposits with igneous rocks and their attendant metamorphic phenomena is notorious. Furthermore, in many of these regions there can be traced a regular gradation in the composition of the ores from the igneous rocks themselves outwards into the surrounding strata. Hence it appears that the nature of the ores is a function of the temperature: the non-metallic minerals also show similar relations, both in the gangue minerals of veins and in the constituents of the country rock. All this really means that certain metals and their accompanying non-metallic minerals are characteristically formed at high temperatures and high pressures; that chemical action is more intense under such conditions and that the country rock is more altered. With increasing distance from the source of heat a different set of minerals are found and chemical changes are less strongly marked, while the deposits found at and near the surface are again different and there may be little or no alteration of the country rock. It must be noted however that the temperature zones are not necessarily parallel to the pressure zones, so that a further set of complications are introduced. This subject is discussed in detail in the chapter on ore zones.

It is found as a matter of empirical observation that certain minerals are found under conditions that lead us to believe that they were formed at high temperatures and pressures while others have been formed under different conditions: hence classifications have been drawn up founded on these criteria. To take some actual examples, tin and tourmaline are believed to be high temperature minerals, formed in a deep zone, copper belonging to a rather shallower zone, zinc and lead being formed at still lower temperatures, and so on. Hence we have a division of ore-deposits into those formed in deep, middle and shallow zones. It is evident that in such a case there can be no hard and fast lines and transitional forms must exist, but the basis seems a satisfactory one and does correspond to real facts in nature. This classification is largely developed by Lindgren[1].

Somewhat similar in principle is the idea founded on the thermal character of the ores and their relation to igneous rocks and water. Here we have the chief subdivisions as follows:

(1) Magmatic segregations: ores actually separated from igneous magmas by processes of crystallization.

(2) Hydrothermal or aqueo-igneous deposits: those formed by highly-heated watery solutions mainly of magmatic origin.

(3) Aqueous deposits: those laid down from water at or near the ordinary temperature, the water often being meteoric.

Some writers have introduced further subdivisions, but the above expresses the underlying idea with sufficient clearness.

In any case there is no doubt that the presence or absence of water of whatever origin is a determining factor in the character of a vast number of ore-deposits and any satisfactory classification ought to take this into account in some way or other. It must of course be clearly understood that the above discussion refers solely to the origin of primary ores. Later changes of oxidation and re-precipitation with their accompanying chemical

[1] Lindgren, *Mineral Deposits*, 2nd edition, New York, 1919, especially chapter XIV.

reactions and physical changes are here left entirely out of account.

But in any complete classification these secondary types have to be considered as well, and here we are confronted with two very different classes of phenomena. There are first of all ores formed by chemical and physical changes in the primary deposits themselves, changes which take place underground and commonly within the limits of the original deposit. This group comprises the products of oxidation, reduction and secondary enrichment in veins and other masses of like origin, by which the original constituents are made to combine differently and become re-distributed at different levels, most commonly below their original position, while some constituents may be removed altogether in solution. The second category includes those instances where ore-minerals of a stable nature are removed bodily from their original position by the action of the ordinary surface agents of denudation, being afterwards re-deposited in a new location as some form of alluvial deposit. In some ways intermediate between these two groups are gossans and all the different kinds of mineral deposits left on the outcrops of lodes, such for example as those to which the term *shoad* is often applied. By many modern geologists these are classed as *residual* and *eluvial* deposits. They are often loosely described by technical writers as alluvials, although not transported. In fact the same term is sometimes applied also to cases where a mineralized rock in place, although in reality entirely undisturbed, is in such a decomposed condition as to be workable by the methods usually applied to true alluvials. Hence a good deal of confusion has arisen. Beyond these is another very important group of deposits, chiefly confined to ores of iron, manganese and aluminium, which are ordinary bedded sediments, but contain enough of the metals either as a primary or secondary constituent, to be workable as ores.

In the investigation of many ore-bodies, especially those of the primary type, great difficulty has often arisen in deciding to what extent actual replacement of older minerals and rocks has taken place: many veins and other masses are undoubtedly simple infillings of open spaces, while in other instances it is equally obvious that rock originally occupying the space has been more or less completely replaced by new material, and both these processes have occurred under similar physical conditions according to the chemical character of the rock and the ore-bearing media. Sometimes again the infilling of a fissure has been accompanied by replacement of its walls extending for some distance on either side. Hence in any scheme for the classification of such ore-bodies we have to include both infillings and replacements in each category. Any attempt to separate them must lead to endless confusion.

Taking into account the considerations briefly outlined above the following scheme of classification has been drawn up:

I. Primary deposits.
   (a) Magmatic segregations within the boundaries of igneous rock-masses.
   (b) Pegmatites and abyssal infillings and replacements formed from material extracted from magmas by differentiation: water of magmatic origin only.
   (c) Infillings and replacements at moderate depths with moderate temperature and pressure: magmatic water or water heated by descent.
   (d) Infillings and replacements near the surface at low temperature and pressure: water of meteoric origin.

II. Secondary deposits.
   (a) Oxidation ores of the superficial zone of active water circulation.
   (b) Reduction and cementation deposits of the zone of secondary enrichment, with stagnant water.

III. Stratified deposits.
   (a) Bedded ores formed primarily as sediments.
   (b) Bedded ores formed by metasomatism of sediments.

IV. Superficial deposits.
   (a) Residual and eluvial deposits, gossan and shoad.
   (b) Transported alluvial deposits.

Referring to the above table, it is obvious that the subdivisions adopted in group I are purely arbitrary, and in fact all form one continuous series, but the sub-groups do in reality correspond to fairly well-marked differences of type, both in regard to metal-content and gangue mineral. In the other groups the distinctions are real and not arbitrary, being founded on purely geological considerations, and therefore of truly genetic value.

It is with much hesitation that the author ventures to add one more to the already innumerable classifications of ore-deposits that have been put forward: none of these are satisfactory, and it is not claimed that this is better than the rest, but it does express in a simple way the general principles set out in the earlier parts of this chapter.

It is important to note that in this scheme replacements and contact-deposits are not specifically distinguished from infillings and vein-deposits, because experience has shown that it is in practice impossible in a great number of cases to decide definitely whether a given ore-body has been formed by the actual filling of an empty space or by gradual removal of soluble or unstable material with concomitant deposition of some other mineral or group of minerals. On this subject opinions differ very widely in individual cases, and the process cannot be seen in actual operation. From most points of view, however, and especially from the practical standpoint, the difference is not of the smallest importance.

CHAPTER VIII

THE RELATION OF ORE-DEPOSIT TO EXTERNAL INFLUENCES

Oxidation and Secondary Enrichment. It is well known to every one concerned with mining that when an ore-body is followed downwards from the surface the character of the ore almost invariably undergoes change. To put it shortly it is very rare to find unaltered primary ores at the surface. Different types of ore vary greatly in this respect, however. A few primary ores, such as native gold and tinstone, are so stable that they can survive without change even a prolonged period of weathering, and in a few instances a mineralized region may have been so recently glaciated that all altered material has been swept away bodily down to entirely unaltered rock. But in the vast majority of cases it is easy to show the existence of an upper layer of altered rocks and correspondingly altered ore-minerals changing either gradually or suddenly in depth to the unaltered ore.

This alteration of the superficial layers is primarily a physico-chemical process, but it is also truly geological, being controlled by many factors depending on climatic conditions, the nature of the surrounding rocks, their permeability by air and water, their susceptibility to solution and chemical change, and of course to a great extent on the character of the ores themselves. In all of these processes the underground circulation of water naturally plays a part of the highest importance. In fact it is hardly too much to say that this is the dominant factor. From this it is evident that the influence of climate must be in the first rank, and in point of fact a great difference is noted in the downward extension of the altered zone in moist and in arid regions respectively, as also in the character and composition of the secondary minerals produced in each case. The nature of the geological processes to which the country rock has been subjected is in this connection also very important. If the rocks are much faulted and shattered the surface water naturally permeates them down to great depths and in a similar way there are great facilities for the rise of solutions of volcanic or other origin from below. Such solutions are often very active chemically and may bring about notable changes in rocks and ores. Thus it is clear that numerous conditions have to be taken into account, and the subject is by no means simple. It is very necessary to bear in mind the fundamental difference that exists between secondary ore-zones produced by such processes and the zones of primary ore-deposit dealt with in another chapter. When one kind of zone is superposed on another great confusion may be brought about and these two sets of zones are not

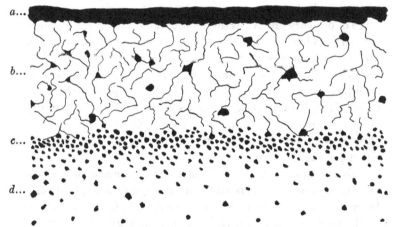

Fig. 26. Longitudinal section of vein with oxidation and secondary enrichment. *a*, gossan; *b*, zone of oxidized ore; *c*, zone of secondary enrichment; *d*, lean primary ore in depth.

necessarily or usually parallel. Furthermore, instances are known where superficial and secondary zones of alteration have been buried deeply below newer sedimentary or volcanic rocks. Of this instances will be given later.

In order to clear the way for a systematic treatment of the subject it may be well to state at the outset that in a great number of cases it is possible to recognize three clearly defined zones from the surface downwards, as follows[1]:

       (*a*) The zone of oxidation.

       (*b*) The zone of secondary enrichment.

       (*c*) The primary ore zone.

[1] The occurrence of a zone of secondary enrichment is by no means universal; the ores of some metals do not lend themselves to this process.

Of these zones the first two will be considered in detail: the nature of primary ores has already been discussed at length.

The Zone of Oxidation. Nearly all ores, as just stated, are unstable in the presence of water and free oxygen, and being themselves generally of complex character, they often react on each other when in contact with air or moisture. As already described in the chapter on water circulation in the rocks, the outermost layers of the crust to varying depths, are as a rule incompletely saturated with water, having cavities filled with air, and the water is in a state of free circulation. The conditions are therefore eminently suitable for processes of oxidation. These are also helped to a large extent by the gases dissolved in the water and by acids derived from various sources, from organic matter and from the decomposition of rocks and minerals. This process of decomposition is a cumulative one, in that the reactions first initiated set free other chemically active substances and the changes probably tend to go on at a constantly increasing rate. The lower limit of this zone of free oxidation is the level of permanent, more or less stagnant ground water, below which there is little or no active circulation, no free oxygen and the water is usually alkaline rather than acid. The conditions here are those of reduction rather than oxidation, and as will be seen in a later section, very notable and characteristic chemical changes take place at this horizon.

Hence it follows that in this uppermost zone the ore-minerals are converted into new compounds which themselves contain oxygen. The number of these is endless, but the more important of the general groups are oxides, hydroxides, hydrates, carbonates, sulphates and silicates, with chlorides and less universally such compounds as phosphates, arsenates, molybdates and many others, according to local circumstances. The occurrence of these compounds in the zone of oxidation is largely controlled by two factors; their own solubility, and the characters of the gangue and country rock as regards their liability to react with the products of oxidation. This subject can be best illustrated by means of a specific example. Take the common case of a vein containing galena and blende in a gangue of quartz in a siliceous country rock. The lead sulphide undergoes oxidation to lead sulphate, which is very insoluble and chiefly remains *in situ* as the mineral anglesite. Blende, on the other hand, oxidizes to the very soluble zinc sulphate, which is readily removed by circulating water. In this simple case there will be no reaction with the gangue or country rock and the result is that the oxidation zone is impoverished in zinc and relatively enriched in lead. The matter is nearly always complicated however by the presence of dissolved carbon dioxide in the water, and most carbonates are rather insoluble: when this is so the lead by preference forms cerussite and zinc also forms an insoluble carbonate, calamine. It is a general rule of chemical equilibrium that when a solid can be formed by the mixing of two solutions this formation always takes place. Thus, taking the simple equation

$$BaCl_2 + H_2SO_4 = BaSO_4 + 2HCl$$

the reaction will always proceed towards the right, forming barytes and free hydrochloric acid. Consequently an insoluble mineral will always be formed when its constituents exist in a solution in the zone of oxidation, and carbonates are very common, either simple carbonates like cerussite and calamine or the so-called basic carbonates like malachite and azurite, which are compounds of carbonate and hydroxide in fixed proportions. With a calcite gangue or a calcareous country rock carbonates are naturally very common, since copper sulphate formed by oxidation of chalcopyrite reacts readily with calcium carbonate. Under laboratory conditions when fragments of calcite or of limestone are put into a solution of copper sulphate the green basic carbonate is formed.

Another possible type of alteration in the oxidizing zone is removal of gangue-minerals by solution, either with or without alteration of the ore. A process of this kind obviously leads to a relative concentration of the ore, and in some instances has led to conspicuous enrichment of the superficial zones and residual surface deposits. This often occurs with gold, tinstone and wolfram.

Gossan or Iron Cap. A phenomenon well known to all miners and prospectors is the formation on the backs or outcrops of veins and other metalliferous bodies of a conspicuous deposit of iron oxide, generally hydrated, rarely in dry climates anhydrous ferric oxide (true haematite). This is generally called *gossan* (from an old Cornish miner's term) or *iron cap* (French, *chapeau de fer*; German, *eiserner Hut*). Gossans are most conspicuously developed where lodes are rich in iron pyrites or copper pyrites, but are found in deposits of any kind rich in iron, whether in the form of sulphide, oxide or carbonate. The most common form is a cellular or spongy mass of hydrated iron oxide of a colour varying from bright red to brown, often best described as limonite, although this so-called mineral has really no definite composition. Scattered through the gossan, or concentrated at its base are often to be found crystals, lumps or grains of specially resistant minerals, such as tinstone or gold. At Mount Morgan in Queensland there was found a very remarkable gossan of great thickness, up to 160 feet in places, extraordinarily rich in gold. In this case the gossan was haematite, and had been formed by the oxidation of auriferous copper-bearing pyrites, which is now worked in depth for copper. It is hardly necessary to give any more examples of the process of gossan formation, which is of almost universal occurrence and is quite similar to the ordinary brown weathering of rocks rich in iron compounds of any kind. The accumulation of the gossan to great thicknesses depends on the very insoluble character of hydrated iron oxides. It is usually very porous and offers little obstacle to the downward percolation of rain water. Hence oxidation and solution may go on actively beneath it and in the next zone below is found the chief development of the other oxidation ores.

The Relation of Ore Zones to Denudation. From what has just been said it is clear that the existence of the zone of oxidation is a direct result of the operation of the ordinary geological agents. Its upper limit is the surface of the ground; its lower limit is approximately the water-table. But it is most important to remember that the position of neither of these limits is fixed. The ground surface is constantly being worn away, and the water-table is being lowered with it, though not necessarily at the same rate. Hence the whole zone must be conceived of as working downwards, so that oxidation and other chemical processes are constantly attacking new material.

Metallic compounds also are constantly trickling downwards in solution and being brought under the influence of different physical conditions. As soon as ground-water level is reached the character of the water is different, reduction prevailing rather than oxidation. Hence further chemical reactions occur: in a general way these are rather complicated, and the principles involved can perhaps be best explained by means of a concrete example. The phenomenon is specially conspicuous in the case of copper deposits and an ideal case will be chosen from the ores of this metal.

Changes in a Copper Lode. In order to simplify the question as much as possible let us suppose a lode of copper pyrites in a quartz gangue with a siliceous or other stable country rock, assuming also that the lode is newly exposed to chemical and physical denudation by removal of some overlying rock which it did not penetrate. The sulphide ore will be attacked at once by atmospheric gases and water and will be decomposed to iron oxide (gossan), copper sulphate and sulphuric acid[1]. The copper sulphate solution will trickle downwards and in the case supposed there is nothing to precipitate it in the zone of oxidation. But when it reaches the region of stagnant water with no free oxygen and comes in contact with unaltered primary sulphide, the sulphate will be reduced to cuprous sulphide, which may be deposited on the surface of the primary chalcopyrite or in cracks, as the minerals chalcocite or covellite, or as often happens it may react with the chalcopyrite to form bornite. In many instances the primary ore also contains small quantities, perhaps mere traces, of other metals, such as antimony, arsenic or bismuth, which are set free along with the copper, and in the zone of reduction and precipitation form various complex minerals, such as tetrahedrite or tennantite. The occurrence in a similar way of gold and silver is reserved for further consideration. The result of these processes is the formation, usually at or just below ground-water level, of a zone of rich copper ore, known as the *zone of secondary enrichment*. Below this comes the lean ore of the depths, which is often too poor to be worked.

[1] At first ferrous sulphate will be formed, but this is an extremely unstable substance and oxidizes at once to ferric oxide and free acid.

Thus in this, the simplest ideal case, we can recognize four zones: the gossan at the top, then the zone of oxidation and leaching, then the zone of reduction and reprecipitation, with the lean ore below. The depths of each of these zones must evidently depend on local circumstances and will vary both absolutely and relatively in each case. But a point of the greatest importance to be borne in mind is that as the surface is worn away by denudation and the general drainage level of the district is lowered the whole system must descend, bringing about further modifications which will be considered presently.

All natural circulating waters contain more or less carbon dioxide, as well as dissolved salts, such as chlorides. These nearly always bring about further changes, especially in the oxidation zone. When carbon dioxide or a dissolved carbonate come in contact with copper sulphate insoluble copper carbonates are formed, usually the basic carbonates malachite and azurite, while dissolved silica precipitates chrysocolla. These minerals produce the well-known green and blue coloration of oxidized copper ores. When such carbonates or other insoluble precipitates are formed there may be a very considerable accumulation of ore in the oxidation zone, and they are formed in great abundance where the solutions have the opportunity of reacting with a carbonate gangue or with a calcareous country rock.

In many examples of copper lodes that are undergoing this process there are likewise found masses of cuprite and native copper. These are generally found at the lower limit of the oxidation zone, at about the level of the permanent water-table or just above it. They must be regarded as indicating the beginnings of reduction, being formed from solutions that have escaped precipitation in the oxidation zone and have come within the influence of the reducing agencies of the ground-water. It is possible that they are formed by a temporary rise of the ground-water into the carbonate zone: thus the precipitating effect of the sulphides is eliminated, and a low oxide (cuprite, Cu_2O), or native metal is the result.

Thus when this process has been in operation for a considerable time the relations will be as shown in Fig. 26. It will be clear from the description that the space between the present ground surface and the lower limit of the zone of secondary enrichment must contain all the metals that were originally in that part of the vein now removed by denudation. Nearly all the iron is concentrated in the gossan, while the copper is distributed according to local circumstances in the other two zones, the ratio between the copper contents of the zones depending on the extent to which precipitation in insoluble forms has taken place in the oxidation zone.

In the above discussion the mineralogy has purposely been simplified as much as possible, but as a matter of fact the total number of minerals that actually occur in zones of each of these types is very large indeed. However no good purpose is usually served by making any subject more complicated than is necessary to illustrate the general principles involved.

The Influence of Climate on Secondary Changes. Since the changes in ore-deposits that lead to oxidation and secondary enrichment are to a very large extent the work of water of atmospheric origin and the substances dissolved in it, it is easy to see that climate must have a very important bearing on the matter, since it controls both the amount of the circulating waters and their composition. In a region of normally or excessively high rainfall water is abundant but contains a low proportion of dissolved matter, owing to dilution. In an arid region, on the other hand, water is scarce but rich in soluble substances. Under such conditions it is possible that oxidation may be slow, but precipitation of secondary ores considerable if insoluble compounds are formed at once. This is notably the case with some silver ores, as will be explained shortly.

If the rocks are much fissured and the total rainfall small the zone of free water circulation and oxidation may extend to very great depths: for example in many of the very deep copper and silver mines of the Andes in Chile and Peru it is doubtful whether the bottom of the oxidation zone has ever been reached, even at 1500 or 2000 feet from the surface. Another point of great importance that is illustrated by the same district is the prevalence in the circulating waters of chlorides, alkaline carbonates and other soluble salts such as are always found under the climatic conditions postulated. The practical result is the formation of a peculiar and characteristic set of minerals, in the case of copper largely basic chlorides, basic carbonates and basic sulphates, in the case of silver compounds with the haloid elements chlorine, bromine and iodine. The importance of these reactions will appear when the changes in silver deposits are discussed. Thus aridity may be regarded as constituting one special type of climatic influence on secondary ore-changes.

In certain regions with an arctic climate oxidation and alteration of ores is greatly hindered or even entirely prevented by the permanently frozen condition of the ground-water in the upper layers of the crust. In such circumstances there can obviously be no circulation and transfer of dissolved material to lower levels, at least to any important extent: such processes can only take place within the depth of a very few feet below the surface to which the summer thaw extends; a permanently frozen zone, if such exists, must form an impassable barrier. At the present time the frozen regions of the world are more noted for alluvial gold-working than for any kind of deep mining, as is only natural, when all their drawbacks are considered.

Secondary Zones in Silver Deposits. In the majority of silver deposits the primary ore of the silver is argentiferous galena, but in some important instances the silver is mainly contained in argentite and other sulphides, stephanite, polybasite and similar complex minerals with arsenic in place of antimony. In some places these complex sulphides appear to be undoubtedly primary, though they occur much more commonly as secondary ores. As a simple case we will assume a vein of argentite with traces of the antimony and arsenic minerals, subjected to the same conditions as we have supposed in the case of copper.

The simplest possible change is the oxidation of argentite to silver sulphate, a compound which is slightly soluble in water and does not exist as a natural mineral, because it is always precipitated from solution as some insoluble compound. The very insoluble character of silver chloride and of the corresponding bromine and iodine compounds is well known; the silver sulphate solution even when saturated only contains a very small amount of the salt and many natural waters contain enough chloride to precipitate the silver as cerargyrite or horn silver. Any soluble carbonate would also have the same effect, as silver carbonate is very much less soluble than the sulphate, in the ratio of about 1 : 180. Hence silver veins in regions with saline waters, that is to say especially in arid regions, often show rich deposits of chlorides, bromides, or iodides of silver (cerargyrite, bromyrite or iodyrite). In regions with a larger volume of circulating water the silver sulphate is certainly carried downwards in quantity, since many of the deposits show notable enrichment in zones of secondary deposition, as with copper. Also many silver lodes some way below the surface are remarkably rich in native silver, often in peculiar platy and wiry forms. This silver is undoubtedly due to reduction of compound minerals, perhaps sulphides, but the chemistry of the process is not clear. When ground-water level is reached the silver still in solution is precipitated, sometimes as a second generation of argentite, often as pyrargyrite and proustite by reaction with antimony and arsenic sulphides, sometimes as stephanite, or polybasite. Thus the relations in a silver lode are nearly the same as in a chalcopyrite lode, the chief difference being the possible absence of any gossan, if no iron is present, and the absence of anything to represent the oxide ore, cuprite, in the case of copper.

Zoning in Mixed Lodes. Most actual lodes are of course not so simple in composition as the ideal examples just described; as a rule they contain two or more metals as sulphides and often a considerable number; it is not at all unusual to find copper, silver, lead and zinc all in one vein and besides the iron that is necessarily contained in chalcopyrite, both pyrite and pyrrhotite are common. Gold is also widely distributed in sulphides and antimony, arsenic and bismuth are often present in varying quantities. Besides all these, the zonary arrangement of primary ores introduces further complications, above all because the primary and secondary zones are not necessarily parallel. Again differences may be introduced by variations in the character of the gangue minerals. It is evident that patches or zones of calcareous gangue may have very notable effects in causing precipitation of carbonates. In fact the possible variations are endless, as may be seen for example by reference to Emmons' masterly monograph on the subject[1]. As before stated, the phenomena are most clearly seen in copper and silver: they are less conspicuous with lead and zinc, though probably of some importance: the occurrence of secondary enrichment in tin deposits is still to some extent a matter of discussion.

The general conclusions reached with regard to copper, silver, lead and zinc in this very brief discussion may be conveniently summed up in the following table; since all these ores actually occur in nature in mixture this should in reality be a solid model and not a flat plane. Since lead and zinc nearly always occur together they are put in the same column. The relative position of the names within any rectangular division does not indicate their actual distribution in space within that zone, nor do all the minerals inserted necessarily occur together. The whole table is generalized as much as possible. Many of the complex secondary sulphides form mixed crystals to an unlimited extent, e.g. tetrahedrite, which often contains both copper and silver.

	Copper	Silver	Lead and Zinc
Zone of Oxidation	Gossan		**Cerussite Anglesite Calamine** Pyromorphite Hemimorphite
	Malachite Azurite Chrysocolla Atacamite	**Horn Silver**	
Upper limit of variable water level	**Cuprite Native Copper**	**Native Silver**	
Upper limit of permanent water level			
Zone of Secondary Enrichment	**Chalcocite Bornite** Tetrahedrite	**Pyrargyrite Proustite**	
Zone of Primary Ore	**Chalcopyrite** Enargite	**Argentite Argentiferous Galena**	**Galena Blende**

[1] Emmons, "The Enrichment of Ore Deposits," Bull. 625, U.S. Geol. Survey, 1917 (513 pp. with very full bibliography).

The Chemistry of Secondary Enrichment of Copper Ores. A large amount of careful experimental work has been carried out on this subject, especially in America, and it is found that many of the observed reactions can be reproduced in the laboratory. At the outset there are certain general facts to be considered. In the first place practically all copper deposits contain iron, either as chalcopyrite, pyrite or pyrrhotite, or in other more complex forms. Now all iron compounds are specially liable to oxidation, all the ferrous salts being very unstable, and it is known that both ferrous and ferric sulphates are specially active chemical agents, having the power of attacking a vast number of other natural compounds. Copper also forms two distinct series of compounds, cuprous and cupric salts, and of these the cuprous series readily undergo oxidation. Hence all the materials for a complicated set of reactions are ready to hand.

The commonest of all copper ores is chalcopyrite, or a mixture of pyrite and chalcopyrite. Field observation shows that when this decomposes in presence of water and oxygen the result is a mixture of hydrated ferric oxide and copper sulphate. It is very difficult or impossible to draw up a simple and convincing chemical equation to represent this change. In all probability, however, the steps in the process are somewhat as follows:

$$CuFeS_2 + 4O_2 = CuSO_4 + FeSO_4$$
$$FeS_2 + H_2O + 7O = FeSO_4 + H_2SO_4$$
$$2FeSO_4 + H_2SO_4 + O = Fe_2(SO_4)_3 + H_2O$$
$$Fe_2(SO_4)_3 + 4H_2O = Fe_2O_3 . H_2O + 3H_2SO_4.$$

The final result therefore is a precipitate of solid iron oxide and an acid solution of copper sulphate. Since, however, there is no definite experimental evidence to support these precise equations it is better to represent them in a general way thus:

$$\left. \begin{array}{c} FeS_2 \\ CuFeS_2 \end{array} \right\} \rightarrow FeSO_4 \rightarrow Fe_2(SO_4)_3 \rightarrow Fe_2O_3 . H_2O.$$
$$\downarrow \qquad\qquad\qquad\qquad \downarrow$$
$$CuSO_4 \qquad\qquad\qquad H_2SO_4$$

The next step in the process, namely the effect of an acid solution of copper sulphate on different sulphides has been investigated experimentally by Zies, Allen and Merwin[1], who obtained chalcocite from pyrite by the following quantitative reaction:

$$5FeS_2 + 14CuSO_4 + 12H_2O = 7Cu_2S + 5FeSO_4 + 12H_2SO_4.$$
pyrite   chalcocite

Under slightly different conditions covellite was also obtained. When chalcopyrite was treated with copper sulphate chalcocite was also deposited:

$$5CuFeS_2 + 11CuSO_4 + 8H_2O = 8Cu_2S + 5FeSO_4 + 8H_2SO_4.$$

Chalcopyrite was also altered to covellite by a much simpler reaction:

$$CuFeS_2 + CuSO_4 = 2CuS + FeSO_4$$

but in nature covellite is much less common than chalcocite, perhaps owing to a further reaction with the ferrous sulphate formed, which would itself undergo oxidation, at the same time reducing the cupric sulphide to Cu_2S, chalcocite. In the same series of experiments sphalerite and galena were found to alter first to covellite and then to chalcocite. In some experiments it was found that $CuSO_4$ eventually changed chalcocite to metallic copper and sulphuric acid.

Some of the equations given above are so complicated that it is not possible to interpret them strictly as representing exactly what happens in nature, but they undoubtedly do illustrate in a general way what occurs when the downward-moving solutions of copper salts come in contact with the primary sulphides of the lower zones.

Metamorphism and Ore-deposits. A considerable amount of doubt exists as to the part played by metamorphism in the genesis of ore-deposits. This doubt arises to a certain extent from a want of precision in the definition of the term metamorphism, which is used in different senses by different authors, often somewhat vaguely and inaccurately. The definition here

[1] Zies, Allen and Merwin, "Some Reactions involved in Secondary Copper Enrichment," *Econ. Geol.* vol. XI, 1916, pp. 407–503. See also Young and Moore, "Laboratory Studies on Secondary Sulphide Enrichment," *ibid.* pp. 349–65 and 574–81.

adopted will be as follows: metamorphism is the production of new minerals or new structures, or both, in pre-existing rocks, by heat or pressure, or both. This amounts in point of fact to a rearrangement of the constituents of the rock, and is framed expressly to exclude those cases where there is addition of new material from outside, with or without removal of material already present. For changes of the class just mentioned we shall adopt the term metasomatism, which means a change of substance. In metamorphism in the strict sense the bulk analysis of the rock is the same before and after the alteration; in metasomatism it is different. It must be admitted that it is sometimes difficult to apply this rule quite strictly: sometimes metamorphic processes involve loss of water or other volatile constituent, but there should be no *gain* of material.

The great majority of the ore-deposits described by most authors as metamorphic and contact deposits are in point of fact metasomatic in origin, in that the metallic constituents have been added during the chemical, mineralogical and physical changes that took place. These are often described as replacements, and in many instances the use of this term is well justified, since gain in one constituent almost necessarily involves loss of another, at any rate on a large scale, owing to the increase of volume otherwise involved. The addition of small quantities of a valuable constituent like gold might however make a payable ore with no appreciable change of volume. Metasomatic deposits of this kind are very common and of enormous practical importance.

There are however a number of ore-deposits in which it is clear that the ores have acquired their present condition and mineralogical character as a result of metamorphic processes in the stricter sense, as defined above. This of course implies that the metallic constituents were present in the rock before the metamorphism and have merely assumed new physical and mineralogical characters, or entered into new chemical combinations. Thus in a sense the genesis of the ore is not due to the metamorphism since its elements were there before, though existing in different combinations: it is the *present condition* of the ore that is to be referred to metamorphism, not its origin. It is obvious however that the state of combination of a metal in an ore may be a matter of great importance to the miner and to the metallurgist, since different minerals vary so much in their qualities with regard to ore-dressing, transport and smelting.

Two classes of Metamorphism. The various changes included under the heading of metamorphism are due, as before stated, either to heat or to pressure, or both combined. In many instances it is difficult to disentangle the effects of the two causes, since increase of temperature in rocks is generally accompanied by increase of pressure, owing to expansion and other causes, and it is almost impossible to conceive of intense pressure without production of heat by friction. Nevertheless there are cases in which one or the other cause is clearly dominant. Many terms have been used to express the character of the two classes, and of these perhaps the most satisfactory are *thermal* and *dynamic*. Broadly speaking then, we may say that the effect of thermal metamorphism is to cause the elements of the rock to enter into new combinations, without necessarily much change of structure, at any rate on the large scale, while dynamic metamorphism produces in addition to new minerals, also new and conspicuous structures in the rock; in other words dynamic metamorphism gives rise to gneisses and schists.

The Causes of Thermal Metamorphism. Without entering into an elaborate discussion of the origin of thermal metamorphism, such as may be found in any text-book of general geology, it will suffice to say that two chief processes are here operative, namely: a general rise in the temperature of the rocks owing to deep burying under a thick blanket of sediment, usually resulting from a general and long-continued subsidence of the region, or the intrusion of large masses of heated rock-magma into already existing rocks. The effects of the first type mentioned are especially liable to be complicated by pressure and the resulting rocks are often very like the crystalline schists due to dynamic metamorphism. In some cases also it appears that there has been complete fusion up to a certain level, so that the fused rock behaves to all intents and purposes as a true igneous magma and may show the phenomena characteristic of normal vein-formation. This is of course an extreme case, but has probably happened with some of the great bathyliths of the pre-Cambrian, as in Canada. More common and more easy to understand are the phenomena accompanying the intrusion of igneous masses; these are much more local in character and on a smaller scale, so that the effects are usually confined to a somewhat narrow shell around the intrusion. The effects of this type are often described as contact-metamorphism. This term is however often used somewhat loosely by writers on ore-deposits to include many of what should more accurately be described as replacements: that is to say, the ore-deposits are metasomatic rather than strictly metamorphic.

The Effects of Thermal Metamorphism. In the cases now to be considered, where there is no addition of material from outside, the changes brought about are mainly of a mineralogical nature: that is to say that the constituents already existing in the rock may undergo a change of physical state and already existing combinations may be broken up and differently combined. The final result will obviously depend on the chemical constitution of the original rocks and the temperature and pressure relations prevailing at the time. The study of metamorphism from this point of view is simply a matter of physical chemistry, and the principal factors to be taken into account are the ranges of stability of the different chemical compounds which we call minerals.

The principles involved can best be illustrated by taking a very simple case. Suppose we have a pure sandstone, consisting of quartz grains with a cement of quartz: here there is only one component, silica, and no chemical reaction can occur; only a physical change is possible and under the influence of high temperature such a rock will crystallize, the distinction between grains and cement will be wiped out, and the result will be a quartzite. Again in the case of a pure limestone, consisting of calcium carbonate alone, either as calcite or aragonite. When limestone is heated in a kiln it breaks up into lime and carbon dioxide, but underground this dissociation cannot take place, and the only possible result is crystallization as calcite, which is the stable form under the given conditions; such a rock is called marble. When there are two components however the conditions become different and chemical reactions may occur. Thus if it were possible to have a rock composed entirely of silica and alumina, these oxides would combine directly thus:

$$Al_2O_3 + SiO_2 = Al_2SiO_5.$$

No natural rocks are actually so simple as this, but this reaction does occur extensively in clay rocks when metamorphosed, forming the minerals andalusite, kyanite or sillimanite, all of which have the composition Al_2SiO_5. The occurrence of each is determined by conditions of temperature and pressure. In a similar way in a siliceous limestone the following reaction occurs:

$$CaCO_3 + SiO_2 = CaSiO_3 + CO_2.$$

It is not clear how the carbon dioxide escapes, but it obviously does do so, and a band of sandy limestone when followed up to the edge of an intrusion is often found to pass into a rock wholly or partly composed of the lime-silicate, wollastonite, $CaSiO_3$. Any excess of calcium carbonate will crystallize as calcite, and the rock will then be a mixture of wollastonite and calcite, or wollastonite-marble.

When three components are present the relations are evidently more complex, but the same general principles hold: to take another concrete case, a rock composed of calcium carbonate, alumina and silica will yield various minerals according to the proportions of the original mixture and the temperature to which it is raised. This case has been worked out experimentally in great detail, owing to its importance in the technology of Portland cement. The whole matter is very complex, and it must suffice to say here that the minerals most commonly formed in nature from impure argillaceous limestones are lime-garnet and lime-felspar. If magnesia is present in addition, i.e. if the original limestone is dolomitic, pyroxenes or amphiboles will also be formed. Thermally metamorphosed limestones and dolomites of this kind are very commonly associated with the so-called contact ore-deposits, but in most instances there has been more or less metasomatism as well. These will be considered separately. It is evident that the processes of metamorphism in the restricted sense as here defined cannot in themselves give rise to ore-deposits, unless the useful metal is already present in the rock. However the study of these processes has thrown considerable light on the present condition of certain valuable ore-bodies; that is to say their mineralogical character, though not their bulk chemical composition, has been determined by pure thermal metamorphism. In the case of one interesting type the principles were first worked out for dolomitic limestones as follows: the bearing of these facts on certain classes of ore-deposits will appear later.

Dedolomitization. This word has been introduced to express the mineralogical changes that take place when a dolomitic limestone is subjected to heat, as for example from an intrusive igneous mass[1]. In some instances dolomite crystallizes as such, forming a dolomite marble, but in other cases the mineral is dissociated as follows:

$$CaMg(CO_3)_2 = CaCO_3 + MgO + CO_2.$$

The calcium carbonate crystallizes as calcite and the magnesia forms the mineral periclase, which however is very hygroscopic and is soon hydrated to brucite, $Mg(OH)_2$. This mixture is called brucite-marble.

When silica is present as sand-grains or chert the results depend largely on the relative proportions of the dolomite and silica. When the former is in excess the reaction may be represented as follows:

$$2CaMg(CO_3)_2 + SiO_2 = 2CaCO_3 + Mg_2SiO_4 + 2CO_2.$$
$$\text{calcite} \qquad \text{forsterite}$$

The mineral forsterite, Mg_2SiO_4, is a member of the olivine group, and is therefore easily converted to serpentine. When quite fresh the mixture of calcite and forsterite is called forsterite-marble; when serpentinized it is often called ophicalcite or Connemara marble, the green colour of the serpentine being due to traces of iron.

When, on the other hand, the silica is in excess the results are

[1] Teall, *Geol. Mag.* 1903, p. 513. Harker, "The Tertiary Igneous Rocks of Skye," *Mem. Geol. Survey*, 1904, pp. 144–51. Hatch and Rastall, *Quart. Journ. Geol. Soc.* vol. LVI, 1910, pp. 507–20.

different and the mineral diopside, a member of the pyroxene group, is formed, thus:

$$CaMg(CO_3)_2 + 2SiO_2 = CaMg(SiO_3)_2 + 2CO_2.$$
$$\text{diopside}$$

The metamorphosed rock will then consist of diopside alone, or a mixture of diopside and quartz, according to circumstances. A reaction of this kind is known to have occurred in many instances, and a good example is afforded by some of the contact-pyroxenites and "skarn" rocks of Canada, Sweden and Finland.

When alumina is also present a large variety of minerals may be formed, according to local circumstances, such as spinel, garnet, idocrase, epidote, zoisite and scapolite. It is hardly necessary to discuss the formation of these in detail, as in many instances there is a certain amount of metasomatism as well. In particular it is quite clear that sometimes there has been a considerable diffusion of material across the boundary between the igneous and the sedimentary rock, especially a diffusion of silica from the intrusion, so that a zone of silicates may be formed along the outer part of the sediment. Similarly in certain cases, lime and magnesia have diffused into the igneous rock, forming a marginal zone of more basic character to a granite, which may take the form of diorite or even gabbro. Some of the Laurentian granites of Canada show this very clearly: the basified margin of the intrusion passes almost imperceptibly into the pyroxenite of the outer portion of the original dolomite. A similar diffusion of silica from a granite into a dolomite has been demonstrated in Natal, where a regular gradation from more acid to less acid minerals can be traced in successive zones, beginning with diopside rock and ending with forsterite-marble. In all these cases of metamorphism of dolomites the guiding principle is that the magnesia molecule is attacked first, and lime silicates are only formed when all the magnesia has been used up.

The Zinc Ores of New Jersey. We may now proceed to apply the principles just stated to certain other similar cases in which the metallic constituents of the rocks are of commercial value. The best instance of this is furnished by the famous zinc deposits of Franklin Furnace, New Jersey, U.S.A. These ores consist of minerals containing zinc, manganese and iron embedded in a gangue of crystalline calcite, and the ore-body behaves in every way structurally like a metamorphosed sedimentary stratum. It is well known that many other deposits of zinc and manganese ores have been formed by metasomatism or replacement of limestone, so that they now consist of masses of zinc ore, either as sulphide or oxidized ores, calamine, smithsonite, etc., as at Malmedy in Belgium. We start then from the assumption of a mass of zinc ore, mainly zinc carbonate, with calcite, with impurities of manganese, iron and silica—the iron and manganese playing the part of the magnesium in the dolomites as just described.

The zinc minerals now found in this ore are zincite, ZnO, gahnite, $ZnAl_2O_4$ (a spinel), franklinite (an iron-zinc-manganese spinel) and willemite, Zn_2SiO_4: since iron and manganese are isomorphous with zinc, we may assume an ideal case and substitute one for the other as required in any proportion. The parallelism between the zinc and magnesium minerals then becomes obvious, as shown in the following table:

brucite, MgO.	zincite, ZnO.
spinel, $MgO.Al_2O_3$.	gahnite, $ZnO.Al_2O_3$.
	franklinite, $(Zn, Mn.Fe)O.(Fe, Mn)_2O_3$.
forsterite, $2MgO.SiO_2$	willemite, $2ZnO.SiO_2$.

Only when we come to the diopside group does the analogy fail; there is no double silicate of calcium and zinc, but since abundance of calcite is present and silica was therefore not in excess, we cannot expect to find a mineral of this type. The parallelism is striking and suggests that this is an instance of the simple thermal metamorphism of a calcareous rock originally containing zinc, manganese, iron and alumina.

Weathering Processes in Rocks. The study of the various changes that take place in rocks as a result of the action of weathering agencies is in some cases of considerable importance to mining geologists, since such changes may bring about great alterations in both the chemical composition and physical characters of ore, gangue, or country rock, and thus may have an important bearing on the mining and dressing of the ores. These changes are very variable in character and often very far-reaching in their effects. Moreover it is important to notice that such changes may be either beneficial or the reverse. The ore may be leached out, or concentrated; the gangue and the country rock may be hardened, or softened: in fact the variations are endless.

One very important class of changes, namely, the oxidation of the ores, essentially a process of weathering, is dealt with in some detail elsewhere. Another somewhat similar process is the concentration of ores on the outcrop of lodes by removal of less stable material. Many such instances will be found in the descriptive sections of this book (see, for example, under the headings gold and tin). It is proposed to deal here in a general way with some of the alterations that take place in the country rock of different types, usually leading to disintegration and removal of some or all of the rock-forming minerals, and thus changing the general appearance and bulk-composition of the rock.

Since weathering is due to meteoric agents, it is naturally largely controlled by climate. Its effects in the hot, damp tropics differ greatly from those seen in the arid regions of the western United States or in the frozen wastes of Siberia. In temperate climates with well-marked seasons they are most complicated, since all the different factors, heat and cold, water, snow and ice, wind, and organic agencies, all act in turn and their effects are superposed and often difficult to disentangle. In the tropics with their hot, moist climate and abundant vegetation, rock-decay is very rapid and often extends to great depths. In desert regions the changes are more superficial and the most notable feature is the production of great masses of sand and loose stones. In arctic regions chemical decomposition is slow and the greatest effects are due to frost and snow water. But in nearly all regions the rocks show more or less superficial decomposition; sometimes a mere skin, sometimes extending downwards for hundreds of feet, so that it is often very difficult to determine, in prospecting, what the original character of the rock may have been. It is of special importance for the mining geologist to make himself familiar with the appearance of rocks when weathered: specimens of a rock from a decomposed surface outcrop may look very different from specimens of the same rock from a deep shaft[1]. A further complication is added by the fact that the mineralized portion of a rock, even when fresh and unweathered, may be of quite different composition and characters to the same rock when not mineralized. In studying the effects of weathering on a given rock, it is necessary to bear this fact carefully in mind, and much useful experience can be gained by the examination of the waste dumps of a mine which is known to be working in one originally uniform rock. Ore-bodies vary very much in their susceptibility to weathering: some, such as quartz-veins are very resistant and tend to stick out as bold features of the landscape (*e.g.* some gold reefs). Others are more easily weathered than the rest of the rock (*e.g.* the conspicuous gossans of many sulphide veins), while others again may outcrop merely as hollows, filled with surface rubbish, owing to more rapid removal of the ore. As a general rule the mineralized portion of a rock weathers at a different rate to the rest, and since mineral veins are often also faults they may weather very rapidly owing to easy access of water.

It is not proposed to describe here in detail the actual mechanism of the weathering processes that occur in rocks under various conditions: full information on this subject may be found in any general text-book of geology.

Weathering of Igneous Rocks. Since the minerals of deep-seated igneous rocks are formed at high temperatures and pressures they are naturally unstable under the very different conditions that prevail at the surface, and tend to break up into simpler and usually more highly oxidized compounds. Of the common rock-forming minerals quartz and white mica are the most stable and usually undergo little or no alteration, thus becoming concentrated in the weathered layer. On the other hand the felspars and the ferromagnesian minerals, such as

---

[1] In this respect most museum collections are of little use. Much trouble is taken by curators to obtain "fresh" specimens of rocks, but these are not what the geologist sees on the ground. A first experience in the field is usually somewhat disconcerting, however well the student has learnt his rocks in the class-room, since the weathered stuff looks entirely different, and in the tropics especially, it is often impossible to dig down to un-weathered rock.

biotite, hornblende, augite and olivine, are easily decomposed. The alkali felspars, orthoclase, microcline and albite, lose their potash or soda, and generally some silica, forming kaolin or sometimes aggregates of white mica, while at the same time the rock crumbles to a powder at the surface. Felspars with more lime give rise to more complex changes, often yielding both white mica and calcite. Biotite changes from brown to green and forms chlorite, while hornblende and augite also give rise to various green or brown chloritic minerals, at the same time setting free a good deal of iron oxide, which forms a rusty coating on the surface. Olivine is generally converted first into serpentine and eventually to carbonates.

It will thus be seen that in weathering igneous rocks may yield a good many new chemical compounds, some of which are very stable under the surface conditions, and accumulate, such as white mica, kaolin and iron oxides, together with the original quartz, if such is present. On the other hand the alkalis set free from felspars and the carbonates are soluble and are carried away. The general result is to break up the rock and form a loose layer, which may also be carried away by mechanical agencies, such as running water, wind and moving ice. The extent of removal of the weathered layer is thus very much a matter of climate. Where water and air are able to penetrate along joints or other open fissures weathering may extend far under ground and of course tends to accentuate the presence of open joints. This also applies to rocks of classes other than igneous.

The Weathering of Sediments. This subject is not quite so easy to deal with as the weathering of the igneous rocks, because the sediments show a much wider range of composition and structure: the effects also depend to a great extent on the degree of cementation and metamorphism that they have undergone. Sediments usually consist of minerals that are stable under atmospheric conditions, in so far as they are not soluble in water or ground-solutions, hence the disintegration is generally physical or mechanical rather than chemical in its nature. The simplest case is that of a sandstone, where the cement is dissolved and the quartz grains set free, thus forming a surface layer of sand. In the same way conglomerates form gravel-beds, which may be difficult to distinguish from newly-deposited river gravels. Such are largely formed from the auriferous and other conglomerates of the southern Transvaal, especially the Elsburg series.

Most of the argillaceous rocks are of so fine a texture that they simply form mud, even from well-cleaved slates. Owing to oxidation of the iron sulphide so abundant in such rocks the weathered portions are usually yellow or brown, and this brown stain often extends far underground where joints or fault-fractures are present.

It is well known that limestones and dolomites are much more soluble than any other rocks, except the salt deposits. Consequently they weather in a special way. The outcrops of limestones are often covered by a red residual deposit of iron oxide and other insoluble impurities, of a clayey nature, and similar material may accumulate in caves and hollows below the surface. In regions of heavy rainfall, where the residue is washed away, limestones often show a perfectly bare surface at the outcrop, since the weathering is all due to solution. It is unnecessary here to enter into the special features of limestone denudation, both above and underground. The results of it are specially favourable to ore-formation of several types and are discussed elsewhere, especially in the chapter on lead and zinc.

Neither is it necessary to consider in detail the processes affecting highly ferruginous rocks, which often give rise to considerable masses of residual ironstones and alter the characters and composition of bedded iron ores near their outcrop.

Weathering and Ore-deposits. The chief point incidental to our present purpose brought out in the last two sections is that the superficial alterations of rocks are bound to have some effect on the ores contained in them, if any are present: disseminated ores, such as tinstone in granites, platinum in serpentines and gold in many rocks, are set free, and may be later concentrated by water or other action. Less stable ores are oxidized and may be leached out altogether, or on the other hand, like compounds of iron and manganese, they may be concentrated at the surface as residual deposits. The ultimate fate of oxidized ores is dealt with at some length in the section on oxidation and secondary enrichment. It is obvious, however, that it is one of the necessary qualifications for a successful prospector to be able to judge from the characters of weathered rock what it would be like in depth. Such an indication as the gossan capping of a sulphide vein is obvious enough, but the widespread surface of a sedimentary or igneous rock is more difficult to deal with, since so many factors are involved whose effects are very variable under different conditions. For this purpose experience is after all the only safe guide, and is worth volumes of text-books. Much may be learnt by careful study of quarries or open workings of any kind, whether metalliferous or not.

In many districts also further complications are introduced by the presence of thick coverings of glacial drift or other transported and alluvial deposits, made up of material entirely different from that of the underlying rock. In such cases, in the absence of natural or artificial exposures, it is obvious that boring or digging pits is the only resource.

Rock-changes accompanying Mineralization. The formation of ore-deposits is as a rule a somewhat complicated process: the effects are not confined to the mere introduction of the metal or metals into their present position, but frequently notable and far-reaching alterations are brought about in the surrounding rocks. It is comparatively rare to find a clean-cut vein, filled with ore and gangue cutting across unaltered rock. Since most ores are deposited from solutions (or vapours) it is only natural that these should penetrate beyond the actual walls of the fissure, if such exists, and when, as is so common, an ore-body is a belt of shattered rock without definite boundaries, or a more or less complete replacement of the substance of the rock itself, it is obvious that notable changes must be set up in its neighbourhood.

In the cases about to be described it is not desired to imply that these processes form a necessary part of the mineralization: in some instances the alteration of the rock and the introduction of the ore are all due to one agency: in other instances it can only be asserted that the two phenomena are often associated, possibly sometimes because a previous change in the rock has rendered it more susceptible to mineralization. It is usually extremely difficult to prove that two processes were exactly contemporaneous, or even that they had a common origin.

The following list comprises the chief alteration processes that occur in rocks, either igneous, sedimentary or metamorphic; they are divided into groups according to their genetic relationships:

    I. Tourmalinization, greisening, kaolinization, scapolitization.

    II. Silicification, sericitization, propylitization, pyritization.

    III. Dolomitization, carbonatization.

These groups must not be interpreted too strictly: they are more a matter of convenience than anything else. The first group comprises some of the phenomena of pneumatolysis, which certainly take place at a fairly high temperature: they are described in Chapter II and need not be discussed further in this place, except to remark that the term kaolinization is often used somewhat loosely to describe some of the effects of weathering in felspars. This is unduly to extend its meaning.

The processes of group II are very common accompaniments of mineralization and must be examined in some detail. They seem to be due for the most part to waters derived from deep-seated sources, often magmatic, and are often contemporaneous with the formation of the ores.

Silicification. In this case, however, it is necessary to draw a distinction between two processes, similar in their general effects, but differing essentially in origin. On the one hand there is a type of silicification which is a perfectly normal process in some classes of sedimentary rocks: for example the cementing of sands by silica and their conversion to quartzites, and the hardening of clays to shales and mudstones; the former is always and the latter often due to deposition of silica from ground water. Again the metasomatic replacement of limestones by silica, with formation of chert, is a common phenomenon which can without doubt go on at low temperatures. In many instances the formation of cherts in limestone has acted as a hindrance to ore-deposit, the distribution of the ores showing that the unaltered limestone has been replaced while the cherts acted as an impervious barrier[1]. On the other hand there is a type of silicification, which is clearly connected with lode-formation and the introduction of ores. The extreme type is the formation of greisen veins, which grade into masses of pure quartz, the amount of mica being very variable. It is often difficult to decide whether masses of quartz in lodes and in altered country rock are intrusive into actual fissures or have been formed by molecular replacement. Definite stringers may generally be considered as deposited in open cracks, but irregular masses of quartz can often be shown to be replacements of other minerals, especially felspars.

Very frequently however the silica replacing other minerals is deposited not as quartz, but as chalcedony or even as some amorphous form of hornstone or opal. Endless names are applied to these forms of silica, such as jasper, carnelian, bloodstone and many others, depending on shades of colour, mostly due to iron. Many specimens of jasper are silicified rhyolites. Flints, agates, onyx, etc., are all closely allied to cherts.

Sericitization. In a general way this process may be described as a common decomposition-reaction in felspars, especially in alkali-felspars; therefore it is characteristic of the alteration of igneous rocks of the more acid types. Essentially it consists of an alteration of the felspar crystals to an aggregate of finely-divided mica. Such a change is a common result of ordinary weathering, but it often takes place in a more exaggerated form under the influence of mineralizing solutions. By a very simple reaction, involving only hydration and leaching out of a certain proportion of the potash, orthoclase and microcline are converted into muscovite. Here there occurs merely a change in the relative proportions of the chemical constituents, potash and silica being set free. The latter is often deposited again immediately as quartz. But in the case of the plagioclase felspars, according to a number of investigations carried out of late years in America and elsewhere, aggregates of potash mica may also result from the action of solutions of magmatic origin. This

[1] Cherts may also be primary sediments, deposited under peculiar conditions, such as the ferruginous cherts of Lake Superior and the "calico rock" of South Africa (see p. 329).

involves a loss of the soda of the original plagioclase, while the lime molecule often gives rise to carbonates, or sometimes to minerals of the epidote group, especially zoisite. The reason for this type of change is by no means clear, but the potash must obviously be derived from the magmatic solution. Secondary potash felspar of the clear glassy variety known as adularia, is also frequently formed. By this process a diorite for example, or a granodiorite or monzonite, or the corresponding dyke-rocks and lavas, are converted into aggregates of quartz and scaly mica in fine flakes, often with some finely divided felspar as well, and a sericitized rock of original intermediate character may come to contain a good deal of free quartz without much change of total composition. The alteration of orthoclase to muscovite may be represented by the following equation:

$$\underset{\text{orthoclase}}{K_2O \cdot Al_2O_3 \cdot 6SiO_2} + xH_2O \rightleftharpoons \underset{\text{muscovite}}{(H, K)_2O \cdot Al_2O_3 \cdot 2SiO_2} + \underset{\text{quartz}}{4SiO_2}$$

Under the conditions that prevail at low and moderate temperatures and pressures, in the later stages of vulcanicity, as well as under the action of weathering agencies, this equation proceeds from left to right. It is probable that under intense dynamic metamorphism the action is reversed, as in the formation of certain felspathic gneisses.

Propylitization. This is a type of alteration brought about in certain igneous rocks mainly by the action of solutions belonging to a late or solfataric stage of vulcanicity. It is specially characteristic of andesites, basalts or other types of basic lavas and their corresponding intrusions. These rocks are always rich in ferromagnesian minerals, hornblende, augite or olivine, and their plagioclase felspars contain a large proportion of the lime molecule. Hence the characteristic secondary minerals are chlorite and epidote, with carbonates of lime and magnesia and sometimes a little sericite derived from the felspar. Often also a good deal of pyrite or marcasite are developed: the iron for these sulphides probably comes from the ferromagnesian minerals and the sulphur from the volcanic solutions or vapours. The altered rock as a rule has a green colour owing to the abundance of chlorite; there is a decrease of alkalis and sometimes of silica, but the total bulk composition is not much altered unless the amount of sulphur introduced is very large. Naturally, as in all changes of this type, there is an increase of water in the altered rock.

Pyritization. This term is here used in a somewhat narrow sense to indicate an exaggerated form of the type of change alluded to in the preceding paragraph: it is not here intended to include all the manifold forms of sulphide mineralization of which pyrite is so characteristic, but only the development of pyrite from the iron content of the original ferromagnesian minerals of igneous rocks by the action of volcanic solutions or vapours rich in sulphur.

In nature it is not usually possible to establish a clear-cut line of demarkation between the different types of rock-alteration just described. Most commonly they are found associated to different degrees in the same rocks. The mineralized portions of such rocks usually consist of an aggregate of sericitic mica, chlorite, epidote, pyrite, iron oxides, carbonates of lime, magnesia and iron, and many other minerals, with more or less secondary silica in the form of quartz or other crystalline, crypto-crystalline or amorphous forms. All of these occur in endless variety.

Changes belonging to this category mainly occur in volcanic rocks and are specially characteristic of the shallow-seated gold-silver mineralization, often telluridic, of the Tertiary eruptives of the western United States and of Hungary and Transylvania. They are most typically developed in rocks of andesitic and basaltic type, including also dacites and some of the less alkaline of the so-called trachytes (*i.e.* monzonitic lavas), as well as the intrusive masses of similar composition associated with each of these types. In most instances the resulting rocks really show a close analogy to some of the so-called pneumatolytic vein-types of the deeper-seated mineralization, allowance being made for differences of original composition and difference of physical and chemical conditions. A propylitized andesite or porphyrite does not really differ in any essential feature from some of the chloritic lode-stuff of the tin and copper deposits of Cornwall for example, except in the absence of certain minerals containing fluorine, boron and other elements characteristic of higher temperature and pressure, and in the character of the ore-minerals, which are in the less deep-seated deposits mainly free gold, auriferous sulphides and tellurides, instead of cassiterite, wolfram, mispickel and chalcopyrite. The difference is in reality an exposition of the zonary principle, with modifications in each case appropriate to the special conditions.

Formation of Alunite. In some regions where lavas rich in potash felspars have been acted on by solfataric vapours near the surface there is often a considerable development of the mineral alunite, a hydrated potassium aluminium sulphate, which unlike common alum, is insoluble and forms a white or pinkish granular aggregate and imparts to the rocks a bleached appearance, which is often confused with kaolinization. The sulphur required for the formation of this mineral may be of direct volcanic (solfataric) origin, or it may be formed by oxidation of pyrite or marcasite. Lavas rich in alunite sometimes contain a workable amount of gold, which is probably derived from auriferous pyrites. Some large deposits of alunite also form a valuable source of potash compounds: such are worked in Italy, Nevada, Utah and New South Wales.

Carbonates in Altered Rocks. In nearly all the cases of alteration as above described and in most rock-changes due to weathering and low-temperature agencies, much carbonate is commonly developed. This may include not only calcite, but also dolomite, chalybite and a variety of isomorphous mixtures of these compounds. This is quite apart from the ordinary deposition of crystalline carbonates as vein-fillings in open spaces: these are probably deposited from solution and are not due to complex reactions among pre-existing minerals, such as have given rise to the carbonates of the altered wall-rocks. The source of the carbon dioxide in the latter case is perhaps to be sought in the volcanic emanations.

Dolomitization and Carbonatization. The alteration of masses of originally more or less pure limestone to magnesian limestone or dolomite rock is a very common process, taking place on a very large scale, but it is not of much importance in the study of ore-deposits, and it is not necessary to enter into a discussion of the theories that have been put forward in explanation of this somewhat obscure subject. So far as the formation of replacement ore-bodies is concerned dolomite rocks behave much the same as ordinary limestones, and their metamorphic changes are also very similar, since lime and magnesia are closely allied from the chemical point of view.

In many processes of ore-deposit however there is to be seen a large development of carbonates of lime, magnesia and iron formed by the alteration of other minerals, including silicates. In such instances the source of the carbon dioxide is not always clear: in many cases it appears to be derived from mineralizing solutions, and it is known that development of carbon dioxide is characteristic of the later stages of vulcanicity, as shown by the common occurrence of effervescing springs and jets of the gas in volcanic regions.

CHAPTER IX

METALLOGENESIS

General Introduction. The highly important and interesting subject of the origin and distribution of the metals throughout the crust of the earth can be approached from two different points of view, which are nevertheless cognate. We have firstly the geographical distribution, the arrangement in space and sequence in time of the different metals, and secondly their association genetically with definite rock-types, or the petrographical view of the subject. Both of these are logically comprised in the general expression *metallogenesis*. From the practical standpoint the former would seem at first sight to be the more important and so it is when we are dealing with known deposits. But the significance of the second becomes more apparent when we consider that it may give us a clue as to where to look for hitherto unknown deposits of a particular character: it is in fact the foundation of scientific prospecting. Just as the search for coal or oil is based on stratigraphical geology, so must prospecting for metals be based on petrology.

In this connection it is mainly with the rarer constituents of the earth's crust that we are concerned[1]. Of the more common elements only iron and aluminium are of first-class importance as metals, though compounds of calcium, magnesium, potassium and sodium are also largely exploited: these however find no constructional applications in the metallic state. All the other metals must be regarded as rare elements from the mineralogical point of view and it is only when they have undergone some natural process of local concentration that they become of economic importance. In the same way, though iron and aluminium are in reality enormously abundant they usually exist in some intractable form, or in material of such low grade as to be unworkable. We must therefore understand by the

---

[1] See table of average composition of the crust of the earth, p. 4, quoted from Clarke, *Data of Geochemistry*.

term metallogenesis the natural processes of concentration as just defined which have given rise to the workable ore-deposits.

In an illuminating paper Washington has discussed the behaviour of the elements as regards the formation of rocks and ore-minerals[1]. He refers them to two main groups of the periodic table: (1) the petrogenic elements, characteristic of and most abundant in the igneous rocks; these are of low atomic weight and occur normally as silicates, oxides, chlorides and fluorides; (2) the metallogenic elements, rare or absent in normal igneous rocks, but existing as ores. They are of high atomic weight and occur in nature as native metals, sulphides, arsenides, but not primarily as oxides or silicates. This scheme of Washington's fits in remarkably well with the periodic classification of the elements and with the facts of mineral paragenesis in ore-deposits, but its author has perhaps not sufficiently taken into account the indubitably close connection that exists between igneous rocks and ores, ignoring the fact that many of the most important deposits of copper, silver, gold, tin, lead, zinc and the minor metals are patently differentiation products of igneous magmas. He has thus made too sharp a distinction between his groups. Nevertheless this publication is of the greatest interest and brings out very clearly the fact that there are in nature two fairly well-defined groups of elements, differing notably in their petrological and mineralogical behaviour. They may also be spoken of in a very general way as the silicate-forming and the non-silicate-forming elements, on the understanding that the term "on a large scale" is implied in the definition.

At this point a very broad and fundamental distinction comes in. An important class of ore-deposits, of which gold-placers and tin-gravels are the best examples, are formed on the surface of the earth by the ordinary operations of the superficial geological processes of denudation and deposition. The valuable constituents of these deposits are derived, and they are therefore properly called secondary ore-deposits. With these we are not at present concerned, since their origin is obvious and their

[1] Washington, "The Chemistry of the Earth's Crust," *Journ. Franklin Inst. Philadelphia*, vol. cxc, 1920, pp. 757–815.

valuable constituents are always obtained at second-hand from deposits of the other class.

The subject of this chapter is the origin of the *primary* ore-deposits; of those that have been formed underground, and there concentrated into payable forms; it is necessary here to take into account also various subsequent changes that have taken place in these deposits and modified their original character for the better or for the worse, since these changes are often of great practical importance. This is not the place to enter into a discussion of the forms assumed by such deposits: this subject is treated elsewhere in this book (see Chapter V). What we are really concerned with is their composition and their origin, and the question of form will only be considered in so far as it has any bearing on the other points.

Since the middle of the nineteenth century there has been a vast amount of discussion as to the origin of the primary ore-deposits. It is not possible to go into this in detail, and it must suffice to sum up very briefly indeed the general trend of thought on this subject. The earlier tendency was to attribute everything to the action of water: in most cases either explicitly or by implication the meteoric waters of the upper layers of the crust. But difficulties soon began to present themselves, especially with regard to the source of the metallic compounds. To explain this, the hypothesis of the so-called lateral secretion was invented, which supposed that the metals already existed disseminated in the surrounding rocks and were dissolved out of them by the ground-water and re-deposited as lodes, veins and other metalliferous deposits. In support of this theory a great number of rock-analyses were made, showing infinitesimal amounts of metals, not more than could be accounted for by impurities in the reagents used and often far below the limits of possible analytical accuracy.

When the inadequacy of this theory in its original form had been sufficiently demonstrated, it was improved upon by supposing that the crust-waters descend far into the interior of the earth and come up again containing soluble metallic compounds which are deposited as veins. Here again an insuperable difficulty arises in the well-known fact that as a rule very deep mines are dry: it is only when open fractures exist that they need pumping. It is abundantly clear that meteoric waters do not normally penetrate more than a few hundred feet, and cannot have been the source of veins which have in some cases been followed down for some thousands of feet. The only logical conclusion is therefore that the material of deep veins came upwards, not downwards.

Furthermore it has now been abundantly demonstrated that no real line of demarcation exists between the so-called "mineral veins" and the dykes which are universally admitted to be of igneous origin: every possible gradation exists between the two, and in many cases a quartz vein has been traced laterally by a gradual mineralogical transition into a pegmatite. Now it is not here maintained that all primary ore-deposits, or even all such ore-deposits having a dyke-like or vein-like form, are of direct igneous origin, but it may safely be considered as proved that a large proportion of them are so. This subject will be again referred to later. There is also an important class generally known collectively as the contact deposits which also need separate consideration, but it may be said at once that these mostly arise from the special type of rock-alteration brought about by the intrusion of igneous masses; in other words they are due to metamorphism in the broadest sense of that term. Here again the introduction of the metallic elements is directly due to the igneous activity, although the ultimate derivation of the metals is not always quite so clear.

For the present however it is advisable to confine our attention to the types of ore-deposit which are believed to be of direct igneous origin, either as segregations within the limits of an intrusive mass, or as offshoots from an intrusion in the shape of dykes, lodes and veins in their many forms. There must also be taken into account all those cases where the mineral deposits are obviously connected with volcanic eruptions, as in Hungary, the western United States, Mexico and the Andes: in many, if not most, of these instances recent research has shown that the mineralization is largely due to processes taking place after the actual eruptions; in fact to an intrusive phase of dykes and veins injected into the lavas after their consolidation, and to disseminations due to vapours and solutions belonging to still later stages of eruptive activity. It is also worth noting that the gangue-minerals of some veins may be wholly or in part of later date than the ore-minerals and may be derived from a quite different source.

The extended study of different types of ore-bodies during the last few years has brought out quite clearly the fact that certain metals tend to be associated with certain types of igneous rock: of this there are now many good examples. Only one or two of the most clearly established cases need be mentioned at this stage, such as the association of tin and tungsten with granites, of nickeliferous sulphides with gabbros and norites, and of chromium and platinum with peridotites. Another class of facts has also to be taken into account here, namely, that although some metals occur in rocks of almost all compositions, nevertheless they enter into different combinations in different rock types. A concrete example will perhaps make this point clearer.

Iron in igneous rocks. As already stated iron is one of the more abundant elements in the accessible part of the earth: according to Clarke[1] the average of a very large number of analyses shows that the content of metallic iron in the igneous rocks amounts to 4·50 per cent. This iron however exists in nature in two forms, as ferrous and ferric iron, the former being slightly the more abundant in the unweathered rocks. With regard to its combination with other elements to form minerals from magmatic solutions the behaviour of iron is peculiar and not fully understood. Its tendency to form silicates appears to be much less strong than with the other important rock-forming elements. Magnesia, lime, alumina, potash and soda are always completely saturated by silica if sufficient of the latter is present, forming silicates up to their full capacity, whereas it is possible for oxides of iron, especially magnetite, to separate from a magma containing an excess of silica. Hence we commonly find some accessory magnetite even in granites, and in some instances enormous masses of magnetite have separated by differentiation from acid intrusives. In basic rocks, on the other hand, ferrous iron is frequently found in combination with titanium, as

[1] *The Data of Geochemistry*, 4th edition, 1920, p. 35.

titaniferous magnetite or ilmenite, while in the ultrabasic rocks it is often combined with chromium as picotite or chromite. Hence not only the proportion but also the state of combination of the iron varies according to the composition of the rock as a whole. No doubt this is controlled to some extent by the relative abundance of such elements as titanium and chromium in the basic rocks. It may be supposed that iron has a strong affinity for these and tends to combine with them if they are present. The concentration of such elements into basic rocks is evidently part of the larger question of the differentiation of the magmas before intrusion.

Thus the most characteristic combinations of iron may be arranged as follows:

acid and intermediate	basic	ultrabasic
magnetite	ilmenite	chromite.

The reason for the separation of magnetite and quartz from the same magma, instead of iron silicates, is difficult to explain: it probably depends on the fact that iron does not readily form simple silicates, but requires the admixture of a certain proportion of the magnesia- or lime-silicate molecules in order to crystallize as an amphibole or a pyroxene: this is perhaps connected with the comparatively low freezing-point of silicates rich in iron.

It would appear therefore that in the ultrabasic and basic rocks the ferrous iron molecule combines with chromium or titanium when these are present to form chromite or ilmenite until these molecules are saturated, and any iron then remaining over unites with magnesia and lime in silicates. In the acid rocks however its behaviour is different, and oxides of iron can separate in large quantities from an acid magma: the affinity of iron for silica appears therefore to be slight, or what is really the same thing, iron oxides are very insoluble in acid silicate solutions.

Iron and Sulphur. We must next consider the behaviour of iron towards sulphur. Ferrous iron can unite with sulphur in two proportions, as FeS and FeS_2. The physico-mineralogical relationships of these two sulphides are obviously complex and as yet little understood, but at any rate it is clear that iron sulphide separates from magmas sometimes as pyrites, FeS_2, and sometimes as pyrrhotite, which is approximately FeS. (It usually seems to contain some amount of FeS_2 in solid solution.) Both these minerals are common in igneous rocks. The widespread occurrence of pyrites in almost all types is a matter of universal knowledge and it has recently been shown that pyrrhotite is a common accessory in granites[1]. But both of these minerals are also found as segregations of great size from igneous rocks, or forming what are often described as contact deposits, which must be regarded as equally of magmatic origin. They are also very commonly associated with notable quantities of other and more valuable metals, of which at present we need only mention copper and nickel.

Magmatic Sulphides. The separation of sulphides from silicate magmas is a subject of the very greatest importance and interest from the metallogenetic point of view. It is of all the more significance in that it affords direct evidence of the occurrence in igneous rocks of a type of differentiation depending on the principle of limited miscibility in liquids, as was long ago clearly explained by Vogt in his great work on slags[2].

Vogt's researches showed that at ordinary furnace temperatures the mutual solubilities of sulphides and silicates are quite small, usually of the order of 2 to 5 per cent., and that the solubility increases with rise of temperature: therefore it is concluded that at some very high temperature the two solutions would become perfectly miscible, though such temperatures have not been attained experimentally. From this it follows that there is a limit to the amount of sulphide which can exist evenly disseminated throughout an igneous rock: that is that primary sulphide ores in igneous rocks must necessarily be of low grade unless differentiation has occurred, since the temperature of crystallization of such rocks is at least as low as that of the ordinary smelting furnace, and probably considerably lower. It also follows that the differentiated fraction, which may be called the matte, must necessarily be poor in silicates, since the solu-

---

[1] Rastall and Wilcockson, *Quart. Journ. Geol. Soc.* vol. LXXI, 1915, p. 617.
[2] Vogt, *Die Silikatschmelzlösungen*, Part I, Kristiania, 1903, p. 96.

bilities at the same temperature are of the same order of magnitude. Hence a silicate-sulphide magma, on cooling, must split into two fractions, one rich in silicates with a little sulphide, the other mainly sulphide with a little silicate. Such are the nickeliferous sulphides of the Sudbury ore-bodies and many others.

There is here also another analogy with matte smelting in that the sulphide fraction is heavy and sinks to the bottom, while the silicate fraction is light and floats on the top, like slag. This feature is clearly shown in the field-relations of the Sudbury ores and of the great Insizwa laccolith in Griqualand East. The latter is of so much interest as to merit a somewhat full description. The Insizwa Mountain rises to a height of about 6000 feet above the sea and consists of a great intrusion injected into strata of the Beaufort series, belonging to the Karroo system. The intrusion is variously described as a laccolith and a basin-shaped sheet and is about 300 feet thick as a maximum. It shows a very interesting example of stratification due to gravity-differentiation, ranging from gabbro or norite without olivine and with a little quartz at the top through olivine-gabbro to picrite. The lowest layer of all is again a gabbro, a chilled margin, showing the average composition of the whole mass before differentiation, but a short distance above the base is a layer extraordinarily rich in sulphides, in places nearly pure sulphide, consisting of a mixture of pyrrhotite, pentlandite and chalcopyrite, carrying an appreciable amount of platinum, palladium and gold. The amount of sulphide seems to increase as the contact is followed down into the interior of the basin. The solid sulphide, so far as yet exploited, is rarely more than 2 feet thick, but there is a well-marked zone of richly mineralized picrite. This is a very clear and well-established example of the differentiation of sulphides by gravity-sinking from a basic (gabbro) magma, and shows many analogies to the Sudbury mass: the proportion of nickel and platinum is a noteworthy feature of resemblance[1]. Here not only do we find a definite

---

[1] Du Toit, "The Geology of Part of the Transkei." *Expl. Sheet* 27, *Geol. Survey of S. Africa*, 1917, pp. 18–27. Goodchild, *Trans. Inst. Min. Met.* vol. XXVI, 1917, pp. 12–58.

subsidence of a sulphide layer to the bottom of an intrusion, but the intrusion itself is also stratified by sinking of the heavy olivine crystals, leaving an impoverished almost acid solution at the top. This again is analogous to the granitic differentiate at Sudbury, though in this instance the variation is not quite so marked.

In Norway there are a good many sulphide masses which are now commonly attributed to an igneous origin: of these we need only mention those of Röros and Sulitjelma as typical of the rest. The Röros deposits, which have long been known, were formerly thought to be of sedimentary origin, but this idea is now abandoned and they are considered to be genetically connected with intrusions of gabbro. They form lenticular masses, enclosed in chlorite schist, and consist of pyrite, chalcopyrite, pyrrhotite and blende. A fact of great significance as indicating a high temperature of formation is the presence of biotite in the sulphide bodies[1]. At Sulitjelma the ore occurs along the contact between gabbro and schist, the latter being brecciated and penetrated by veins and seams of the sulphide ore.

One of the most important occurrences of sulphide ore in the whole world is the nickeliferous pyrrhotite of Sudbury, Ontario. As this is described in detail elsewhere, it need not be further discussed here (see p. 366).

Primary Low-temperature Sulphides. Hitherto reference has been made mainly to sulphide ores lying either within or closely adjacent to large masses of igneous rock, and therefore formed at or near the temperature of solidification of the intrusion itself, which in the case of basic rocks in particular, is undoubtedly high. But there remains another class of sulphides, typically represented by galena and zinc-blende, formed at lower temperatures, in some instances perhaps at or near atmospheric temperatures. The mode of origin of these ores and the precise conditions of their formation present some difficulties, and it appears that they can actually be formed within a wide range of temperature and pressure. Both lead and zinc sulphides are common accompaniments of the types of sulphide deposits

[1] Ries and Somers, *Trans. Amer. Inst. Min. Eng.* vol. LVIII, 1918, pp. 244–64.

already described, though often in subordinate quantities: these must necessarily have been formed at high temperatures, but most of the important lead-zinc deposits of the world, often rich in silver, show rather different characteristics, since they are found in areas where there is little or no evidence of igneous activity: such are the lead-zinc ores of northern England, of Moresnet and of the Joplin district of Missouri. It is deposits of this type that have lent the strongest support to theories of ore-formation by meteoric waters, either by lateral secretion or by descent to great depths in the crust, followed by ascension as ore-bearing solutions. It must be admitted that in many ways these theories offer the easiest explanations, but there are difficulties in the way of their acceptance. Some of these difficulties have already been mentioned (see p. 66). Besides the objections founded on chemical grounds to lateral secretion and the absence of water in depth we have to take into account also the character and distribution of the ores themselves, and also the composition of the gangue minerals. If the metals are derived from the surrounding rocks it is difficult to see why they are so limited in number: it might be expected that all metals or at any rate a large number would be present, but this is not the case. As a rule only one or two metals are present in quantity. But the presence of such gangue minerals as barytes and fluorite in immense masses is still harder to account for on the lateral secretion theory. Nevertheless something very like it is accepted by a large number of American geologists to explain the great zinc-lead deposits of the Mississippi valley and other similar cases. Almost alone among American writers, Pirsson has attributed them to "the quiet upward movement of volatile magmatic material[1]." The genesis of the lead-zinc deposits of the Lower Palaeozoic and Carboniferous formations of Great Britain is discussed at some length in another section of this book (see p. 301), and it is unnecessary to repeat it here. It is admitted that the evidence for a direct magmatic origin for these ores is not strong, but on the whole this idea seems to present fewer difficulties than the others: at any rate it is significant that the mineralization of the Carboniferous rocks certainly took place

[1] Pirsson, *Econ. Geol.* vol. x, 1915, pp. 180–6.

about the time of the Armorican-Pennine uplifts and disturbances which were clearly accompanied by so much igneous
activity and ore-formation in Cornwall and in France and
Germany[1].

However this may be, it is at any rate clear that great masses
of such sulphides now exist under conditions that point to
deposition under low pressure and low temperature: the solutions
that carried the ore, if derived either directly or indirectly from
magmas, deposited them far from their source and obviously in
cool regions of the crust, since the limestones that enclose them
do not show the slightest sign of thermal metamorphism: no
changes more far-reaching than silicification and dolomitization
are known, whereas these are common and characteristic.

In all discussions of the origin of such ores, however, so far
as the writer is aware, little or no attention has been paid to the
possibility of a history in two stages; first, deposit from magmatic sources in some rocks perhaps not now visible under a
cover of newer rocks, or removed by denudation; and secondly,
transfer to their present position by circulating waters, which
may have been of meteoric origin. This idea removes the prevailing difficulty as to the source of the metallic compounds,
which has been felt, even if not expressed, by all upholders of
the downward displacement and lateral secretion theories. It
has never yet been proved in any specific case that such ores
are truly primary. Such an origin as above sketched would really
form a sort of connecting link between lateral secretion and
secondary enrichment, with both of which processes it has
analogies.

Arsenides, Selenides and Tellurides. Arsenic is an element
of widespread occurrence in ore-deposits, very commonly in
close association with sulphur. The simple primary arsenides,
for example those of nickel and cobalt, occur in much the
same way as the sulphides of the same metals, and often in
association with them. The silver-cobalt veins of Cobalt, Ontario,
described on p. 485, afford an excellent example of this type.
Arsenopyrite or mispickel, FeSAa, iron sulpharsenide, is one of
the commonest minerals of the high-temperature vein-deposits,

---

[1] Finlayson, *Quart. Journ. Geol. Soc.* vol. LXVI, 1910, pp. 281–327.

and löllingite, $FeAs_2$, is not uncommon. Among the secondary ore-minerals of the zone of enrichment again arsenic compounds are very abundant and in these the arsenic must be derived from primary ores. In minerals of all these classes arsenic, though in itself almost a metal, nevertheless behaves as an electronegative element like sulphur, taking the place of oxygen and forming compounds analogous to the oxides of the higher zones. The elements selenium and tellurium are in their chemical behaviour very similar to sulphur. Tellurium possesses the unusual property of combining with gold to form tellurides, which are valuable ores in Western Australia, Hungary and Colorado. It is believed that tellurides do not form directly from igneous magmas like the sulphides, but at any rate they seem to be primary minerals. Similarly selenium combines with several metals to form selenides; in Sumatra it is associated with gold, although the existence of gold selenide as a definite mineral has never been established. Selenium is also recovered in considerable quantity in the electrolytic refining of copper and must therefore exist in the cupriferous pyrite deposits. Both selenium and tellurium occur in traces along with a considerable amount of arsenic in the pyrite deposits of Rio Tinto in Spain. From these and many other instances it is evident that all these elements are primary constituents of the metalliferous magmas from which these sulphide deposits are derived.

Non-ferrous Oxide Segregations. As already pointed out many of the important magmatic iron ores, especially those in the acid rocks, are oxides, chiefly magnetite. The non-ferrous oxide segregations also constitute a valuable class of primary ore-deposits, of which the most important is undoubtedly cassiterite. It is necessary to state however that cassiterite is almost universally associated with other ores which are not oxides, especially wolfram, mispickel, molybdenite and many others, so that it is almost impossible to treat of the genesis of tin alone without the other metals of its paragenesis, namely: tungsten, molybdenum, arsenic, bismuth and copper. These will be mentioned incidentally when necessary without further explanation.

The first important point is that tin is invariably found in association with granite[1] and nearly always with granite showing special peculiarities, namely the presence of a group of minerals containing fluorine, boron, lithium, beryllium and other elements of low atomic weight, highly volatile, and for the most part chemically very active. These give rise to tourmaline, topaz, fluorite, lithia mica and beryl, with many rarer species. Tin occurs in disseminations in the granite itself apparently as an original constituent; in many varieties of dykes, especially in pegmatites; in quartz veins and lodes, traversing both the granite and the country rock in its neighbourhood, and it is often associated with peculiar rock types formed by alteration of the granite during the later stages of its cooling. This is equivalent to saying that tin ore is a mineral of pneumatolysis. The pneumatolytic changes are beyond all dispute brought about by agents derived from the granite magma, including highly heated water, and there is no doubt that the tin came from the same source, probably as volatile tin fluoride, SnF_4. In the same way tungsten travelled as fluoride, WF_6, which boils at a much lower temperature than tin fluoride: the volatility of arsenic compounds is also well known. It appears that uranium, with its train of radio-active descendants, also belongs to this group.

It is evident therefore that the formation of ore-deposits of this class is simply a special phase of a group of phenomena attendant on the latest stages of the cooling and crystallization of an igneous magma. For these phenomena there is in reality no satisfactory name in general use. They are generally expressed by the term pneumatolysis, which is not altogether a fortunate one, since by its etymology it conveys the idea of the action of vapours only. Vapours undoubtedly do play an important part in these processes, but it is necessary to make provision for the inclusion of the effects of solutions as well. The best known and most characteristic of these processes is in some ways the formation of pegmatites, which are often important ore-carriers, especially in respect of the non-sulphide ores. For a detailed discussion of the origin of pegmatites see p. 35.

[1] In Mexico tin occurs in rhyolites, but this does not really form an exception to the rule, since rhyolite and granite have the same composition.

Although in times past attempts have been made to show that pegmatites are of hydrothermal origin and not truly igneous, no one now disputes that they are dyke-rocks, just as much as the quartz-porphyries or the dolerites. It follows therefore that the ore-minerals contained in them are likewise of igneous origin, since these are obviously primary constituents, along with the quartz, felspar, and other minerals. In fact the ore-minerals are often demonstrably of earlier formation than the latter. Furthermore, it is now generally recognized that many quartz-veins, both metalliferous and otherwise, are merely continuations of pegmatites, being in fact a highly siliceous facies of them, and in many instances a gradual transition has been observed in the field from a normal pegmatite of granitic composition to a pure quartz vein.

To sum up this part of the subject, without needless repetition of detail, we may say that the oxide ores in the broad sense are specially characteristic of the pegmatitic and pneumatolytic phases of acid intrusions, usually large deep-seated masses of granites, less commonly, as in Bolivia, accompanying hypabyssal injections of a less acid and less deep-seated character. No occurrence of tin and tungsten is known in connection with basic rocks. It is evident that the tin, tungsten and molybdenum ores are high-temperature minerals, formed by differentiation during the cooling of acid magmas rich in water and other volatile constituents.

Native Metals. A considerable number of metals occur naturally in the native state: in a few cases, such as gold and platinum, this is the principal method of occurrence, while with other metals it is a mineralogical rarity, or at any rate of little commercial importance. With other metals again, especially silver and copper, the native condition is the result of secondary processes, usually reduction following oxidation and solution, as detailed in Chapter VIII. We have therefore two distinct and well-defined categories of native metals, the primary and the secondary. It is with the first category only that we are now concerned.

Platinum. The greater part of the platinum supply of the world has hitherto been obtained from alluvial deposits in the

Ural Mountains. It is known however that this platinum is derived from the denudation of serpentines. These serpentines are alteration products of peridotites that consist chiefly of olivine and chromite and the platinum is an original constituent of these rocks, having separated direct from the magma, as a primary mineral. Quite lately a similar association of metallic platinum with peridotite has been discovered in the Serranía de Ronda, in the extreme south of Spain. This association of chromium and platinum with ultrabasic rock-types extremely rich in magnesium is one of the clearest cases of a definite metal-paragenesis yet known. It would appear that in the primary deep-seated differentiation of igneous magmas, to whatever cause this may be due, the chromium and platinum tend to follow the fraction having the lowest silica-percentage and the highest magnesia content.

Nickel-iron. Although of no commercial importance the occurrences of the natural alloys of iron and nickel are of great theoretical interest. It has long been known that many meteorites consist almost entirely of iron alloyed with nickel and a similar composition is established in the case of some remarkable masses of native iron found on the west coast of Greenland. For long these were also supposed to be of meteoric origin, but they are now explained as being due to the reduction of the iron-content of basic igneous rock on its upward passage through seams of coal. In a similar way minute grains of native iron have been observed in basalts in north-eastern Ireland.

Gold. By far the greater part of the gold of the world is found in the native state. Gold forms tellurides and apparently also selenides, but it does not form natural sulphides. The telluride ores, though certainly primary, are described in sufficient detail elsewhere. We may say that all the free-milling gold is in the metallic state: the actual condition of some of the low-grade gold ores from which gold can be extracted by cyanidation or other processes is somewhat uncertain: it is probably some form of solid solution or isomorphous combination in sulphides, arsenides and similar minerals. We are here dealing only with the metallic, usually visible, gold associated with igneous rocks of all kinds and with the different forms of veins and reefs which are of such wide distribution and so variable in form. With respect to its relation to igneous rocks gold is singularly cosmopolitan: it has been found in quantity in association with types of all degrees of acidity and with both alkaline and subalkaline families, plutonic, hypabyssal and volcanic. Perhaps the most characteristic manner of occurrence of native gold is in quartz veins and reefs in all their innumerable varieties of form: as previously explained many of these are undoubtedly merely an extremely siliceous type of pegmatite, and these pass by imperceptible transitions into veins formed by thermal waters, which are of course ultimately of magmatic origin, though from their manner of occurrence this is not always quite obvious. If a generalization on this subject can be made at all, it may perhaps be said that gold occurs most commonly in association with rocks of intermediate type, neither very acid nor very basic, neither very alkaline nor very rich in lime: in America at any rate many of the primary gold deposits are associated with monzonites, granodiorites and andesites, all intermediate types. Nevertheless there are many exceptions to this rule.

Paragenesis of Metals. It has long been a commonplace of mineralogical chemistry that certain pairs and groups of metals tend to occur in association. Examples of this are innumerable. In many instances the chemical resemblances between different metals are so strong that great difficulty has been found in separating them, as for example in the platinum group and those commonly known as the metals of the rare earths. Then again we have many cases of isomorphism, where the metals concerned habitually form mixed crystals, like magnesium, iron, manganese, zinc and chromium in the spinels (magnetite, picotite, franklinite and chromite); tantalum and niobium in columbite; cobalt and nickel in sulphides and arsenides; and magnesium, iron and calcium in many groups of the silicates. In each of these two classes the association is natural and what might be expected to occur on general chemical and physical grounds. There are however other instances of metal associations, many of them of great economic importance, for which no reason can as yet be assigned.

In the case of the first group the reason for the close association in nature is fairly obvious. Since the chemist finds so much difficulty in effecting a separation of these metals by methods specially adapted for the purpose, often of great refinement, it is likely that in the somewhat rough and ready large-scale metallurgical processes of nature these elements will tend to segregate into groups, rather than separately. Thus native platinum, so called, is usually in reality a complex alloy of platinum, palladium, iridium, osmium, ruthenium and perhaps other metals of the group. All of these have very similar physical properties and are similarly affected by the processes of differentiation. Consequently they all migrate together into the most basic magma-fraction, along with magnesium, iron and chromium, separating out along with chromite and picotite in the peridotites.

The mineralogical behaviour of the rare-earth group, though entirely different in detail, is controlled by similar principles. The numerous metals of this class, of which thorium and cerium are of the most technical importance, mainly associate themselves with siliceous and aluminous magmas. The immense variety of minerals containing them are chiefly known as constituents of the pegmatites connected with granites and syenites, where they are exploited as a source of the necessary materials for incandescent gas mantles and Nernst lamps[1]. The mineral monazite, which is now the chief source of thorium and cerium, is found as an accessory mineral in certain granites, while the monazite sands of commerce are natural concentrations of the heavy residues from the denudation of such granites.

The study of isomorphism in minerals presents many features of chemical and physical interest, and is also of considerable technical importance, since many metals occur as constituents of mixed crystals, or members of isomorphous series. In the case of native metals this is specially liable to occur, since most of them crystallize in the cubic system in which the crystal angles are fixed once for all, and in consequence a minimum of

---

[1] Brögger, *Zeits. für Kryst.* vol. XVI, 1890. Hidden and Warren, *Am. Journ. Sci.* vol. XXII, 1906, p. 515. Hess, Bull. 340, *U.S. Geol. Surv.* 1908, pp. 286-94. Johnstone, *The Rare Earth Industry*, London, 1915, pp. 1-33.

strain is developed by the mixing of different molecules. The case of the platinum group has already been mentioned; gold nearly always contains a considerable amount of silver as an alloy or in solid solution. Native iron is nearly always nickeliferous, and so on. In many compounds also this phenomenon is well-developed and is even more striking.

Isomorphism in the Sulphides and Arsenides. There is hardly any group of minerals of greater technical importance collectively than the sulphides and many of them show in a very high degree the class of phenomena just indicated. Among many examples it is only practicable to select a few. Perhaps the most instructive of all cases is that of the disulphides and diarsenides of iron, cobalt and nickel, generally known as the pyrite-marcasite group. The peculiar crystal form of pyrite is well known: it commonly occurs either in cubes or as a remarkable figure known as the pentagonal dodecahedron. But besides pyrite a whole group of disulphides and diarsenides of nickel and cobalt show the same peculiarity. Thus we have the minerals smaltite, $CoAs_2$, and chloanthite, $NiAs_2$. Furthermore neither of these are ever pure, but every possible gradation exists between them, so that the general formula may be written,

$$\text{Smaltite-chloanthite} = x CoAs_2 + y NiAs_2.$$

Curiously enough the simple compounds NiS_2 and CoS_2 do not seem to exist in nature. There are also the rather more rare minerals hauerite, MnS_2, and sperrylite, $PtAs_2$, which show the same crystallographic characters. Then again we have cobaltite, $CoAsS$, and gersdorffite, $NiAsS$, and also, rarely, ullmannite, $NiSbS$: the last shows that here, as elsewhere, antimony can replace arsenic (see *postea*).

Now it is obvious that $CoAsS$ can also be written $CoAs_2 + CoS_2$ and similarly for the nickel compound.

Hence we are justified in regarding all these minerals as mixtures of the isomorphous molecules, FeS_2, $CoAs_2$, $NiAs_2$, MnS_2, CoS_2, NiS_2, $NiSb_2$ and so forth, all crystallizing in the same class of the cubic system and capable of forming mixed crystals.

But there is a further point. There are known also a number

of minerals of similar composition to the foregoing, crystallizing in the orthorhombic system, such as marcasite, FeS_2, löllingite, $FeAs_2$, arsenopyrite, $FeSAs$[1], safflorite, $CoAs_2$ and so on. These form another series parallel to the first, and in some cases supply the missing members. The whole may be arranged in a tabular form as follows:

Comosition	Cubic system	Orthorhombic system
FeS_2	pyrite	marcasite
$FeAs_2$	—	löllingite
$CoAs_2$	smaltite	safflorite
$NiAs_2$	chloanthite	rammelsbergite
$FeAsS$	—	arsenopyrite
$CoAsS$	cobaltite	—
$NiAsS$	gersdorffite	—
$NiSbS$	ullmannite	—

Thus we see that not only do these compounds show clear evidence of isomorphism and the formation of mixed crystals, but several of them are also dimorphous, existing in two perfectly distinct crystalline forms.

Sulphides of Lead, Zinc and Silver. The sulphides of lead and zinc, galena and blende, are commonly found in close association in primary ores, but in separate and distinct crystals: although both belonging to the cubic system, they crystallize in different classes and are not isomorphous, hence they cannot form true mixed crystals. On the other hand galena usually contains a certain amount of silver as sulphide, argentite, Ag_2S, which appears to belong to the same class of the cubic system and can therefore mix readily with galena. Zinc-blende usually contains cadmium, sometimes up to 5 per cent., and this raises a point of some interest as to the relations of the two metals. Zinc-blende is cubic, while cadmium sulphide, greenockite, is hexagonal. But zinc also possesses a second dimorphous form of sulphide, wurtzite, having the same composition as blende, but belonging to the same class of the hexagonal system as greenockite. Hence zinc and cadmium sulphides show crystallographic isomorphism in one form at any rate: the cubic cadmium sulphide is not known to exist. The massive forms of zinc

[1] $FeSAs = FeS_2 + FeAs_2$.

sulphide, known in Germany as Schalenblende, consist of wurtzite and not of blende.

In their chemical behaviour the three elements sulphur, selenium and tellurium are very similar, consequently, as might be expected, they combine with various groups of metals to form series of sulphides, selenides and tellurides. The tellurides of the noble metals are of great commercial importance. Closely similar to galena in their physical properties are the sulphide, selenide and telluride of silver, argentite, aguilarite and hessite respectively. Argentite, Ag_2S, is apparently the most important primary of silver and the original source of the rich silver minerals of the zones of oxidation and secondary enrichment. Although the corresponding selenide and telluride are of much less importance, aguilarite being apparently only known from Mexico, yet they are of interest from the theoretical point of view; hessite often contains gold and graduates into the gold telluride ores. Mercuric sulphide, HgS, crystallizes in two forms, as the hexagonal cinnabar, the commoner form, and metacinnabarite, which is cubic and isomorphous with blende, as also are the rare corresponding minerals tiemannite, $HgSe$ and, coloradoite, $HgTe$. From these examples it is clear that sulphur, selenium and tellurium form series of isomorphous compounds.

Compounds of Gold. Among the gold ores, as distinguished from placers, the metal occurs in a considerable variety of ways. In the endless varieties of veins, visible gold is usually native. But there are many gold-bearing ores from which the gold can only be extracted by some special process. In many instances it is known that gold exists in some form in the primary sulphides, and it is still more common in the secondary sulphides of the zone of enrichment and in oxidized ores. Although no definite mineral consisting of gold sulphide is known, yet it is certain that the metal exists somehow in the sulphides, and we may say that gold with sulphur is one of the important metallogenetic types of gold occurrence, having a very wide distribution. Gold with selenium is also known, though only from two localities in any quantity, namely Redjang Lebong in Sumatra, and Republic, Washington, U.S.A. Although it is still uncertain whether a gold selenide actually exists, a black amorphous substance found in these deposits is believed to be such a mineral. The ores of Tonopah, Nevada, show some affinity to this type.

On the other hand, the telluride ores are of first-class commercial importance: they occur in large quantities, for example, in Western Australia, Colorado and Hungary, and include a large number of closely-related minerals, most of which contain silver as well as gold. The most important of them are calaverite, krennerite, sylvanite and petzite[1]. Two Australian minerals, kalgoorlite and coolgardite, also contain mercury[2]. Although some difference of opinion exists as to the exact composition of the telluride minerals, there is no doubt that they are true chemical compounds, namely, tellurides of gold, silver and mercury and they afford an excellent example of a metallo-genetic facies. By analogy the corresponding sulphur and selenium facies of gold deposit can also be recognized.

Isomorphism in the Secondary Sulphides. Among the minerals of the zone of secondary enrichment there are to be found a great number of examples of isomorphism and mixed crystals. This would naturally be expected, since in this zone many elements are present, having been set free by weathering processes; they tend to rearrange themselves in accordance with their chemical affinities, and sort themselves out into groups of related molecules. It is obviously impossible to deal with all the mineral combinations that can arise in these circumstances and only a few typical examples can be chosen, chiefly from the copper and silver minerals. In this zone, among the acid-forming oxides we find not only sulphur, but to a very large extent antimony and arsenic. Very many of the ore minerals belong to the classes of compounds known as sulphantimonides and sulpharsenides, and in some instances bismuth seems to play a similar part: most of the antimony and arsenic compounds seem to have rare bismuth analogues. Among the basic constituents in a similar way one atom of divalent lead often replaces two atoms of monovalent silver or copper.

---

[1] Kemp, *The Mineral Industry*, vol. VI, 1898, p. 295.
[2] Spencer, *Min. Mag.* vol. XIII, 1903, p. 268.

Perhaps the most striking case is that of the ruby silver ores, proustite and pyrargyrite, which are most conveniently written $3Ag_2S . As_2S_3$ and $3Ag_2S . Sb_2S_3$ respectively. Both of these belong to the same class of the hexagonal system and show very similar characters. Another pair of minerals related in the same way are tetrahedrite (fahlerz), $4Cu_2S . Sb_2S_3$, and tennantite, $4Cu_2S . As_2S_3$: these belong to the tetrahedral class of the cubic system and appear to form mixed crystals in all proportions.

A generalized study of the large number of minerals that have been described as occurring under the conditions specified shows that most of them can be reduced to a few general types, as follows:

(1) $$R''S . R'''_2S_3$$

in which R'' may be Pb, Ag_2 or Cu_2. R''' may be, Sb, As, Bi.

Example   zinkenite $= PbS . Sb_2S_3$.

(2) $$2R''S . R'''_2S_3$$

in which R'' may be Pb or Ag_2. R''' may be Sb, As, Bi.

Example   jamesonite $= 2PbS . Sb_2S_3$.

(3) The pyrargyrite-proustite group
$$3R''S . R'''_2S_3.$$

(4) The tetrahedrite-tennantite group
$$4R''S . R'''_2S_3.$$

(5) A series of still more basic salts of rather ill-defined character, including stephanite, $5Ag_2S . Sb_2S_3$, polybasite, $9Ag_2S . Sb_2S_3$, and polyargyrite, $12Ag_2S . Sb_2S_3$.

The existence of the first four groups is well-established, but the fifth group includes a large number of rare minerals, many of which may possibly be explained as solid solutions. The evidence for their existence as definite mineral species is in many cases unsatisfactory. The one fact which emerges clearly is that silver, lead and copper are capable of combining under reducing conditions with sulphur and antimony, arsenic and bismuth in a large number of different ways and these compounds or pseudocompounds afford an excellent example of a characteristic mineral paragenesis, namely, that belonging to the secondary enrichment zone derived from primary ores of the metals concerned. Under such conditions it would appear that antimony, arsenic and bismuth lose their basic character and take on the functions of acid-forming elements with sulphur instead of oxygen.

Metallogenetic Provinces. It is a fact well known to every mining geologist that in different parts of the world any given metallic ore may occur with very various mineral associates, both of metalliferous ores and gangue minerals. Thus if we take the occurrences of any one common metal which is found in many localities, we can subdivide these localities into groups characterized by the presence of some secondary constituent and thus map out the world into a series of provinces, so far as that particular metal is concerned. But it is possible to go further than this. Some parts of the world are strongly mineralized, whereas in other areas ore-deposits are rare or absent. The reason for these differences is mainly geological and depends chiefly on the prevalence or otherwise in the area concerned of igneous rocks accompanying crust-disturbances. Detailed study shows that each mineralized area is commonly characterized by the occurrence of a particular set of metals often in particular states of combination with other elements.

We must now turn to the detailed consideration of a cognate and highly interesting part of the same subject, namely, the distribution in space and time of the metals or groups of metals, and the correlation between the principal epochs of mineralization and periods of earth-movement and igneous activity. It is now a commonplace of petrology that igneous activity is as a rule synchronous with the major phases of crustal disturbance[1]. This may now be accepted as axiomatic. But it is only of late years that it has been realized by geologists that the same rule applies to the formation of at any rate most of the important primary ore-deposits. In the case of tin the connection is obvious: with copper it is rather less so, as likewise in the case of gold and silver; with lead and zinc more doubt may arise, while for the

[1] Bertrand, *Bull. Soc. Géol. France*, vol. XVI, 1888, p. 573. Harker, *The Natural History of Igneous Rocks*, London, 1909, especially Chap. IV. Lake and Rastall, *Textbook of Geology*, 3rd edition, 1920, Chap. XXXI.

sedimentary and metamorphic iron ores, economically the most important, quite another set of causes must be invoked.

Besides their definite distribution in time, igneous activity and the accompanying mineralization must also be considered from the geographical point of view, not alone in connection with the geography of the present, but also with that of the past, which is in fact the same thing as stratigraphical geology. We cannot fix the age of ore-deposits in years; we can only refer them to their position with reference to the geological time-scale as revealed by the succession of the stratified formations and their accompanying igneous rocks. Hence it appears that we can recognize metallogenetic regions in space and metallogenetic epochs in time.

A metallogenetic province may be defined as an area wherein at a given time a particular metal or groups of metals are deposited as primary ores.

When mineralized areas are studied in detail it soon becomes clear that a very great range in type is possible. It is obviously impracticable to describe all known variations in metalliferous deposits: we can only study a few particular cases. The general principles involved can be worked out most clearly by selecting one particular metal or a small group of metals and studying the distribution in space and time of their occurrences. For a variety of reasons, unnecessary to discuss in detail, the ores of tin offer special advantages for this purpose. Most of the leading facts concerning the distribution of tin and its associates are dealt with elsewhere in this book (see Chapter XIII), and will not be repeated here: only the most general conclusions have any bearing on the present subject.

Tin and its Associates. The study of the distribution of tin ores is rendered unusually simple by two facts: (a) the very small number of tin-bearing minerals; (b) the stability of the more important of these minerals, so that complications due to oxidation and secondary enrichment are almost wholly excluded. The highly resistant and stable character of cassiterite is well known, while stannine, the only other tin mineral of any importance, though not stable, is limited in its distribution.

The chief tin-producing areas of the world are as follows: Cornwall; Saxony and Bohemia; Spain and Portugal; Lower Burma and the Malay Peninsula; China; Eastern Australia; Northern Nigeria; South Africa; Bolivia. From the commercial point of view these regions are of very unequal value, and some, such as Saxony and Bohemia, mainly of historic interest, but all of them illustrate some geological point. Other regions which are not commercial producers also possess scientific significance and will be referred to when necessary. The geological conditions in China are very imperfectly known, so that no general conclusions can be formulated, but most of the others have been worked out in considerable detail. In most areas the greater part of the actual output comes from some form of superficial deposit, but the conditions of lode-occurrence are generally known.

Cornwall and Brittany form detached portions of a single geological unit, the Armorican Highlands, while the tin-field of Saxony and Bohemia, in the Erzgebirge, is of very similar type, belonging to the great Hercynian mountain system of central Europe. The tin-bearing area of north-western Spain and northern Portugal forms part of the ancient plateau known as the Iberian Meseta, which consists of Palaeozoic and older rocks, strongly folded by the same series of earth-movements as gave rise to the Armorican and Hercynian mountain chains. All these regions have likewise been invaded by granite intrusions accompanying the folding and specially characterized by a high development of pneumatolytic phenomena. To this pneumatolytic phase the mineralization is mainly due. It is clear that in all these areas these phenomena were contemporaneous, or nearly so, and stratigraphical considerations show that they must be assigned to the Permo-Carboniferous period; at any rate they are post-Carboniferous and for the most part pre-Permian, though some of the German intrusions appear to be of Permian age.

The distribution of metals in the lodes in Cornwall is described elsewhere (see Chapter XIII). It there appears that tin is found mainly in the lower parts of the lodes, being succeeded upwards by copper. The other metals that predominate in the tin-bearing

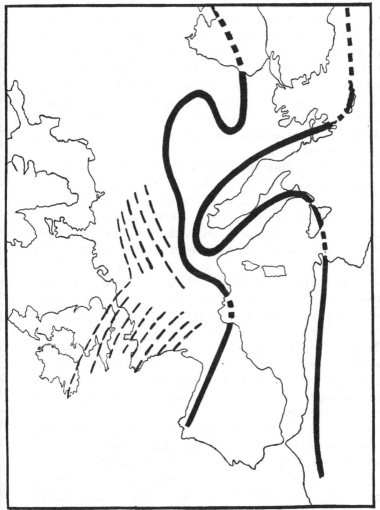

Fig. 27. The post-Carboniferous fold-systems of western Europe. Thin broken lines are the trend-lines of the Permo-Carboniferous (Armorican and Hercynian) folds: thick black lines are folds of the Alpine mountain system.

portions of the lodes are tungsten and arsenic, some lodes being rich in one or both of these, and some, in the higher portions carrying mainly wolfram without tin[1]. Although the copper is found at a higher level it is clear that it came from the same source at the same time, hence we may say that the characteristic paragenetic metals here are tin, tungsten, copper and arsenic. Of the non-metallic elements the most interesting are fluorine and boron, which gave rise to a great development of tourmaline, with smaller quantities of topaz and fluorspar. The occurrence of uranium in Cornwall is also worthy of note, since it forms a point of resemblance to the Erzgebirge: molybdenum is also fairly common.

The mineral assemblage in Brittany is very much like that of Cornwall and a detailed description is scarcely needed.

The tin-lodes of the Erzgebirge belong essentially to the same type. They contain in addition a considerable amount of wolfram and sometimes chalcopyrite, while the lead-zinc-silver lodes contain a little tin and are obviously in genetic connection with them. This region is specially characterized by abundance of lithium in the form of lepidolite and the occurrence of uranium minerals has just been mentioned. The pneumatolysis is more particularly characterized by abundance of topaz, indicating an excess of fluorine in the magmatic residues.

In Spain and Portugal the tin is accompanied by wolfram and in some localities there are important wolfram lodes without tin, similar to the lodes of Castle an Dinas in Cornwall and Hemerdon in Devon. Autunite and other uranium minerals are common accompaniments.

Summarizing these facts in the most general way possible we may say that the Permo-Carboniferous metallogenetic province of western Europe is marked by an earlier phase with tin, tungsten and copper as main constituents, with uranium as a subsidiary local characteristic. The lead-zinc veins of these areas appear to be for the most part of somewhat later date than the tin veins, though genetically related. The non-metallic constituents include both fluorine and boron, varying locally in relative importance.

[1] Davison, *Geol. Mag.* vol. LVIII, 1921, p. 505. *Cornish Institute of Engineers,* vol. VIII, 1920.

Lower Burma and the Malay Peninsula, with an intervening stretch of Siamese territory and the islands of Banka and Billiton, obviously form a single metallogenetic province. The exact age of the mineralization is not known, but is probably either Permo-Carboniferous or Mesozoic. Here the most interesting feature is the gradual change in the relations of tin and tungsten from north to south. In the north, in Tavoy, the lodes contain mainly wolfram, in the south mainly tinstone. Where wolfram is abundant, bismuth is also characteristic, and in the same area tourmaline is rare. This region possesses features in common with the mineralized belt of eastern Australia, especially in the occurrence of bismuth.

Throughout the greater part of the folded belt of the eastern side of the Australian continent, from northern Queensland to Tasmania granitic intrusions of Permo-Carboniferous age have given rise to a high degree of mineralization, the chief metals being tin, tungsten, molybdenum and bismuth. These vary somewhat in relative amount in different parts of the area: on the whole tungsten and bismuth are most abundant in northern Queensland, molybdenum in New South Wales and tin in Tasmania. The manner of occurrence of the minerals, in pipes, also forms a local peculiarity, though similar pipes occur elsewhere[1].

Northern Nigeria presents some interesting features, since it appears that here there were two epochs of tin formation, one Archaean and one Mesozoic. Wolfram is present in the same lodes, but no other metals are prominent. Topaz is very common, while tourmaline is local in its distribution and is not found along with the tin lodes of the Ningi Hills at any rate[2].

The tin deposits of Bolivia are in many ways the most remarkable and possess certain features not seen elsewhere. Here we have tin and tungsten associated with great quantities of silver ores; bismuth is also present in considerable quantity and

[1] Mineralization belonging to earlier geological periods is common in eastern Australia: thus the great Broken Hill deposit appears to be pre-Cambrian, while there are Devonian ore-deposits in the N.E. of New South Wales.

[2] Williams, *Geol. Mag.* vol. LVII, 1920, pp. 434–447.

a very interesting feature is the presence of minerals containing the exceedingly rare metal germanium, hardly known elsewhere. These deposits also are much younger than the rest of the important tin ores of the world, since they are connected with the intrusive phase of the great Tertiary vulcanicity of the Andes.

The tin-tungsten lodes of Alaska are of some interest, so far as known, since they seem to be characterized by certain peculiar minerals containing boron to the exclusion of fluorine.

From the facts here stated it may be concluded that tungsten always accompanies tin in some part of each large field, although the two metals are not necessarily found in the same lodes. On the whole the distribution of tungsten is probably less wide than that of tin. The accompanying metals show more variation: thus bismuth is found in three chief areas, Burma-Malaya, Australia and Bolivia, while molybdenum is really abundant only in Australia. The Bolivian type with silver and germanium is exceptional.

These data can be conveniently summarized in the following table.

Province	Chief metals	Subsidiary metals	Rare accompaniments	Dominant non-metallic elements
Cornwall	tin, tungsten	copper, arsenic	uranium	{ boron, { fluorine
Erzgebirge	tin, tungsten	...	uranium	fluorine
Malaya	tin, tungsten	bismuth	...	fluorine
Australia	tin, tungsten	molybdenum, bismuth	...	...
Nigeria	tin, tungsten	...	...	fluorine
Bolivia	tin, tungsten	silver, bismuth	germanium	...
Alaska	tin, tungsten	...	...	boron

Thus it will be seen that each province possesses some distinguishing characteristic, differentiating it from all the others: no two are exactly alike in all respects.

The Distribution of Silver. The element silver occurs in nature in a considerable number of different associations: its study is not quite so simple as that of tin, for example, since silver ores are peculiarly liable to oxidation and various forms of secondary deposition. Nevertheless, when these factors have been eliminated it is possible to recognize several distinct types of primary deposit. In all probability the most widely distributed primary ore of silver is the sulphide, argentite, from which most of the secondary sulpharsenide and sulphantimonide ores seem to be derived. Ore-deposits of this type are particularly common in the American continent, in the Cordilleran region of South, Central and North America. Of this the silver-tin deposits of Bolivia form a peculiar and local facies. In the rest of this vast stretch of country gold and copper are the more common associates. The very widely-spread occurrence of silver in lead ores is in reality only a variety of this type, since it has been shown that the silver in galena exists as strings and films of argentite[1]. There are many varieties of lead-silver veins, distinguished mainly by their gangue-minerals, such as the lead-silver-tourmaline veins of the Boulder bathylith in Montana (especially the Alta vein); the Broken Hill type with garnet; the galena-tetrahedrite-siderite veins of Wood River, and the Coeur d'Alene district, both in Idaho; the pyritic galena-quartz veins of Freiberg, Saxony; the quartz-carbonate veins of Przibram, Bohemia; and the barytes-siderite veins of Clausthal in the Harz Mountains; all of these appear to be of high temperature origin and undoubtedly primary. Silver is also abundant in most of the lead-zinc deposits of low-temperature type, which need not be enumerated here.

Native silver of primary origin also occurs in various ways; of these perhaps the most widely distributed is as a natural alloy with gold. This is of almost universal occurrence, native gold never being quite free from silver and some varieties containing a large proportion.

Of more interest is the vein-type characterized by silver, cobalt and nickel. Such veins have long been known in Germany, but the most important occurrence at the present time is in the Cobalt district in Ontario. For a description of this area see p. 485.

Finally we find silver occurring along with gold in the telluride ores. Most of the telluride minerals are in point of fact isomorphous mixtures of gold and silver tellurides, in any proportion, conforming to two principal types, namely, calaverite $(Au, Ag)Te_2$, and petzite, $(Au, Ag)_2Te$. These minerals

[1] Nissen and Hoyt, *Econ. Geol.* vol. x, 1915, pp. 172–9.

seem to be formed at moderate temperatures and are easily decomposed in the oxidation zone. As before mentioned the principal telluride regions are Hungary, Colorado and Western Australia.

Metallogenetic Epochs. Just as petrologists have found it possible to recognize clearly-marked periods of igneous activity in definite relation to the stratigraphical succession, so also has it now become possible in many cases to fix age-limits to the different periods of mineralization in various parts of the globe. The methods employed are naturally identical with those used in the case of the igneous rocks. When the ore-deposits are in actual and visible connection with igneous rocks, the matter is exceedingly simple, but in some instances no absolute criterion of this kind is available. Let us take as an example some of the ore-deposits of the British Isles: the tin-lodes of Cornwall are obviously connected with the granites, which are clearly post-Carboniferous and pre-Permian, so that their age is fixed within narrow stratigraphical limits. The lead-zinc ores of northern England are certainly post-Carboniferous, or at any rate later than the Lower Carboniferous: they also appear to be pre-Permian, although there is perhaps no absolute proof of this: it is however inferred that they belong to the same general sequence as the tin-copper-lead-zinc deposits of the south-west. But when we come to consider the lead-zinc veins of central Wales, the Lake District and the Leadhills, our information is even less precise: they are certainly post-Silurian, but there is no means of ascertaining definitely whether they belong to the Caledonian epoch of igneous activity (Devonian) or to the post-Carboniferous (Armorican-Pennine) epoch. If the latter is the case they should be similar to the lead-zinc veins in the Lower Carboniferous. The common presence of copper however forms a distinguishing feature and suggests a different date and origin, presumably Caledonian. It is possible, however, as suggested elsewhere, that these veins, as now seen in the Ordovician and Silurian rocks, represent lower levels in the same series of deposits as are found in the Carboniferous limestone in North Wales and the Pennines, the upper portions, with lead and zinc only, having been removed by denudation. If this is so they must belong to the Armorican series and almost the whole of the mineralization of the British Isles must be assigned to the same date, namely, Permo-Carboniferous.

An excellent example of repeated epochs of mineralization is afforded by the pre-Cambrian ore-deposits of the province of Ontario, Canada[1]. In this region chemical deposition of iron formations has taken place at three different periods, in the Grenville, Timiskamian and Animikian, while metalliferous vein deposits are connected with the basic intrusions of post-Timiskamian age, with the Algoman granite intrusions, and with the Keweenawan intrusions.

The following condensed table summarizes the main conclusions of Miller and Knight.

> KEWEENAWAN. Basic intrusions with some acid facies.
>> Silver, cobalt, nickel and arsenic at Cobalt.
>> Nickel and copper at Sudbury.
>> Gold in various localities.
>
> ANIMIKIAN. Chemical deposition of iron formations.
> ALGOMAN. Granite intrusions.
>> Gold at Porcupine and elsewhere.
>> Galena, zinc-blende and fluorspar.
> Basic intrusions of post-Timiskamian age.
>> Nickel and chromite.
>> Magnetite and titaniferous magnetite.
>
> TIMISKAMIAN. Chemical deposition of iron formations.
> LAURENTIAN. Granite intrusions.
> GRENVILLE. Chemical deposition of iron formations.
> KEEWATIN. Basic volcanic eruptions.

There are thus three cycles of mineralization, each giving rise to vein deposits and connected with igneous intrusions, all within the limits of the pre-Cambrian.

Metallogenetic Epochs in the United States. The relations of the principal ore-deposits of the United States to the geological history of the country have been well summarized by Lindgren[2]. From this point of view the United States can be conveniently divided into two entirely different and contrasted

---

[1] Miller and Knight, *Trans. Roy. Soc. Canada*, Ser. III, vol. IX, 1915, pp. 241–9. Summarized in *Geol. Mag.* 1916, p. 573.

[2] Lindgren, "Metallogenetic Epochs," *Econ. Geol.* vol. IV, 1909, p. 409.

regions, as follows: first we have the Eastern States, broadly comprised in the Appalachian region from Nova Scotia to Alabama, where the principal periods of folding and eruptivity are pre-Cambrian and Palaeozoic. In this region the earliest and most important period of mineralization is connected with the pre-Cambrian intrusives. The basic intrusions gave rise to deposits of iron ores (largely titaniferous), copper and nickel, while the granites carry gold and silver. Lead, zinc, mercury and antimony are comparatively rare and unimportant. At a later date there was a feeble development of iron and copper ores in connection with the so-called Triassic Traps. Since then no mineralization has taken place in the Eastern States.

In the Western States mineralization has been much more important and of quite different date. Of pre-Cambrian age are a certain number of quartz-veins in schists in the Rockies and in the Great Basin, carrying gold and copper with tourmaline and garnet, evidently of deep-seated magmatic origin, and not of much importance. In the Trias and Jurassic some copper deposits were formed in connection with eruptions of andesite and basalt. In the late Mesozoic, about the middle of the Cretaceous, there began a period of intense mineralization, which may indeed be said to be still in progress. The first event was the intrusion of the great series of bathyliths of quartz-monzonite or granodiorite which are the most marked feature of the geology of the Pacific coast region. There are two principal bathyliths, one of which extends almost throughout California, and the other from Washington to Alaska, besides many relatively smaller though still very large ones. Along the margins of these bathyliths intense mineralization occurred, characterized specially by deposits of gold and copper. Lead and zinc are not important in California but are found in a similar way in Nevada and Idaho. The other metals are not found in any quantity. The early part of the Tertiary period, particularly the Eocene, was marked by the intrusion of innumerable laccoliths and dykes of acid and intermediate rocks, with deposits of gold and silver, with much lead and zinc, and some copper and iron. Arsenic and antimony are fairly common. The late Tertiary ore-deposits are mostly associated with volcanic eruptions, chiefly andesites and rhyolites. They also carry gold and silver, while antimony and tellurides are abundant. Copper, lead and zinc are not important. The latest basalt eruptions of post-Pliocene date with hot springs are chiefly characterized by deposition of mercury, and in some places this is still in progress.

Thus in this area it is possible to trace a regular sequence of mineralization, correlated with a gradual change in the character of the igneous activity. Gold and silver occur almost throughout, but as we progress from the vast deep-seated bathyliths of the early phases to the last surface eruptions there is a fairly well-defined sequence of metals. Thus:

gold and silver	basic surface eruptions ...	mercury.
	intermediate surface eruptions	antimony, tellurium.
	intermediate laccoliths ...	lead, zinc, antimony, arsenic.
	acid bathyliths	copper.

The character of the mineralization and the species of metal is thus clearly correlated with temperature and pressure, as indicated by depth, with the acidity of the magma, and with the time factor. Here as elsewhere gold is the ubiquitous metal, appearing to be independent of time and space.

The Ore-deposits of Montana. One of the most complete and satisfactory detailed studies of a metallogenetic epoch as contrasted with the broad generalizations just expounded was recently carried out by Billingsley and Grimes in connection with the copper deposits of Montana[1]. In this region the general sequence of events was as follows: the main folding of the Rocky Mountain chains took place in the Middle Cretaceous, followed by andesite eruptions in the Upper Cretaceous. After this there was a considerable amount of overthrust faulting. The intrusion of the main Boulder bathylith and its accompanying minor phases was probably Eocene, while rhyolite flows and normal faulting continued from Oligocene to Pliocene.

---

[1] Billingsley and Grimes, *Trans. Amer. Inst. Min. Eng.* vol. LVIII, 1918, pp. 284–361. Summary in *Geol. Mag.* vol. LVI, 1919, p. 280.

The intrusive phenomena of the Eocene can be divided into four stages:

(1) Preliminary injections of basic diorite and pyroxenite.
(2) The main mass of the bathylith, consisting of quartz-monzonite.
(3) Dykes of aplite and pegmatite with tourmaline and other pneumatolytic minerals.
(4) Quartz-porphyry.

As early as the Upper Cretaceous andesite stage deposits of native copper were formed, with magnetite and galena carrying some gold and silver, but the greater part of the mineralization is closely connected with the plutonic and hypabyssal intrusions. The vein-formation of the monzonite stage was mostly confined to outlying bosses and cupolas: it consisted mainly of disseminations of pyrrhotite, pyrite and magnetite, with copper and bismuth, together with fissure-veins containing copper, lead and zinc. The aplite-pegmatite stage was accompanied by deposition of chalcopyrite, galena, blende, pyrite, arsenopyrite and rhodochrosite. Much more important was the mineralization accompanying the intrusion of the quartz-porphyries, the deposits usually lying rather deep in the bathylith. They carry extraordinarily rich copper deposits, including enargite, tennantite, tetrahedrite, bornite, covellite and chalcocite. Accompanying the rhyolite flows of the middle and upper Tertiary are a few auriferous sulphide veins. It is calculated that perhaps 99 per cent. of all the ore minerals of the district belong to the intrusions and especially to their later stages. The ores are clearly magmatic and have been concentrated by differentiation. In each intrusive phase there seems to have been a regular sequence of (1) contact and border deposits, (2) internal segregations and (3) fissure and fault veins. It is also noted that in the successive stages the horizon of ore-deposition migrates steadily downwards, doubtless in consequence of falling temperature. This is a remarkably clear and instructive example of the close connection between earth-movements, vulcanicity and ore-formation.

Generalizations. In the preceding sections of this chapter we have considered the mode of occurrence and mineral associations of most of the principal metals and in the chapter on the igneous rocks an account has been given of the physical and chemical laws controlling these occurrences. It now remains to sum up in a general way the conclusions arrived at from the facts of observation and from the theories based on those facts. In the first place it is clear that there is a very marked tendency for the association of certain metals with particular types of igneous rock: the facts need not be again enumerated. From this it follows that the metallic compounds in the magmas are subject to differentiation, though not necessarily under exactly the same laws as the silicate minerals. Furthermore, and this is a point of the utmost importance, the metallic compounds

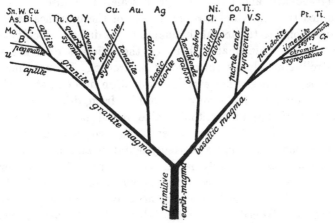

Fig. 28. Genealogical tree showing connection of metals with igneous rocks.

separating from magmas of different composition behave in a way characteristic for each. To put the matter as briefly as possible, the metallic differentiates from basic magmas separate in the *earliest* stages of cooling and tend to *sink to the bottom*, while those derived from acid magmas separate in the *latest* stages of cooling and tend to *rise to the top*[1]. As an example of the first group we have the nickeliferous sulphides associated with norites and gabbros of the Sudbury and Insizwa types; of the second group the pegmatitic lode-deposits of the granite bathyliths and laccoliths.

The next point of importance is the fact that certain metals

[1] Rastall, *Journ. Geol.* vol. XXIX, 1921, pp. 487–501.

and groups of metals show a genetic connection with the associated igneous rock types. Some are rather strictly confined to a narrow range of silica-percentage, while others show a wider distribution. This is probably connected with their natural affinity for other non-metallic elements, and the presence or absence of fluxes, including water.

It is on the basis of the principle set out in the last paragraph that the genealogical tree presented in Fig. 28 has been constructed. The evidence on which it is based is to be found in the earlier chapters of this book. The figure is of course purely diagrammatic and is not drawn to scale. It is also to be remarked that the thickness and length of the branches are not proportional to the actual abundance of the different rock-types. The angles of inclination between the different branches are controlled merely by convenience of drawing. The crossing of various branches is however intentional, in order to show that similar final products may be formed from different partial magmas: that is to say, that similar rock-types and ore-types may have different lines of descent from a common ancestor.

This diagram, though admittedly tentative and incomplete, is in fact a basis for a truly genetic classification of primary ore-deposits. It takes no account of secondary ores of any kind or of the phenomena of local enrichment by supergene processes. It may be claimed however that this systematic arrangement does throw light on the origin and relationships of the primary ores, which are of fundamental importance, as being the original source of all the workable metalliferous deposits of the globe.

CHAPTER X

METALLOGENETIC ZONES

NOTE. The substance of a considerable part of this chapter has already appeared as a special article in *Economic Geology* for March, 1923, vol. XVIII, No. 2, pp. 105–21.

The General Problem. One of the most important problems of metallogenesis at the present day is that of the existence of a definite arrangement in space of the ore-deposits of metalliferous regions. In many localities it is known that some sequence exists in the minerals of economic value. There are several instances where the general character of the ores of a particular mine or of a whole region has been found to undergo changes in depth, and in other instances lateral variations can be traced either along the courses of single lodes or in the metal content of all the lodes when followed in certain directions. There are thus many isolated facts available, but hitherto few attempts have been made to generalize from them, or to build up a coherent account of the whole subject. It may be doubted whether at the present time sufficient data are yet available, and the investigation is attended with peculiar difficulties of a practical nature, since frequently when the product of a mine or of a mining field undergoes a marked change in any given direction work forthwith stops in that direction. However there are cases, and important cases, where continued exploitation of changed ores has been economically possible and these have yielded evidence of much value. The best known of these is perhaps the mining field of Cornwall, where copper in the upper levels in many mines gave place to tin in depth and the latter has been mined for many years with successful results. Other examples will appear in the course of this discussion.

The practical bearing of this subject is obvious and does not need elaboration. It is clear that it would be of value to a mining geologist if he were able to predict with a reasonable amount of certainty changes in the character of ore likely to be encountered in the deeper levels of a mine, or to form an opinion as to the probable horizontal extent of a run of ore. This problem is however not quite of the same nature as that involved in the study of ore-shoots in the ordinary sense of the word. This last expression involves the idea of variations in the concentration of ore of constant qualitative composition, whereas the point now under consideration is rather a vertical or lateral variation in the mineralogical character and metal content of the deposits, an alteration in the *kind* of metal rather than in the *quantity* of metal.

It is evident that in a problem of this kind there are many factors to be taken into account, since the possible causes of variation are numerous. In the first place it is necessary to eliminate here all the features connected with oxidation and secondary enrichment of primary ores; although this is the simplest and best known cause of ore-zones, it is not the subject now under discussion. What we are here concerned with is variations in the character of *primary* ores, a much more fundamental matter. Sometimes however the two sets of phenomena may be somewhat difficult to disentangle; account must always be taken of the simple fact that primary and secondary ore-zones are not necessarily parallel, although they may be so. Secondary ore-zones must conform more or less accurately to the contour of the country as existing at the time of their formation, and their depth and general disposition depends largely on the ground-water relations of the district. The primary zones on the other hand have no necessary relation to the land surface either at the time of their formation or at any other time, and may undergo any amount of geological disturbance by tilting or otherwise at any time after their formation. Likewise they are entirely dependent on the cause to which they owe their formation, whatever this may be. In a great number of cases primary ore-zones can be shown to be genetically related to igneous activity, especially to the intrusion of plutonic bathyliths and laccoliths. Such zones are obviously for the most part to be found in the types of ore-deposits occurring as veins and lodes, which often have great vertical and horizontal extent; hence they are more likely to exist in connection with acid intrusions. The limited and localized character of the ore-deposits of the basic plutonic rocks does not favour their development, except perhaps on a very small scale.

Although as above stated there is frequently to be observed a change in the metals present in ores in depth in any one vein, nevertheless this is by no means universal, and there are many cases of persistence to great depths. One of the most striking is the Morro Velho mine in Brazil, belonging to the St John del Rey Co. This is now the deepest mine in the world, having reached, in 1923, a vertical depth of 6726 feet below the surface. The ore-body is a vertical vein of lenticular cross section, pitching at 45°, and has a maximum thickness of about 30 feet. All reports on the mine agree in the remarkable fact that there is no material change in the tenor of the ore at the greatest depth yet reached. The ore body consists of a mixed gangue of quartz, chalybite, dolomite and calcite, with pyrite, pyrrhotite, arsenopyrite and a little chalcopyrite. Most of the gold is in the arsenopyrite and free gold is rare. The great gold reefs of Mysore, India, are also now worked to a depth of about 6000 feet. These are certainly extreme cases, but other instances could be quoted in which ores have persisted without change to great depths. More common, however, are instances in which the amount of payable ore, or rather more correctly where the tenor of the ore falls off in depth, without much change in its character. Investors in mines are only too well acquainted with this phenomenon, and it is hardly necessary to particularize further on a somewhat painful subject.

Variation in depth. To most British mining geologists the best known case of a definite change in the chief ore-production of a large district is that of western Cornwall. A hundred years ago most of the lode-mining in Cornwall was copper-mining and the chief product of many mines that still exist was copper ore, tin taking quite a subsidiary place, especially in the Camborne-Redruth area. As time went on however and the mines became deeper copper gradually gave place to tin, and the production of the former metal is now negligible in quantity. Such mines as Dolcoath, East Pool and South Crofty are now regarded simply as tin-mines, with tungsten and arsenic as subsidiary products. The numerous abandoned shallow mines of this area were mainly copper mines and when the supply of this metal fell off in depth, most of them were closed down. The distribution of the two metals in this area is obviously in close connection with the form of the granite masses. These take the form of bosses and cupolas, probably connected at no great depth and forming the upper portions of a great bathylith: the junctions of the granite and slate are as a rule fairly steep. Since the metalliferous zones are apparently on the whole parallel to these junctions they are necessarily oblique to the present surface and probably follow in depth, approximately at any rate, the undulations of the granite-slate contact. Nevertheless, it is possible that the richer mineralization is confined to the neighbourhood of the domes and cupolas, and is less developed where the junction lay originally deeper. The rich tin zone lies partly within the granite and partly in the slate, but the greater part of the copper-bearing portion of the lodes lies in the slate. Near the upper levels of the tin zone is a sub-zone of wolfram, and arsenopyrite is also abundant here. In the upper part of the copper zone, zinc, lead, silver and other metals begin to come in. It must be borne in mind that this brief statement is highly generalized and many local exceptions could doubtless be brought forward, but in the main the sequence holds, and the general succession of ores can be shown diagrammatically as in Figs. 30 and 47.

Besides the lodes running parallel to the granite-slate contact, often called the *Champion Lodes*, there are others having a general direction nearly at right angles to them, called *Cross Courses*; these are demonstrably newer than the former set and some of them are faults, displacing the main lodes. Some of the cross courses, especially at a considerable distance from the granite contact, contain ores of lead and zinc, and still further away are some large iron lodes, mainly siderite, where not oxidised. Hence it appears that the general succession in Cornwall from below upwards may be generalized as tin, tungsten, copper, zinc, lead, iron[1]. Whether anything lies below the tin is not known, but it is unlikely: the tin content of the lodes at the

[1] Davison, "The Primary Zones of Cornish Lodes," *Geol. Mag.* vol. LVIII, 1921, pp. 505–12.

deepest level yet reached in the New Shaft at Dolcoath shows a distinct falling-off, and appears to be unpayable below about 3000 feet from the surface. (Further details on the ore-deposits of this area will be found in other sections of this book; see p. 190.)

In one respect this is a particularly easy case to deal with, since the space-relations of the different types are simple and well known. Only two rock-types here exist in any large amount, granite and slate. The greenstones and elvans (quartz-porphyries) are hardly present in sufficient quantity to affect the general question, although they certainly do cause local variations and interruptions of lodes. Both granite and slate are well-defined and homogeneous masses, arranged on a simple plan, and both apparently favourable to lode-formation. We are not here confronted with a complication such as occurs in the Shelve district in Shropshire, where the metalliferous veins are strictly confined to one type of coarse-grained sediment, the Mytton Grits, and die out at the margin of the overlying and finer-grained Hope Shales[1]. The suitability of the country rock for vein-formation is a factor of considerable importance in a question of this sort. It is quite within the bounds of chemical theory that from a mixed solution of metallic compounds one or more might be precipitated on passing through a rock of a particular composition, leaving others in solution. In such a case the ore-zone would have to conform to the extent of the precipitating rock. This may in some instances be the origin of ore-shoots. Apart from this, however, is the simple fact that different metals may be deposited from solution at different temperatures in a homogeneous rock without the interference of actual chemical reactions, such deposition being purely a physical process. In the case of mineralization due to intrusion this would naturally lead to successive zones concentric to the intrusion, in conformity with the distribution of the isotherms. But here again there are various complications to be taken into

---

[1] "Special Reports on the Mineral Resources of Great Britain," vol. XXIII, "Lead and Zinc Ores in the Pre-Carboniferous Rocks of West Shropshire and North Wales," by B. Smith and H. Dewey, *Mem. Geol. Survey*, 1921. Hall, *Mining Mag.*, vol. XXVII, 1922, p. 201.

account, one of which is the time factor. In most instances we have no guarantee that, to put it in simple language, all the metals started from the intrusion at the same time. If they did so, the matter is simple: the metals should be deposited in zones around the intrusion according to the temperature at which they became insoluble in the carrying solutions. In discussions of this subject it is generally assumed, either consciously or unconsciously, that ore-deposition takes place during the *cooling* of the rocks, but in the case of veins in sediments at any rate this is by no means necessarily the case. As is well known rocks are bad conductors of heat, and the temperature of the whole region around the intrusion may actually rise by conduction after the arrival of the mineralizing solutions. In many instances also it is clear from field evidence that earth-movements have taken place during deposition, veins having been reopened and often brecciated, and then partly or wholly filled up again and again by deposition of similar or different ore minerals, sometimes apparently at long subsequent periods. In some instances also there may have been more than one period of mineralization owing to different intrusive phases in the same region. Hence the whole problem is evidently very complex.

An ideal case. Let us however consider briefly from the theoretical point of view an ideal case, simplified as much as possible. The conditions assumed are the intrusion of a metalliferous igneous rock, granite, for instance, into a series of homogeneous stratified rocks of a character favourable to lode-formation, that is, capable of allowing the formation of open fissures under stress. The metallic compounds are to be supposed to concentrate in the last residue of the cooling magma as vapours or solutions, with water and other more or less volatile substances; that is to say, pneumatolysis is at any rate not excluded, though not essential to the argument. During cooling jointing is developed in the granite and the dynamic force of the intrusion gives rise to planes of fracture in the surrounding rocks, usually parallel or at right angles to the contact. The latter class may be continuous with similar fissures in the granite itself. Along these fissures the ore-bearing solutions pass outwards from the granite, often altering and mineralizing the walls of the fissures themselves, both in the granite and in the sediments. From the centre of the granite outwards there must be a gradual decrease of temperature, and there may be an abrupt change of gradient at the contact, though it is unlikely that this will be very marked, at any rate in the later stages of cooling. Thus the solution, travelling outwards, will eventually come to a region where it is no longer in equilibrium and will begin to deposit one or more of its components in the solid form. If the solution is a complex one, the points of supersaturation for the different components will be reached at different temperatures, or what comes to the same thing, at different distances from the point of departure. This in the simplest case must necessarily lead to a zonary arrangement of the solid products of crystallization, and the metals will be deposited in the inverse order of their solubility in the complex magmatic solution. Here, however, a further point of importance arises. The solubility of any one metal in the solution will be affected by the presence of all the other metals and the presence of an additional one may vary the normal order of the others. An example will make this clearer. Let us suppose that in one case three metals, A, B, C, are present and are deposited in that order. In another instance there are four metals, A, B, C, D. Here the presence of D may modify the relative solubilities of A, B, C so that they may be deposited in a different order, such as B, A, C or A, C, B; furthermore the additional metal may come in any position in the series either in space or in time, according to its own solubility in the presence of the other three. Thus it is evidently difficult to predict the behaviour of a given case on the basis of data derived from other cases. Empirical observation alone can afford sure ground until sufficient experimental facts have been accumulated to cover every case. On physico-chemical grounds however it seems safe to assume that the same assemblage of metals under similar conditions will be deposited in the same order.

Variations in lead-zinc lodes. It certainly is a fact that in some groups of metals a definite order in space is well known to exist. For example, a comparison of a great number of reports on silver-lead-zinc mines shows that commonly in depth zinc increases at the expense of lead and often entirely replaces it, while silver is more abundant in the upper levels of the lead zone. It has already been mentioned that in Cornwall lead and zinc occur further from the granites than copper, and it is a very interesting and important subject of enquiry whether there is any evidence of the occurrence of an important zone of copper ores below the lead-zinc veins of the north of England. As to this point there is actually some evidence. It is well known that copper ores occur in small quantities in many localities in the Carboniferous Limestone and a very interesting case is that of the once famous Ecton mine on the borders of Staffordshire and Derbyshire. This mine has apparently been forgotten by the present generation of writers on mining geology, but affords some useful lessons. The following description is taken from Phillips-Louis, *Ore Deposits*, 1896, p. 284: "The principal deposit, which as early as the year 1778 had been worked to a depth of 200 fathoms, is a pipe-vein, piercing the highly-contorted limestone beds almost vertically. There are eight main lodes coursing E. and W., and the same number of N. and S. veins, together with many smaller branches. The upper portions of the lodes contain lead, poor in silver, and blende with copper ores, the latter predominating on the lower levels. The principal ores are chalcopyrite and erubescite, and with these occur oxides and carbonates.... The veins sometimes attain a great thickness; being in one case as much as seventy yards from side to side." This mine in 40 years yielded copper to the value of £670,000 and the ore is stated to have as a rule contained about 15 per cent. of copper. Although the account here quoted is not very clear as to the relations of the "lodes" to the "pipe vein" one thing is quite definite, namely, that the copper was more abundant in depth in this, for the period, unusually deep mine.

As regards this question, there is another consideration that may be of assistance. In the Lower Palaeozoic rocks of central and north Wales, the Lake District and the Southern Uplands of Scotland are many metalliferous veins, often rich, containing in very varying proportions lead, zinc and copper, with silver and occasionally other metals, such as the arsenic, tungsten and molybdenum of the Carrock Fell area in Cumberland. An ad mirable series of memoirs descriptive of the lead-zinc ores of the British Isles is now in course of publication by the Geological Survey, in which full details of all known occurrences will be found. In innumerable instances it is clearly shown that on the whole the richest zinc deposits are at lower levels than the richest lead deposits, and silver increases in quantity in the upper part of the lead zone. The time-relations of these veins to those of the Carboniferous Limestone, as well as the exact date of formation of the latter, are unknown. It is clear, however, that some are at least post-Triassic, since lead is found in the Dolomitic Conglomerate of the Mendips and lead and copper in the Keuper sandstones of Alderley Edge, Cheshire, while copper ores occur in the Peckforton Hills and at various other points in Cheshire and Shropshire. All these presumably belong to a late phase of the Armorican crust-movements.

It is quite clear that on the whole copper is more abundant in the veins in the Lower Palaeozoic rocks in the localities before mentioned than in those of the Carboniferous Limestone, and the suggestion at once arises that the Lower Palaeozoic veins may represent the lower zones of veins once existing, whose upper portions were similar to those of the Pennine area or of the Carboniferous limestone of Flint and Denbigh, these upper portions having been since removed by denudation[1]. It might be suggested that the veins in the Lower Palaeozoic rocks were related to the Caledonian earth-movements and intrusions and at first sight this view might appear more probable, but Finlayson[2], in his admirable discussion of the whole subject, has brought forward evidence to show that, with the probable exception of Anglesey, all of these are related to the Armorican movements. Why Anglesey is excepted is not quite clear. It may be suggested therefore that if the lead-zinc veins of the Carboniferous could be followed down into the underlying Lower Palaeozoic rocks of the Pennine area, in all probability a copper zone would be found. This conclusion is admittedly highly

---

[1] Throughout the north of England and North Wales the Devonian system is absent, the Carboniferous resting with a marked unconformity on Silurian and older rocks.

[2] Finlayson, *Quart. Journ. Geol. Soc.* vol. LXVI, 1910, p. 284.

speculative, but it may at some future time be worthy of practical investigation. It must however be admitted that the latest description of the lead-zinc deposits of the northern Pennines, quoted on p. 301, does not lend much, if any, support to this hypothesis, since the ores are seen clearly to end off downwards, the lowest visible parts of the veins being usually barren. At any rate it is significant that the order in which the ores are found, namely copper, zinc, lead, from below upwards, is the same as that prevailing in Cornwall.

In many other districts in the British Isles and elsewhere there is evidence that blende is on the whole earlier and deeper than galena. Thus in the Shelve district in Shropshire Mr T. C. F. Hall states that blende increases in depth and there is no appreciable quantity of silver in the galena. In Flintshire there are two sets of veins of which the older contain blende, while the later ones have practically no blende[1]. In Wales again blende increases in depth, and many of these veins carry copper, as is also the case in the Isle of Man. Considerable quantities of pyrite are also found in many veins in the Lower Palaeozoic rocks.

The Ore Zones of Cornwall. Among British examples this question has been more completely studied in Cornwall than anywhere else, and no apology is needed for returning once more to this often-mentioned area, since some work of great importance has lately appeared[2]. From a detailed microscopic study of veinstones Davison has shown that as a general rule, with slight variations, the order of arrival of the minerals in the lodes is as follows:

1. Quartz, mica, tourmaline, topaz, cassiterite, wolfram, stannite and molybdenite.

2. Quartz, chlorite, fluor, chalcopyrite, mispickel, pyrite and blende.

3. Lead and silver ores.

Wolfram is often earlier than cassiterite and the latter is

[1] Hall, "Lead Ores," *Imperial Institute Monographs*, 1921, pp. 49, 51.

[2] E. H. Davison, *Geol. Mag.* vol. LVIII, 1921, pp. 505–12. H. B. Cronshaw, *Trans. Inst. Min. Met.* vol. XXX, 1921, p. 408, with discussion and author's reply.

usually before tourmaline. Chalcopyrite is often before mispickel and pyrite is usually later than both.

The following are the chief metalliferous zones recognized in ascending order, which is equivalent to an order of decreasing age.

4. Silver, lead and zinc minerals.
3. Dominant copper.
2. Tin and copper in fair proportions.
1. Dominant tin.

The ores of zone 4 often occupy fissures belonging to a later series of earth movements, but the other three can all be seen in the same lodes. The latter also sometimes carry lead, zinc and silver in their highest levels along with some copper. In the neighbourhood of ridges or cupolas in the granite copper is often replaced by wolfram, and there is then a wolfram zone just above or just inside the granite. The zones run approximately parallel to the granite-slate contact, which is undulating in depth, patches of metamorphosed slate being seen in places far from any visible granite outcrop. As a rule the base of the copper zone nearly coincides with the contact.

Cronshaw has also investigated many typical lodes in great detail and has drawn up the following scheme showing the relative ages of the different minerals, the newest being at the top.

10 Calcite.
9. Pyrite, haematite and chalybite, with quartz and fluorspar.
8. Galena, blende and chalcopyrite with quartz and fluorspar.
7. Chalcopyrite, with quartz and fluorspar, accessory tourmaline, chlorite and cassiterite.
6. Chlorite, with accessory tourmaline, cassiterite, mispickel, quartz and fluorspar.
5. Fluorspar with accessory tourmaline, cassiterite, mispickel, quartz and chlorite.
4. Blue tourmaline and cassiterite with mispickel and accessory quartz and fluorspar.
3. Brown tourmaline with accessory quartz.
2. Quartz and cassiterite with tourmaline and mispickel.
1. Quartz, with wolfram, mispickel and felspar and local developments of tourmaline, cassiterite and topaz.

Of these groups, numbers 2 to 6 correspond to Davison's tin zone, group 7 to his copper zone and group 8 to his lead-zinc-silver zone. Group 9 is the iron lodes, such as the great Perran lode near Perranporth. The calcite deposits may be of quite different origin, and much later.

From the data here given the following further generalization is drawn, the mineralization being divided into periods corresponding to the order of arrival of the different constituents.

Stages	Gangue	Ore		
10 8, 9 } 7 }	Calcite Quartz	Iron ores { Zinc-blende, galena { Chalcopyrite		
6 5 3, 4 1, 2	Chlorite Fluorspar Tourmaline Quartz	Cassiterite	Mispickel	Wolfram, chalcopyrite

It will be observed that quartz comes at two very distinct periods, while cassiterite and mispickel cover a long range: they both reach a maximum in stages 3 and 4.

From the above summaries it is clear that in Cornwall there is a close correlation between distribution in space and order in time of the different minerals, both ore and gangue, the later minerals being the more distant from their source in the granite. The correlation between the schemes as set forth by Davison and Cronshaw and the primary zones of Spurr[1] is clear and need not be discussed in detail at this point.

Zones in the Lead–Zinc Deposits of the British Isles. Reference has already been made to the evidence that in some cases at any rate the lead-zinc lodes of Britain pass down into copper deposits. A survey of the recent literature of the subject, which is fairly extensive, brings out clearly the fact that in innumerable instances it is possible to establish a general change in the proportional relations of both the ores and the gangue minerals within the limits of the lead-zinc zone itself. In innumerable cases we find casual remarks, without ulterior object,

[1] See next section.

to the effect that the proportion of blende increases downwards, while the galena decreases. In fact it is hardly too much to say that some of the veins which are mainly lead-bearing near the surface, become almost exclusively zinc producers in depth. This is true of the deposits lying in the Lower Palaeozoic rocks as well as those in the Carboniferous limestone of North Wales and the north of England.

A particularly instructive example is afforded by the mining district of west Shropshire, which has recently been exhaustively described[1]. It is quite clear that in this region the veins show a threefold division of minerals, with zinc below, then lead, and finally a capping of barytes near the surface. All these lie in the Mytton series of the Lower Ordovician. The pre-Cambrian rocks of the area to the east contain copper, but the junction is a faulted one, and the depth relation of these to the Palaeozoic is not definitely known: still the fact is highly suggestive. This area is described more fully in the chapter on lead and zinc deposits (p. 289).

In the second part of the *Memoir of the Geological Survey* just quoted Messrs Smith and Dewey give some interesting facts relative to the zonary distribution of ores in the veins of North Wales: in Merionethshire, where gold also occurs in workable quantities, it appears to come above the galena zone, and rapidly dies out downwards. The presence of an appreciable amount of gold is very exceptional for the British Isles. Again in Caernarvonshire we find galena above blende, with sometimes pyrites in the highest levels, as at the well-known pyrites mine of Cae Côch, Trefriw, in the Conway valley.

The Generalized Succession of Ore Zones. From comprehensive studies of the mineral content of ore-bodies, both metallic ores and gangues, and their relation to different rock-

---

[1] Bernard Smith, "Lead and Zinc Ores in the Pre-Carboniferous Rocks of Shropshire and North Wales, Part I. Special Reports on the Mineral Resources of Great Britain," *Mem. Geol. Survey*, 1922. T. C. F. Hall, "Lead Ores," *Imperial Institute Monograph*, 1921, pp. 49–51; *Mining Magazine*, vol. XXVII, 1922, pp. 201–9. An admirable account of the geology and mineral resources of this region is also contained in a privately printed report on the mines of the district drawn up by Mr Hall for Shropshire Mines Ltd., to which the writer has had access by the kindness of the Managing Director of the Company.

types and structures it has been found possible to build up a coherent scheme of ore-formation in relation to depth-zones, which is of course equivalent to saying, in relation to temperature and pressure. This subject engaged the attention of many of the early writers, who were however somewhat handicapped by the paucity of data at their disposal as regards extra-European regions, and by limitations imposed by the rather nebulous ideas prevailing as to petrogenesis. Perhaps the earliest of the generalizations on modern lines was that put forward by Lindgren[1] in 1907. In this he divided ore-formation into five chief types: (1) magmatic segregations, (2) pegmatites, (3) deep-seated veins, (4) shallow veins, (5) surface deposits. Furthermore he gave lists of minerals, both metallic and non-metallic, characteristic of each zone. These subdivisions were laid out on very broad lines: nevertheless the root of the recent ideas of a zonary arrangement of metals is there.

This paper was followed very shortly (in the same year) by another by Spurr, whose classification is more definitely mineralogical, or metallogenetic, and is reproduced here. It is as follows[2]:

6. The zone of earthy gangues, barren of valuable metals.
5. The zone of silver and much gold, with antimony, arsenic, bismuth, selenium and tellurium.
4. The galena-blende zone (galena usually argentiferous).
3. The cupriferous pyrite zone.
2. The free gold-auriferous pyrite zone.
1. The pegmatite zone with tin, tungsten and molybdenum.

In a later paper[3] zone 4 was broken up into two, with lead above and zinc below: this is in entire conformity with British experience. In Britain 1, 3, 4 and 6 are fully developed, but for some reason the rich gold and silver zones are missing.

In 1908 Emmons[4] dealt with the same subject and gave elabo-

---

[1] Lindgren, "The Relation of Ore-Deposition to Physical Conditions," *Econ. Geol.* vol. II, 1907, pp. 103–27.

[2] Spurr, "A Theory of Ore-Deposition," *ibid.* pp. 781–95.

[3] Spurr, *ibid.* vol. VII, 1912, p. 489.

[4] Emmons, "A Genetic Classification of Minerals," *ibid.* vol. III, 1908, pp. 611–27.

rate tables of the zonary distribution of minerals in eight groups, which are very similar to those of Lindgren and Spurr, but showing rather more subdivision in detail.

However, in all these cases the divisions into zones, or rather into types of ore-deposit, are laid out on very broad lines, dealing for the most part with zones measurable by thousands of feet; depths greater than are usually covered by individual mine-shafts, and unlikely to be encountered within limited areas. What is more immediately the subject of investigation is rather the problem of whether it is possible to work out a definite succession of metals within the metallogenetic region of a single phase of mineralization, such as the intrusion of a bathylith or laccolith. On this side of the question Spurr's subdivision of his zone 4 is a step in the right direction, and a similar succession may be regarded as established in connection with certain British ore-provinces, and especially Cornwall, as already set forth in this chapter.

Quite recently the whole subject has been revived and amplified by Kemp[1], who has brought forward certain concrete examples on a comparatively small scale from Bingham Canyon, Utah, and elsewhere, and has quoted and summarized much of the recent literature of the subject.

In both Germany and Austria similar phenomena have been described. In the well-known mining district of the Harz Mts, around Clausthal, experience has shown that in depth lead gives place to zinc, and in the Bleiberg district, near Villach in Kärnten, which formerly supplied most of the lead production of Austria, as the mines got deeper the proportion of zinc increased and this is now the chief product.

American examples. Among American occurrences of ore-deposits showing a zonary arrangement special mention may be made of the Butte district in Montana, U.S.A., one of the most important copper producers of the world. Here the genetic relations of the mineralization and igneous intrusions have been worked out with special clearness and certainty and the metal-

---

[1] Kemp, "The Zonal Distribution of Ores concentrically around an Intrusive Mass or Igneous Center," *ibid.* vol. XVI, 1921, p. 474 and vol. XVII, 1922, p. 46.

logeny shows some interesting features. One point is that it is generally believed that here chalcocite is a primary mineral, and much of the copper is in primary combination with arsenic, as enargite. Thus it would appear that iron is less abundant here than in other copper regions, where the primary ore is chalcopyrite. According to the most recent descriptions there are three more or less well-defined ore-zones[1]. The central one carries mainly copper and is characteristically free from zinc and manganese. An intermediate zone of irregular width carries mainly copper but with a considerable amount of zinc, while near its margin manganese becomes fairly common. The outermost zone mainly carries zinc and manganese with a large amount of silver and some gold. It appears therefore that in this area the sequence is copper, zinc, manganese, silver, gold. Lead appears to be comparatively rare, its place being taken by manganese. In this instance the order copper-zinc, etc., does not seem to have been changed by the presence of manganese, which merely takes the place of lead, and is closely associated with the silver.

In many of the mining areas of western America there appear to be as yet insufficient data for generalization and in both North and South America the problem is greatly complicated by the depth of the oxidation zone, owing to the dryness of the climate and the prevalence of complicated forms of secondary enrichment.

A particularly interesting case is that of the tin-silver deposits of Bolivia. These form so far as is known a unique type and for a long time much doubt existed as to their true character, mainly from want of precise local knowledge, most of the generalizations having been evolved by German geologists who had not visited the country. Quite lately however the whole subject has been carefully investigated by competent American authorities, especially by W. M. Davy[2]. According to the observations of this author the deposits in question really extend over a large tract of country and show a regular gradation in

---

[1] Billingsley and Grimes, *Trans. Amer. Inst. Min. Eng.* vol. LVIII, 1918, pp. 284–361; abstract in *Geol. Mag.* vol. LVI, 1919, p. 280. Sales, Bull. 80, *Amer. Inst. Min. Eng.* 1912, pp. 1523–1616.

[2] Davy, "Ore Deposition in the Bolivian Tin-Silver Deposits," *Econ. Geol.* vol. XV, 1920, pp. 463–96.

character from north to south. In the north the veins are in obvious connection with plutonic intrusions and are characterized by the presence of tin, tungsten, bismuth, tourmaline and topaz, thus belonging to the deep-seated vein-type of pneumatolytic affinities. In the middle region, where the veins are connected with hypabyssal intrusions, tin is less abundant, occurs mainly as stannite and other complex sulphides instead of cassiterite, and is associated with copper and zinc. Finally, the richest silver zone is associated with effusive lavas, much of the silver occurring in tetrahedrite, which here appears to be a primary mineral. Here there seems to be no doubt that the different types of vein-deposit are closely correlated with the depth of the igneous rocks from the surface, or, what comes to the same thing, that they are true temperature zones. It may be taken as established that tourmaline is always a high-temperature, deep-seated mineral, and its presence along with ores certainly shows plutonic conditions. Here then is an example of a more or less normal cassiterite-tourmaline zone of plutonic character grading upwards into a zone with tin sulphide and some copper and this again into a zone with silver, copper, zinc and lead; the last three metals have not apparently been differentiated, at least so far as current descriptions show. It is obvious that the general sequence is the same as in Cornwall and elsewhere, the chief local features being the abundance of tin sulphides and silver.

China. Turning now to another very distant part of the world, a remarkably interesting occurrence of metalliferous zones has recently been established in China[1]. It is well known that this great country is highly mineralized, being specially rich in tin, tungsten and antimony. Of the two last-named metals China is now the largest producer. The mining of antimony is an old-established industry, and during the latest stages of the war an immense production of tungsten ore was obtained from some of the southern provinces. In the memoir just quoted a sketch-map is given of China south of latitude 32° N., showing an area about 400 miles broad and parallel to the coast divided into a double

[1] W. H. Wong, "Les Provinces Metallogéniques de Chine," *Bulletin Geol. Survey of China*, No. 2, 1920.

series of concentric zones, each characterized by the preponder-
ance of a particular metal, each series being related to a belt
of intrusive granite and dioritic rocks. The point of greatest
interest however is that the sequence here established in each
case is precisely the same as that found in England and in parts
of the United States and elsewhere. The Chinese succession of
zones, when followed away from the region of granitic intrusions
is as follows:

tin, copper, zinc, lead, antimony, mercury.

In China there appear to have been two principal periods of
mineralization; the first in the latest Algonkian, which yielded
largely gold in irregular distribution, with a little disseminated
copper, tungsten and molybdenum. The second and principal
period which gave rise to the zonary arrangement just described
occurred between the end of the Palaeozoic and the Jurassic, in
connection with great granodioritic intrusions. We have here
therefore an excellent example of metallogenetic epochs, as well
as of zonary distribution of ores.

Zones of rock-alteration. In this connection it is of
interest to note that, as pointed out by Emmons[1], a zonary
distribution can sometimes be recognized in the types of rock-
alteration that accompany the formation of ore-deposits. Thus
Emmons says "in many metalliferous districts three principal
types of alteration are recognized—chloritic alteration in a zone
extending far out from the deposits, extensive sericitization
within this zone and silication near the master fractures. Where
the fractures are closely spaced the area of sericitic alteration
may be hundreds of feet wide and the chloritic alteration may
extend over thousands of feet." This generalization refers chiefly
to the types of alteration brought about by the great copper-
bearing acid bathyliths of western America, but similar ideas are
applicable in many other regions. In Cornwall something of the
same sort can be observed: thus at Dolcoath, where copper is
predominant from the surface to about 900 feet, with tin below
the vein-stuff of the upper zone is mainly a green quartz-chlorite
rock, while in the deeper zone this is replaced by a blue rock

[1] Emmons, *The Principles of Economic Geology*, New York, 1918, p. 248.

composed of quartz and tourmaline, showing a higher degree of alteration and more far-reaching chemical changes.

Causes of primary zoning. In this discussion it has hitherto been assumed, either explicitly or tacitly, that the mineralization is due directly to the intrusion of igneous rocks. Now the argument is capable of application in another way, in the converse sense. If it is regarded as established that a definite generalized succession of mineral zones exists, the occurrence or otherwise of this sequence can be applied as a test of the igneous origin of any particular set of ore-deposits on a large scale. It is well known, for example, that there has been much discussion as to the primary or secondary origin of certain lead-zinc deposits, especially those of the Pennine area in England and of the Mississippi valley in America. It is not proposed here to discuss either of these specific cases. They are merely brought forward as examples of the possible application of the method. It is however evident that if in any instance there is no zoning at all, or zones occur in an order inconsistent with a general and well-established law, the formation cannot have been due to the causes to which the law is applicable. Further generalizations are obviously needed before this can be considered a workable proposition. Nevertheless the fact remains that if in a large number of areas where the mineralization is known to be of igneous origin the ores occur in a regular order, then, if in any other area the same ores occur in a different order their formation is in all probability due to a different cause.

We will now proceed to consider from the theoretical point of view and on general grounds some of the problems here presented. It may be assumed as an axiom that in any chemical and physical processes involved in ore-formation the controlling factors are temperature and pressure, and that under varying conditions of temperature and pressure the same substances may behave in different ways and form compounds possessing different mineralogical characters, or to put it in a still simpler form, under varying conditions the same *metals* may form different *minerals*.

For the sake of simplicity we will deal first with the case where the mineralization is of igneous origin.

When a mass of heated igneous material is injected into a series of comparatively cool rocks at a considerable depth the lines of equal pressure will, apart from variations due to small local causes, run approximately parallel to the earth's surface[1], since the pressure is due simply to the weight of the over-lying rocks. On the other hand the lines of equal temperature, or isotherms, are the resultant of two independent factors. First there is the gradual increase of temperature downwards, which is of universal occurrence, though the rate of increase varies locally: secondly there is the temperature gradient set up by the heated intrusion. If this last was the sole cause the isotherms would be parallel to the contact surface. But the combination of the two causes must give the isotherms a more complex form, varying with the shape of each individual intrusion. A simple

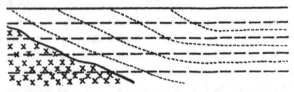

Fig. 29. Isobars (broken lines) and isotherms (dotted lines) in relation to an intrusion. *By permission of the Editor of Economic Geology.*

instance is shown diagrammatically in Fig. 29. The isobars and isotherms therefore cut one another obliquely, marking off regions which may be regarded as the stability fields of certain possible chemical compounds[2].

In this ideal case the intrusion is assumed to give off a number of different products, *A, B, C ...*, which travel outwards to distances controlled by the temperature, and form compounds controlled jointly by temperature and pressure: hence there is evidently much room for variation. The same considerations apply to changes set up in the country rock by the heat of the intrusion. The minerals found at any point are therefore a

[1] As a matter of convenience lines of equal pressure will here be called *isobars*; this term is commonly associated with barometric pressure, but there is no etymological or other reason why it should be so limited.

[2] Since we are here dealing with a mass in three dimensions we should in strictness always speak of isothermal and isobaric *surfaces*, and the stability fields are in reality solid figures.

function of the chemical composition, the temperature and the pressure, and the whole problem is evidently one of great complexity. When we take into account possible original local variations in the composition of the country rock it becomes still more so, and such cases cannot be profitably considered, at any rate in the present state of our knowledge.

However, taking the simplest possible case and reducing to a purely diagrammatic form the relations may be expressed as in Fig. 30. Here A, B, C, D represent the metals given off from

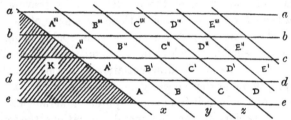

Fig. 30. Diagrammatic representation of stability fields of compoundsc deposited from an intrusion. *By permission of the Editor of Economic Geology.*

the intrusion K; the lines a, b, c... are the isobars and x, y, z the isotherms, here considered for simplicity as straight lines; within the spaces bounded by these lines the metals A, B, C will be deposited as modifications A', B', A'', B'' and so forth, with different gangue minerals and different alterations in the country rock. From this diagram it appears that sinkings in any position will always encounter some kind of change in the character of the ore. Probably the nearest approach that is yet known to this ideal development is to be found in the mineralization of Bolivia, as previously referred to and described more fully in Chapters XIII and XVII. When such an area has undergone deep denudation the effects observed will evidently vary according to the direction of the section formed by the ground surface.

CHAPTER XI

MINERAL FORMATION

The Physical Chemistry of Mineral Formation. Since the great majority of ore-minerals and gangue-minerals are definitely crystalline substances their formation is controlled by the physical and chemical laws that govern the behaviour of this class of matter under the conditions that prevail at the time of their formation. Some of these laws have already been mentioned in the chapter on the igneous rocks, being there employed to elucidate rock-structures and more especially the phenomena of magmatic differentiation. In this chapter we shall apply some of these laws again, as well as others, to a consideration of ore-deposit from a more purely mineralogical standpoint, with special reference to mineral associations, the sequence in time and space of different mineral species and certain changes that may take place after the primary formation of the minerals.

For the sake of simplicity it will be assumed that minerals are deposited from solution. In a foregoing chapter it has been shown that igneous magmas are essentially solutions, though at times they may give rise to vaporous phases. However, as is well known, at high temperatures and especially at high pressures, the distinction between a liquid and a gas disappears, and many ore-deposits have probably been formed under such conditions, *i.e.* above the critical point for water (about 375° C.). Again some minerals, especially in volcanic regions, have formed by sublimation, or direct passage from the gaseous to the solid form. But undoubtedly in the vast majority of cases water played the most important part in deposition: that is to say, most primary ores belong to the hydrothermal phase of igneous activity, while a considerable proportion and especially those of secondary origin are formed under conditions approaching those of normal temperature and pressure.

Hence it follows that the laws of solutions are applicable to problems of mineral genesis. These laws are now known both theoretically from the standpoint of thermodynamics, as also

experimentally. The fundamental idea underlying the study of
the formation and alteration of minerals is that of equilibrium:
just as the permanent existence of ice and water in contact
is impossible with variable temperature and pressure, so also
minerals crystallizing from a fused state must either increase in
quantity or disappear if the temperature is lowered or raised.
But in nearly all instances the formation of minerals is complicated by the fact that the liquids in question do not consist
of one substance only, like water or fused salt, or molten metal.
More than one substance is present and the mass becomes a
solution as distinguished from a simple melt. Among artificial
products fused blast-furnace slag is an example of such a solution
and from it various minerals may crystallize. In ordinary
language a solution means something dissolved in water, but in
the technical sense a solution is any liquid mixture, while of
late years the term has been extended to include solid solutions,
or homogeneous mixtures of solids. Furthermore there is in
reality no distinction between solvent and solute. A mixture of
salt and water must be regarded as a mutual solution or homogeneous mixture of the two substances. Thus in a slag likewise
there is in reality no solvent, but the whole is a homogeneous
fused mass or *melt*. In studying melts or solutions therefore we
soon encounter the conception of what are known as components. A system of ice, water and water-vapour contains only
one component, the compound H_2O: partially cooled molten
lead has only one component, the element Pb, and so on. But
a solution of salt and water contains two different chemical
substances, NaCl and H_2O. This is a system of two components
and most natural solutions are complex, often having many
components.

The Phase Rule. Ice, water, and steam, or molten and
crystallized lead are different forms of the same substance: such
different forms are known as *phases*. In a solution of a given
composition the number and amount of the components are fixed,
but there are two possible external variable factors, temperature
and pressure. These external variables have a controlling influence on the possible number of phases. Thus if the temperature
is lowered part or all of the water may freeze to ice: if the

solution is in a closed vessel and the pressure is reduced, more water will pass into the form of vapour and so on. Hence not only the relative amounts of the phases, but also their number may vary. These variations work according to regular laws. Thus, as is well known, at great heights water boils at a lower temperature than at sea-level: that is, the equilibrium between water and vapour varies according to pressure. This is expressed by saying that the system possesses one degree of freedom. Similarly, if we half-fill a flask with water, exhaust the air, and seal it up the vacuum will be filled by water vapour whose pressure varies with the temperature, but water will still exist as such in the flask through a wide range of both temperature and pressure. A block of metallic lead will also continue to exist under widely different conditions of temperature and pressure, which are here independent variables and there are thus two degrees of freedom.

By a consideration of all possible cases of this kind, aided by the general principles of thermodynamics it has been found possible to formulate a general law, known as the Phase Rule, which is in reality the foundation of the study of the formation of minerals under complex conditions[1]. This rule may be conveniently stated as follows: a system of n components can exist in $n + 2$ phases only when the temperature, pressure and concentration are fixed; if there are n components in $n + 1$ phases, equilibrium can exist while one of these factors varies; if there are only n phases (*i.e.* equal to the number of components), two of the factors may vary. This may be written as the equation

$$F = C + 2 - P,$$

where F = degrees of freedom, C = number of components and P = number of phases.

In the cases hitherto considered we have dealt only with systems of one component, so that the question of concentration has not come into account. It must be remembered however that the absolute amount of the components does not affect equilibrium: thus a saturated solution of salt in water in contact with solid salt has the same percentage composition whatever

---

[1] Findlay, *The Phase Rule and its Applications*, 4th edition, London, 1922.

may be the amount of solid (temperature and pressure remaining constant).

In dealing with the formation of minerals the point usually sought to be determined is the number of phases that can be obtained from a system of certain components under given conditions of temperature and pressure: hence the equation can conveniently be written in the form

$$P = C - F + 2.$$

This law is the foundation of all quantitative studies of mineral genesis, since it shows what must happen to solutions of given composition under given conditions.

Solidification of a single component. This is the simplest possible case and may be exemplified for instance by the process of crystallization of a magmatic sulphide segregation, which has settled to the bottom of a differentiated intrusion or has been injected as an independent body. Here the solidification is entirely a matter of temperature and pressure. The effect of pressure is no doubt important in deep-seated intrusions and the manner in which it will work may be predicted as follows: if the solid crystals are denser than the liquid the freezing point will be raised by pressure. This is the commonest case. There are however some grounds for supposing that at enormously high pressures, such as must prevail deep within the earth, the effect is diminished or even reversed. In the usual case, when the molten mass reaches a definite temperature, varying with the pressure, solid crystals begin to form and continue to grow as the temperature falls, the crystals usually being larger the slower the cooling. However, a phenomenon known as under-cooling is possible, where a melt remains liquid below its real freezing point, so long as it is undisturbed and no solid is introduced from outside. Under-cooling with its attendant metastable and labile conditions is an important factor in the formation of igneous rocks and probably of ore-deposits as well, but its theoretical side cannot be discussed here. The size of the crystals formed also depends to a great extent on the viscosity of the liquid just above the point of solidification. The more mobile the liquid the larger will the crystals be, since free diffusion of molecules is then possible. But the most important condition in any case is rate of cooling. This is a fact known to every metallurgist and chemist. Quick chilling always produces a finely-granular structure and slow cooling favours coarse structure.

In some cases, more especially in silicate melts, the effect is so marked that a very quickly cooled lava, for example, may not have time to permit the individualization of crystals, and the whole solidifies without any discontinuous change of state as a perfectly homogeneous mass, a glass, indistinguishable from artificial glasses. Such a system has no definite freezing point and is in reality an under-cooled liquid. In the sulphides however and apparently in most other metallic minerals nothing analogous to a glass is known, and even the chilled margins of sulphide ore-bodies are always crystalline, though often very finely granular.

This type of solidification is not of much importance in our present study, since the vast majority of primary ores are undoubtedly formed from more or less complex solutions or mixtures of fused material—that is, from systems of more than one component.

Crystallization from Solutions. The simplest case that we have to consider is the crystallization of a single mineral from a solution in which no other mineral substance is present, or to put it into more correct language, the separation of a solid component from a liquid homogeneous mixture of two components, a soluble salt and water, for example. Here the relations can be expressed by a simple curve on a temperature-concentration diagram and such a curve is usually a more or less straight line, since solubility commonly varies directly as the temperature, in nearly all cases increasing with rise of temperature[1]. Thus we obtain a diagram like Fig. 31. The real meaning of this diagram is that at all points above the solubility curve a homogeneous mixture of the two substances is possible; two components forming one phase: below the same line solid crystals can exist in contact with solution, while any point actually on the line

---

[1] In this discussion, for the sake of simplicity, pressure is neglected: as we are dealing here with separation of solids from liquids the effect of pressure is in reality very small.

represents a concentration and temperature at which solid and liquid can co-exist in equilibrium, any variation of temperature necessarily causing a change in the concentration. This diagram is the correct expression of the common fact that when a solution of a given composition is cooled to some definite temperature it begins to deposit solid, not however necessarily though usually a crystalline solid.

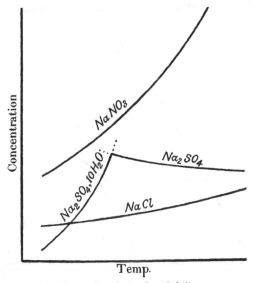

Fig. 31. Examples of simple solubility curves.
Not to scale.

Frequently however complications are introduced into such a system by the fact that many substances crystallize in different forms at different temperatures and that even in the solid form physical changes may take place on change of temperature. The temperatures corresponding to such changes are known as *inversion points*, and usually at such a point there is a change in the run of the solubility curve: the solubility may either increase or decrease at an inversion point. Hence the solubility curve may become either steeper or flatter, or its gradient may even be reversed, if the new inverted form is less soluble with rise of temperature. This happens with sodium sulphate (Glauber's salt) and perhaps with calcium sulphate (gypsum and anhydrite): it is common among salts of calcium in general, especially its compounds with organic acids.

Hitherto in discussing this type of simple solution it has been tacitly assumed that only one of the substances concerned separates in the solid form, *e.g.* a soluble salt from water above the freezing point of the latter; but of much more importance is the case where both components can exist in the solid phase, while in the liquid state they form a homogeneous mixture, or

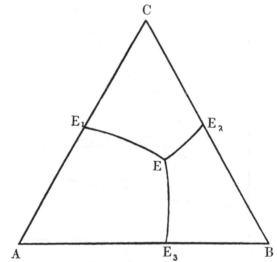

Fig. 32. Eutectics and crystallization curves of a ternary system projected on the base of the solid model.

solution. This is the general case and must now be considered. Each component in the pure state is assumed to have a definite freezing point, and it is a general law that the addition of another substance soluble in a liquid lowers the freezing point of that liquid. Under the conditions supposed this is precisely the eutectic case discussed in Chapter II, under the heading of igneous rocks (see p. 15), and the argument need not be repeated. The general effect is that on complete crystallization crystals of the component in excess of a certain proportion are embedded in a mixture of crystals of both components in a certain definite ratio, the eutectic ratio, which is constant for that particular pair of substances, whatever the original composition of the solution; the only variable quantity in the result is the kind and amount of the first crop of pure crystals, since this depends only on the initial concentration. With three components, A, B, C, there will be first a crop of A or B or C, then a mixture of A and B, B and C, or C and A, and finally a ground mass of $A + B + C$, in the proportions of the ternary eutectic for these three substances. This can be represented by a tri-

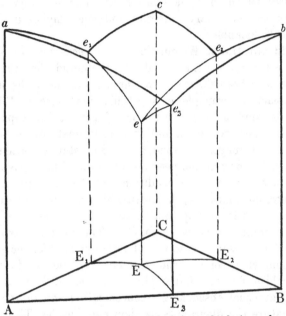

Fig. 33. View of space-model of thermal relations of a three-component system.

angular diagram in which the corners represent pure A, B, C respectively, and any point within it a mixture of the three proportional to the distances of the point from the three corner points: the points E_1, E_2, E_3 are the binary eutectics for each pair AB, BC, AC, while E represents the composition of the ternary eutectic ABC. This figure is in reality a projection on the horizontal plane of a solid model, the temperatures being given by verticals, whose upper limits form a surface. (Fig. 33.)

For four components the model has to take the form of a tetrahedron, while for more than four graphic representation in three dimensional space becomes impossible. Most mineral solutions are really complex systems of this kind and while it is quite certain that there must be a eutectic for the whole number, its position cannot be determined except by observing the freezing point of the last liquid residue, usually a quite impracticable proceeding, as is also analysis of all the individual products. However, fusible metals are commercially useful examples of mixtures which in some cases must approach very near to ternary or quaternary eutectics and even higher grades are practically possible.

Solid Solutions. We must now turn to cases where mixtures are possible in the solid as well as in the liquid phase: one such has already been described in Chapter II, under the heading "the reaction principle in crystallization" (see p. 17). This principle applies to all instances of perfect isomorphism in crystals, where two (or more) chemically analogous substances can form a homogeneous solid, a mixed crystal or so-called solid solution. The description need not be repeated here. Even more commonly do we find the condition that the isomorphism is not complete, but limited, each constituent being able to absorb the other in crystals only up to a certain definite proportion, intermediate mixtures being impossible. The diagram expressing these relations is in a sense a combination of that for the eutectic condition, where solid solution is impossible, and that for perfect isomorphism or unlimited solid solution, as will be seen below. Taking as usual abscissae for concentration and ordinates for temperature the diagram is as shown in Fig. 34. This can be explained as follows: a and b are the freezing points of the pure substances A and B, ae and be give the composition of liquids in equilibrium with points at the same temperature on the solidus curves ad and bf. The process of consolidation can be most easily followed by considering a particular instance. Starting with a solution having the composition and temperature k, and cooling, when the temperature reaches h, the liquid is then in equilibrium with a solid represented by g, of the same temperature, but of different composition, containing a larger proportion of A. This solid will then begin to crystallize, forming a mixed crystal of A and B and altering the composition of the remaining solution. The course of crystallization then proceeds along he for the liquid and gd for the solid. At the point e both components d and f are in equilibrium with the solution, and both will crystallize together, like a eutectic, except that the constituents of the eutectic are not pure A and B, but mixed crystals having the compositions d and f. The divergence of the lines below d and f means that miscibility in the solid form also decreases as the temperature falls, and there takes place a mechanical rearrangement of molecules, such that the mixed crystals are no longer homogeneous, but made up of a visible

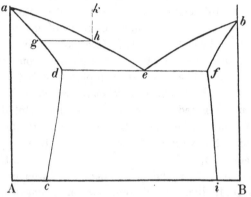

Fig. 34. Thermal relations, liquidus and solidus curves for a system of two components with limited formation of mixed crystals.

intergrowth of the two components, often taking on a graphic form, as may be seen in many microphotographs of alloys and ore minerals.

Mixed Crystals. Both of the types of crystallization just described are of the highest importance in ore-formation. There is no doubt that mixed crystals are much more common than pure compounds in nature; indeed it is somewhat doubtful whether the latter ever occur in any quantity. But it must be remembered that in many natural compounds the mutual solubility in the solid state is very small, often amounting to a mere trace.

On the other hand, the following consideration is of much importance: namely, that this miscibility is not confined to a pair of substances, but may extend to any number. A very good instance of this is afforded by the minerals of the spinel group, which includes magnetite, picotite and chromite, and zinc, iron and manganese spinels (gahnite and franklinite). So far as we know, the metallic molecules in this group of minerals are interchangeable in any proportion. Hence a general formula for the group can be written

$$R''O . R'''_2O_3,$$

in which R'' may be Fe'', Mg, Mn'', Zn, while $R''' = Fe'''$, Cr''' or Al'''. This is one of the most perfect instances of isomorphism or mixed crystal series known. The so-called titaniferous magnetite is probably a mechanical mixture of magnetite, Fe_3O_4, and ilmenite, $FeTiO_3$, and therefore not an instance of isomorphism. Among the silicates the garnet group shows very similar relations. The classical example of perfect isomorphism is afforded by the plagioclase felspars, which have been very carefully investigated. Some of the metallic sulphide minerals also come in here, while the native metals and amalgams almost certainly do. Native gold always contains silver, apparently in homogeneous mixture in any proportion, and native copper usually carries silver and *vice versa*. The large group of tellurides of gold, silver and mercury are also considered to form a continuous series, or rather, more than one continuous series, since we know gold tellurides of the two forms Au_2Te and $AuTe_2$. A very good example of what appears to be a continuous series of mixed crystals is afforded by the minerals of the tetrahedrite-tennantite group, which show a regular gradation from $4Cu_2S . Sb_2S_3$ (tetrahedrite) to $4Cu_2S . As_2S_3$ (tennantite).

On the other hand, a vast number of ore-minerals habitually carry only a small proportion, often a very small proportion, of one or more subsidiary metals. Such, for example, is the case with argentiferous galena. This subject has been investigated experimentally and it is found that the solubility of silver sulphide in solid lead sulphide is very small. If the silver content is more than about 0·2 per cent. the ore is demonstrably a mechanical mixture of galena and argentite as shown by examination of polished surfaces by metallographic methods. In this instance therefore the eutectic point must lie very near the lead end of the series[1]. Similar relations undoubtedly hold for very many other pairs of sulphides both of the precious and base metals, but in very few cases has the subject yet been treated quantitatively, so that the actual limits of miscibility are for the most part quite unknown. Although gold does not combine directly with sulphur either under laboratory conditions or as a natural mineral, nevertheless considerable quantities of gold are obtained from auriferous pyrites and from many other sulphide ores, and it may be suggested that this gold is held in solid solution in the sulphide, possibly as metal, but much more probably as a sulphide with very limited miscibility. The same is true of the silver content of many copper ores: silver and copper are able to replace each other in many compounds, though whether this replacement is unlimited is not certain. As an example may be mentioned the very common occurrence of silver in minerals of the tetrahedrite-tennantite group: further varieties are known containing mercury up to 17 per cent. Thus copper, silver and mercury form an isomorphous series with respect to their sulphantimonides and sulpharsenides. It would be possible to go on multiplying examples of this kind of thing to an indefinite extent, but no good purpose would be served by so doing. It must however be remarked here that a similar series of phenomena also occur when minerals are deposited from watery solutions, as well as from direct fusion; the latter being the type of formation hitherto considered. The presence of the third component, water, in the system does not seem to make any difference in the general principles involved, although it makes the representation of the effects by graphic methods more complicated, since a third coordinate would have to be introduced.

Dimorphism and Polymorphism. Many natural substances are known to exist in two or more distinct modifications, possessing different properties. Such, for example, is the element carbon, which can exist as diamond, graphite or amorphous charcoal according to the conditions of its formation. Silica is known to have seven different crystalline forms, besides other

[1] Nissen and Hoyt, *Econ. Geol.* vol. x, 1915, pp. 172–9.

and perhaps an indefinite number of amorphous or colloidal hydrates. The study of the crystalline forms of silica is of great significance in work on refractories, but does not come within the scope of this book. Among ore-minerals such relationships are also known and a good many examples are mentioned in Chapter IX. These need not be repeated here, but a few words may be said on one or two special cases, which are of practical interest in ore-geology.

One of the commonest substances that habitually occurs in two forms is ferrous sulphide, as pyrite and marcasite. These have the same chemical composition, FeS_2, but different physical properties: most important of all, marcasite is much less stable than pyrite and is very easily oxidized to ferrous sulphate and ultimately to ferric hydrate and free sulphuric acid. However, the thermal relations and conditions of formation of these two minerals are little known, although they are so abundant. Another interesting case is zinc sulphide, which can crystallize as zinc-blende and wurtzite, both of which are found in natural deposits. Crystallized zinc sulphide can be made by the action of sulphuretted hydrogen on an acid solution of zinc sulphate under pressure, and it has been shown that the product is blende or wurtzite or a mixture of both according to the temperature and the concentration of the acid. At a given temperature zinc-blende is formed in a weakly acid solution, wurtzite at a higher degree of acidity, and wurtzite formed at a lower temperature changes on heating with constant acid-concentration into zinc-blende. Thus zinc-blende is normally the crystalline form produced in neutral slightly acid or alkaline solutions, the common condition in primary mineral veins, while it may be converted into wurtzite by the more acid solutions that prevail in the oxidation zone. Wurtzite in ore-deposits is therefore to be regarded as of secondary origin. The relations of pyrite and marcasite are probably similar, the latter forming also in acid solutions. The same relations certainly hold for cinnabar and metacinnabarite.

Formation of Secondary Minerals. The formation of secondary minerals in the zones of oxidation and enrichment is in most cases to be referred to processes more complicated than mere crystallization from watery solutions. Very often the separation of the minerals in the solid form is due to precipitation; that is, the mixing of two solutions, or the mixing of a solution and a gas. It is a general law that if two solutions contain molecules which are able to combine to an insoluble compound, that compound will be formed, and the same rule applies to the passage of a gas through a solution. Thus copper sulphate formed by oxidation of a copper sulphide is a soluble salt, but if a soluble carbonate is added to the solution, or carbon dioxide gas is passed through it, solid copper carbonate will be precipitated. Barium chloride is soluble, but a soluble sulphate or free sulphuric acid will precipitate solid barium sulphate (barytes). This is undoubtedly the origin of many deposits of crystalline and massive barytes.

Another type of reaction here also encountered is observed when certain metallic solutions come in contact with easily decomposed solids: thus a solution of copper sulphate easily decomposes solid calcium carbonate in the form of limestone, forming a deposit of solid copper carbonate and setting free the somewhat soluble calcium sulphate (gypsum) which is easily removed by percolating water. This is an example of replacement, or metasomatism. As a matter of fact naturally formed copper carbonates are always combined with copper hydroxide as malachite or azurite, rather than pure copper carbonate, but this does not affect the argument. We can illustrate the formation of malachite by the following equation:

$$2CuSO_4 + 2CaCO_3 + 5H_2O$$
$$\text{calcite}$$
$$= CuCO_3 . Cu(OH)_2 + 2(CaSO_4 . 2H_2O) + CO_2.$$
$$\text{malachite} \qquad \text{gypsum}$$

Reactions of this kind are extremely common in all kinds of ore-formation, and according to many writers, especially in America, they have in fact played an extremely important part in the deposition of ore-minerals in veins, even among the highly insoluble and comparatively stable sulphides. These conclusions are largely founded on extensive studies of polished ore-specimens by metallographic methods. It may, however, be suggested that some of these apparent replacements are in reality to be explained as due to separation of originally homogeneous solid solutions into two or more phases in the crystalline state, owing to fall of temperature after solidification. Such changes occur very largely in metallic alloys, to which sulphide ores show many close analogies.

We now come to a class of phenomena of which the explanation is still somewhat obscure: the fundamental fact of observation here, again taking the zone of secondary enrichment in copper or silver lodes as a concrete example, is that when solutions of soluble metallic salts in their descent come in contact with solid sulphides, the metal is extracted from the solutions and precipitated on the primary sulphide, either as the same or another mineral. Thus copper sulphate is often precipitated by chalcopyrite as chalcocite or bornite. This is obviously in part a process of reduction, since the cupric sulphate of the solution is commonly reduced to the cuprous sulphide, chalcocite, or to bornite, in which the copper also exists in the cuprous state. Beyond this statement, however, it is difficult to go, as we know little or nothing of the actual chemical, physical and thermal relations of processes of this type. It is nevertheless an important factor in enrichment of ore-bodies, and is worthy of closer study than has yet been bestowed upon it: we know that the reactions occur, but we do not know *why* they occur[1].

The Stability of Minerals. It is a matter of everyday observation that the common minerals of the ore-deposits vary enormously in their power of resistance to the action of the ordinary geological agents, such as the solvent action of water and of natural acids, oxidation, hydration, changes of temperature and so on. These geological agents as we see them at work at and near the surface act within a very limited range of physical conditions. In this brief discussion it is necessary to exclude the very marked effects produced by hot, acid or alkaline springs, solfataras, and other volcanic gas-jets and so on, and to confine our attention to the everyday effects of meteoric agencies, which can be seen in operation anywhere.

---

[1] Young and Moore, *Econ. Geol.* vol. XI, 1916, pp. 349–65 and 574–81. Zies, Allen and Merwin, *ibid.* pp. 407–503. Graton and Murdoch, *Trans. Amer. Inst. Min. Eng.* vol. XLV, 1913, p. 26.

These effects may be summarized as those processes that tend to take place at and near the ordinary atmospheric temperature and pressure. From this point of view minerals can be divided into two broad categories, those with a tendency to form and those with a tendency to decompose. As a generalization it may be said that most primary ores are unstable under such conditions and undergo alteration, changing to secondary minerals. Some primary minerals however possess a high degree of stability within a wide range of conditions: such are gold, cassiterite, tourmaline and quartz, to make a haphazard selection from a wide field of choice. On the other hand many, perhaps the majority, are more or less unstable and are specially prone to changes by hydration and oxidation. Such, for example, are most of the simpler sulphides and nearly all of the very numerous minerals containing ferrous iron.

An important factor in controlling the stability of minerals is the condition of their environment with regard to the presence or absence of water and the character of the substances dissolved in the water. Broadly speaking the water of the higher zones is oxygenated and usually acid in reaction, while the stagnant water of the deeper zones contains no oxygen and is either neutral or even slightly alkaline. Hence in the deeper zone are chiefly found compounds without oxygen, such as the sulphides and arsenides, usually of a simple character: here also are to be found a few ore-minerals with oxygen, of which cassiterite and wolfram are the most characteristic. It is to be noted that the silicates of the deepest zones yet reached contain abundance of oxygen, but there is reason to believe that the heavy core of the earth consists of native metals and sulphides.

On the other hand, in the uppermost zones where there is comparatively free access of air and dissolved gases and acids are abundant nearly all the characteristic minerals contain oxygen, and we find in quantity such compounds as carbonates, sulphates, arsenates, phosphates, silicates, together with many hydrated oxides and occasionally chlorides, bromides and iodides, which though not oxygen salts are stable under similar conditions. Thus in the simplest possible case galena is oxidized to lead sulphate, anglesite, which can often be seen as a crust on galena, although the carbonate, cerussite, is more common. Zinc sulphate is a very soluble salt and does not occur as a mineral. In the case of the copper compounds carbon dioxide is a powerful agent of precipitation. The sulphide is first oxidized to sulphate, which is a soluble salt, but the copper is very easily precipitated again as an insoluble carbonate, malachite or azurite. When fragments of limestone are placed in a solution of copper sulphate they soon become covered with a green coating of basic copper carbonate, malachite, but copper carbonate precipitated from a solution of copper sulphate by means of sodium carbonate is pale blue. Under natural conditions the deep blue basic carbonate azurite is common.

The colours of the oxidized metallic compounds are often very characteristic: thus besides the blue and green copper compounds, we have the green oxidized nickel minerals and the pink tint of cobalt minerals; most of the common phosphates and arsenates are yellow, orange or greenish, the anhydrous and hydrated iron oxides are red, brown or yellow, chromium salts are green or orange, and so on. All these colours are in sharp contrast to the prevailing dark or metallic appearances of the primary minerals. Almost the only brightly coloured primary minerals are the sulphides of arsenic and mercury, and it is doubtful whether the arsenic sulphides are ever truly primary. As to cinnabar there is no question: it is certainly deposited as such from thermal waters.

The Separation of Ore-minerals from Silicate Magmas. It has been shown in an earlier chapter that sulphides possess only very limited solubility in silicates: this question needs no further discussion here. But there are certain ore-minerals belonging to other groups that need a brief consideration at this point. The most important of these groups comprises certain oxides, such as cassiterite and the members of the spinel group, more especially magnetite and chromite, as well as ilmenite. Evidence as to the relation of such minerals is to be obtained from a study of their order of separation during the crystallization of the molten magma, especially in those cases where they occur disseminated as primary constituents of a normal igneous rock. In all cases such minerals are found to separate at a very early stage and usually to form well-shaped crystals of definite outline. In granites for example magnetite is always one of the very first minerals to form, being usually only preceded by apatite and sometimes zircon. This is commonly expressed by saying that magnetite has a high freezing point, but from the physico-chemical standpoint this is an inadequate statement. What is really meant is that magnetite is very insoluble in a silicate melt, or as it may be otherwise expressed, the silicate-magnetite eutectic point lies very near the silica end of the concentration axis in the thermal diagram. Precisely the same argument applies to cassiterite, which when primary in a granite always crystallizes at a very early stage. In the basic rocks ilmenite behaves in a precisely similar way, often forming beautiful skeleton crystals, such as are so often seen in microscope sections of gabbros and dolerites. Again the chromite of the peridotites and serpentines is also as a rule the first mineral to form. The oxides therefore as a group are evidently very insoluble in silicate melts. Similar considerations apply to the common accessory minerals of granites; zircon, apatite, sphene and certain members of the epidote group, which always crystallize early.

Hitherto we have considered only the crystallization of minerals from a normal magma, where they form primary disseminations. The effects are somewhat different when their constituents become concentrated into partial and residual magmas, for example, of the pegmatitic type. Here fluxes are more abundant, and sometimes remarkable effects are produced in the way of coarse texture and large crystals. This subject, however, has been discussed elsewhere (see p. 41).

In one respect there must be important differences in the separation of crystals from silicate magmas and from watery solutions, these differences depending on the much greater viscosity of the silicate melts. The more mobile the solvent, the smaller the number of centres of crystallization that will be set up, and consequently the larger will be the individual crystals. This statement also applies to segregations of the pegmatitic type, which are kept mobile by fluxes. Hence in disseminations the crystals will usually be small and numerous, while in vein-formations deposited from water they may be very large. The best-developed of all crystals are those formed in vugs and geodes, where crystallization is undisturbed and probably very slow, owing to uniformity of conditions for a long period.

The well-known banded structure of mineral veins, with their frequent alternations of ores and gangues, both often in several varieties, shows clearly that the character and composition of the solutions that fill the fissure undergo frequent changes. This variability suggests that vein-filling is often a very long and slow process, and indeed in the case of the formation of well-developed crystals of minerals of low solubility it must necessarily be so. Rapid precipitation from strong solutions always produces either extremely minute crystals or apparently amorphous masses, and very slow cooling or evaporation of artificial solutions is required to give good crystals. In nature the relations must be similar.

To sum up the whole subject, crystals may be formed by three chief processes acting on solutions, using this term in its broadest sense. These processes are: (1) cooling, (2) concentration, (3) precipitation. Cooling and concentration are applicable to the simplest possible systems of one or two components, as well as to those of more complex constitution, concentration being brought about by removal of the solvent by evaporation or otherwise. Precipitation, on the other hand, involves the presence of a larger number of components, usually, in ordinary language, implying two dissolved components in addition to the one regarded as solvent. Further there is the important process of accretion, or precipitation of a dissolved substance on the surface of a solid. By these processes most of the crystalline and amorphous minerals of the ore-deposits have been formed under the infinite variations of conditions that prevail within the earth's crust.

Colloids and Mineral Formation. So far we have dealt, either explicitly or implicitly, only with crystalline substances and normal liquids. It is necessary however to consider, even if only very briefly, the bearing of the colloidal condition of matter on mineral formation. The subject is as yet only in its infancy so far as deposition of minerals is concerned, and therefore the following can only be regarded as of a tentative nature and liable to correction in the light of fuller knowledge. It has long been known that silica, for example, can be obtained in a jelly-like condition, possessing properties very different from those of a true solution, and many other substances, *e.g.* hydroxides of iron and aluminium can apparently exist in similar forms. According to present ideas such are to be regarded as two-phase systems, consisting either of suspensions of very finely divided solid particles in a liquid (suspensoids) or of a dispersion of minute drops of one liquid in another (emulsoids). Most of such systems, now usually called sols, can be coagulated by addition of an electrolyte to more or less solid forms (gels). Thus, for example, addition of a small quantity of hydrochloric acid causes a silica sol to set to a hard glassy-looking gel, which cannot again be got into the liquid form; suspensions of ferric hydrate are also easily precipitated in a solid amorphous form.

From the mineralogical point of view the chief importance of this class of substances is that they solidify in certain well-marked and characteristic forms, either botryoidal, stalactitic or mammillated, very often with a conspicuous concentric and radial structure, such as is so often seen in limonite or in some of the common manganese minerals, pyrolusite and psilomelane. The well-known "kidney" form of haematite is another example. In many of these concretions of all kinds the radial fibres are demonstrably crystalline, but they do not show true crystal faces. Even zinc sulphide, tin dioxide and barytes often occur in similar forms. Such minerals are always demonstrably of secondary origin and have always been formed from aqueous solutions. For example, the concretionary form of tin dioxide, known as wood tin, is believed to be formed by oxidation and redeposition of stannite.

In all questions connected with the deposition of minerals from colloidal solutions it is important to bear in mind the fact that the suspended material of such solutions, if of a mineral nature (that is, inorganic), is easily precipitated by the addition of a very small quantity of almost any electrolyte. Now practically all natural waters contain sufficient dissolved salts to act as electrolytes, hence in this case it is not necessary that there should be any specific substance in solution to act as a precipitant, as in ordinary chemical reactions: almost anything will do. Hence it follows that precipitation of colloidal material should be specially easy: in fact it is difficult to see how any natural colloidal system could continue to exist as such for any length of time. It must inevitably encounter an electrolyte and be precipitated as a gel. This consideration no doubt accounts for the very wide distribution in nature of amorphous forms of silica, hydrated oxides of iron, manganese and aluminium, and many other structureless aggregates.

Formation of Minerals from Gases. There is every reason to believe that in some instances of ore-formation at very high temperatures and under deep-seated conditions minerals have been deposited in the solid form by reactions between substances in the gaseous condition. A case of this kind was long ago determined experimentally by Daubrée, who obtained tin dioxide (cassiterite) by a reaction between tin fluoride and steam, thus:

$$SnF_4 + 2H_2O = SnO_2 + 4HF.$$

Tin fluoride is a volatile substance, which boils at about 700° C. under atmospheric pressure and the reaction gives rise to solid tin dioxide, setting free hydrofluoric acid, which can then go on to produce further changes, especially formation of tourmaline, topaz and fluorspar in rocks. Rutile has also been prepared by a similar type of reaction, and there is every probability that some deposits of specular haematite have been formed by the decomposition of ferric chloride by water, thus:

$$Fe_2Cl_6 + 3H_2O = Fe_2O_3 + 6HCl.$$

It is well known that ferric chloride is a common constituent of volcanic emanations.

It must be remembered however that many processes of mineral formation have gone on at great depths, perhaps under 10 miles or more of overlying rock, where the pressure is very high, and that under such conditions it may be impossible to determine whether a given fluid is to be regarded as a gas or a liquid. The critical point for most of the substances in question is still quite unknown.

Minerals of Complex Composition. We must now consider very briefly the class of substances commonly called molecular compounds, or in certain cases, double salts. These are not mixed crystals, that is, isomorphous mixtures of varying composition, either indefinite or limited: they are definite compounds formed by the union of two end-products in fixed proportions. As a simple example we may take the common gangue-mineral dolomite, which is represented by the formula $CaMg(CO_3)_2$, or as it may also be written, $CaCO_3.MgCO_3$. In the case of an isomorphous mixture of two substances A and B, the physical properties are in general linear functions of the composition, varying regularly from A to B: but in such a compound as dolomite this is not so. It crystallizes in a class of the rhombohedral system different from that in which calcite, $CaCO_3$, and magnesite, $MgCO_3$, crystallize and its density is said not to lie on a straight line joining those of the last-named minerals. It is therefore a true molecular compound.

It is a matter of great difficulty to decide to what extent true molecular compounds exist among ore-minerals: however, a certain amount of experimental work has been done which throws light on the constitution of one or two well-known groups of minerals. Let us take for example pyrargyrite, one of the ruby silver ores, which is so common in enriched zones in silver lodes. This has the composition Ag_3SbS_3, which may also be written $3Ag_2S.Sb_2S_3$. Both of the compounds here separated exist in nature as well-known minerals, Ag_2S being argentite and Sb_2S_3 stibnite (antimonite). The thermal and other relations of these two simple sulphides have been studied from the metallographic standpoint[1], and it is found that the freezing point curve for $Ag_2S - Sb_2S_3$ shows two maxima, corresponding in position with $3Ag_2S.Sb_2S_3$ and $Ag_2S.Sb_2S_3$, the latter being the less common mineral miargyrite, which is known from some localities in Saxony, Bohemia, Mexico and Bolivia. These were the only two compounds actually indicated in this investigation. But there are known a considerable number of other minerals also composed of Ag_2S and Sb_2S_3 in definite proportions, accord-

[1] Jaeger and van Klooster, *Zeits. anorg. Chem.* vol. LXXVIII, 1912, p. 245.

ing to analysis. These are shown in the following table, along with the corresponding compounds of copper and lead, so far as these are known to exist.

Chalcostibite, $Cu_2S.Sb_2S_3$	Miargyrite, $Ag_2S.Sb_2S_3$	Zinkenite, $PbS.Sb_2S_3$
— —	— —	Jamesonite, $2PbS.Sb_2S_3$
Bournonite, $3(Cu_2, Pb)S.Sb_2S_3$	Pyrargyrite, $3Ag_2S.Sb_2S_3$	— —
Tetrahedrite, $4Cu_2S.Sb_2S_3$	— —	— —
— —	Stephanite, $5Ag_2S.Sb_2S_3$	Geocronite, $5PbS.Sb_2S_3$
— —	Polybasite, $9Ag_2S.Sb_2S_3$	— —
— —	Polyargyrite, $12Ag_2S.Sb_2S_3$	— —

It will be seen that in the earlier members the series is fairly complete, but that in the silver group the number of probable molecular compounds is much greater than with copper or lead. Possibly future and more detailed synthetic researches may yield more definite proofs of the existence of such molecular compounds. In precisely the same way there are known to exist many minerals of a similar kind in which antimony is replaced by arsenic: thus proustite is $3Ag_2S.As_2S_3$ and tennantite is $4Cu_2S.As_2S_3$. The last-named appears to form a complete series of mixed crystals with tetrahedrite, while some members of the same group also contain silver and mercury. The miscibility of pyrargyrite and proustite however seems to be very limited. We thus have here examples of double isomorphism, in which both constituents can be replaced by analogous chemical elements, in one group silver, copper, lead and mercury, and in the other antimony and arsenic. In many rather rare minerals also bismuth plays a part similar to the last-named elements.

Such compounds have hitherto as a rule been regarded as salts of complex acids, as sulphantimonides and sulpharsenides, but the point of view just outlined is much more in harmony with what we now know as to the physical properties of these substances. It is moreover difficult to believe in the existence of so great a number of complex acids as would be required to explain the existence of all the silver-antimony-sulphur compounds as above enumerated.

In this connection it is of interest to note that in many respects the sulphides, arsenides, selenides and tellurides of the heavy metals show affinities with metallic alloys. Of these groups the sulphides are the least metallic, e.g. zincblende, the selenides and tellurides being decidedly more so[1]: some of the simple arsenides and antimonides may perhaps be regarded as intermetallic compounds, since arsenic in many of its properties behaves as a metal, while antimony undoubtedly is one. Among such compounds are niccolite, NiAs, chloanthite, $NiAs_2$, smaltite, $CoAs_2$, sperrylite, $PtAs_2$, and löllingite, $FeAs_2$, while even pyrite has many metallic properties. Dyscrasite, Ag_3Sb, is certainly an intermetallic compound, as its existence has been demonstrated metallographically in an investigation of the silver-antimony alloys[2]. It is formed by a reaction between solid silver and the fused liquid at 560° C. In a similar way the selenides and tellurides of silver have been investigated metallographically, and Ag_2Se and Ag_2Te (hessite) have been found to possess the characters of intermetallic compounds, as has also the gold telluride $AuTe_2$, which has two dimorphous forms, corresponding mineralogically to krennerite and sylvanite.

The study of complex minerals is of considerable practical importance, since such minerals naturally give rise to complex ores, many of which are difficult to deal with both in ore-dressing and in the subsequent metallurgical processes. However, many of them possess this redeeming feature in that they often yield as by-products notable quantities of the precious metals, the profit from which may help to render such ores payable in spite of their complexity. Numerous examples of this will be found in the descriptive chapters of this book.

[1] Desch, *Intermetallic Compounds*, London, 1914.
[2] Petrenko, *Zeits. anorg. Chem.* vol. I, 1906, p. 133, quoted by Desch, *Metallography*, 3rd edition, 1922, p. 406.

PART II

DESCRIPTIVE

CHAPTER XII

COPPER

FROM the researches of archaeologists in many quarters of the globe it appears that copper was probably the first metal utilized by man for the manufacture of arms, implements and tools. The earliest known metallic articles consist of nearly pure copper, but at an early date the valuable properties possessed by an alloy of copper and tin were discovered and the Bronze Age forms a commonly recognized stage of human culture, in most parts of the world preceding the age of iron. In later times the technical applications of copper acquired a continually increasing importance for innumerable purposes and its use was greatly extended by the discovery of a large number of alloys, such as gun-metal, bell-metal, brass and the modern refinements of phosphor-bronze and manganese-bronze. But the greatest impetus to the technical applications of copper was given by the development of electricity in an immense variety of ways With the exception of silver, copper possesses the best electrical conductivity of any metal, and is now utilized in enormous and ever-increasing quantities in the construction of all kinds of electrical apparatus. From this it follows that the rapid rise in copper production is one of the most striking features of the mining and metallurgy of the last few decades. The amount of copper produced in the last ten years is approximately equal to that mined in the whole of the nineteenth century, and the output is still increasing at a rapid rate. More than 60 per cent. of the world's yield of copper is now obtained from the United States. One important consequence of the great demand and high price of the metal is that it is now profitable to work great masses of ore of very low grade, which exist in great abundance, so that the world's supply of copper seems secure for at any rate a good many years to come[1].

Ores of Copper. The number of minerals containing copper is very large and those constituting commercially workable ores are numerous. They may be divided roughly into three groups as follows:

> (a) Native copper.
> (b) Sulphide ores.
> (c) Oxidized ores.

The second and third groups must however be interpreted in a somewhat liberal sense, since some of the minerals are complex in composition, and a strictly chemical classification would be very cumbrous. The copper ores show in a striking degree the distinction between primary and secondary minerals, so strongly insisted on in most recent writings on ore-deposits, and they also display very well the phenomena of secondary enrichment, a matter of the highest practical importance, as well as of great theoretical interest. In fact it may be said with truth that our knowledge of this subject is very largely derived from the study of the copper deposits.

The following are the more important ores of copper occurring in sufficient quantity to be worked on a commercial scale.

(a) *Native Copper.* This crystallizes in the cubic system, but often forms thin plates and arborescent growths, or occurs in a massive form in crevices in rocks, in amygdaloidal cavities in lavas, or as the cement of conglomerates. Good crystals are often found in the spongy secondarily enriched portions of veins, frequently in association with cuprite.

(b) *Chalcopyrite* or *Copper Pyrites*, $CuFeS_2$. This is undoubtedly the most important primary ore of copper and is to be regarded as the ultimate source of many of the other copper-bearing minerals, which are derived from it by oxidation and subsequent reduction. It is a very common constituent of primary metalliferous veins of igneous origin and is also found exceptionally as a constituent of sediments, as in the German Kupferschiefer. It crystallizes in the tetragonal system, forming peculiar crystals

---

[1] Hatch, *Geol. Mag.* vol. LVIII, 1921, pp. 32–40.

of sphenoidal shape, but is commonly massive, and often carrying a brilliant tarnish of iridescent colours, whence it is sometimes called peacock copper. Chalcopyrite can be distinguished from the somewhat similar pyrite by its crystal-form and by its deeper yellow colour and tarnish. It is also much softer.

Chalcocite, Redruthite or *Copper Glance,* Cu_2S, is an orthorhombic mineral of lead-grey colour, sometimes also found in massive form. It is common in the upper parts of veins along with cuprite and other reduced ores and is apparently sometimes primary.

Covellite, CuS, is a peculiar mineral of indigo-blue colour, usually found as an incrustation in the zone of secondary enrichment.

Bornite or *Erubescite* is a double sulphide of copper and iron, approximating to Cu_3FeS_3 or $3Cu_2S.Fe_2S_3$. It most commonly occurs massive, showing a peculiar purplish or brownish metallic lustre, with characteristic tarnish. It is also a mineral of the zone of secondary enrichment.

Tetrahedrite, Fahlerz, or *Grey Copper Ore,* $4Cu_2S.Sb_2S_3$. This is perhaps the most typical member of a large group of sulphantimonides and sulpharsenides, very characteristic of the zone of secondary enrichment. It often contains in addition iron, zinc, mercury or silver, replacing part of the copper, while when arsenic replaces antimony it grades into tennantite, $4Cu_2S.As_2S_3$. These two minerals appear to form an isomorphous series of mixed crystals. The amount of silver sometimes rises to 30 per cent. The crystal form is very characteristic, often taking the form of large tetrahedra.

Bournonite or *Wheel Ore* is a sulphide of copper, lead and antimony of rather doubtful constitution, generally represented as $3(Pb, Cu)_2S.Sb_2S_3$. It often forms peculiar and characteristic twin crystals, rather like a cog-wheel.

Enargite is a sulpharsenide of copper, usually written as $3Cu_2S.As_2S_5$: it often contains some antimony replacing arsenic. In some localities this appears to be a primary ore.

(c) *Cuprite,* Cu_2O. Belongs to the cubic system and commonly occurs as octahedra and rhombic dodecahedra, often very well developed. When massive or earthy it is known as tile ore, and when in needle-like crystals as chalcotrichite. It is easily distinguished by its submetallic lustre and ruby-red colour, which sometimes becomes black and dull on exposure to light.

Tenorite, CuO. This is much less common and occurs mostly as a black powder and in concretions, when it is sometimes known as melaconite.

Malachite, a hydrated copper carbonate, $CuCO_3.Cu(OH)_2$, commonly occurs in botryoidal, stalactitic or reniform concretionary masses of a bright green colour: it is only rarely crystalline.

Azurite or *Chessylite*, $2CuCO_3.Cu(OH)_2$, commonly occurs in deep blue monoclinic crystals, often very well developed.

Both these minerals are highly characteristic of the zone of oxidation, where they sometimes occur in enormous quantities and afford valuable ores.

Atacamite, $CuCl_2.3Cu(OH)_2$. A basic copper chloride, usually in dark green masses, rarely crystalline. It is very characteristic of the oxidation zone in regions where for any reason, such as deficient rainfall, the percolating waters contain chlorides, especially in the arid region in Chile.

Chrysocolla is essentially a hydrated copper silicate of a blue or green colour, commonly occurring as incrustations in the oxidation zone.

The characters of the principal copper ores are brought together in a convenient form in the following table:

Name	Composition	Crystal-form or habit	Density	Percent. of Metallic Copper
Native copper	Cu	cubic or massive	8·9	100
Cuprite	Cu_2O	cubic	5·8–6·1	88·8
Chalcopyrite	$CuFeS_2$	tetragonal (commonly massive)	4·2	34·5
Chalcocite	Cu_2S	orthorhombic	5·6	79·8
Covellite	CuS	massive or incrustations	4·6	66·3
Tetrahedrite	$Cu_8Sb_2S_7$	cubic (tetrahedral)	4·8	52·1
Bornite	Cu_3FeS_3	cubic (usually massive)	5·2	55·5
Enargite	Cu_3AsS_4	orthorhombic	4·4	48·3
Bournonite	$(Pb, Cu)_2S.Sb_2S_5$	orthorhombic	5·8	13·0
Malachite	$CuCO_3.Cu(OH)_2$	massive and botryoidal	3·8	57·4
Azurite	$2CuCO_3.Cu(OH)_2$	monoclinic	3·8	55·2

A very large proportion of the world's output of copper is obtained from the so-called cupriferous pyrite and cupriferous pyrrhotite. These are simply pyrite and pyrrhotite containing copper sulphide, either as a mechanical intermixture of chalcopyrite or in solid solution, in sufficient quantity to be workable at a profit as ores of copper, often as a by-product in the manufacture of sulphuric acid. The type is dealt with in detail later (see p. 246).

From the foregoing considerations it is clear that the copper-bearing minerals belong to at least three different types that have originated under fairly well-defined sets of conditions. In the first category we have the primary ores which are either of direct magmatic origin as segregations in igneous intrusions, or deposited as lodes, veins or disseminations by solutions derived directly from igneous magmas. Here also are to be included some of the so-called contact-metamorphic deposits, in which unstable rock-masses have been replaced bodily by material of magmatic origin. The primary group includes chalcopyrite, cupriferous pyrite and pyrrhotite, and apparently in some cases chalcocite, enargite, tetrahedrite and bornite; the primary origin of the last four minerals appears to be well established in certain cases, though sometimes disputed. The native copper of the lavas and conglomerates must also be regarded as primary, as it is derived directly from magmatic solutions. Secondly, in genetic order, we have the oxidized ores, formed by the action of weathering agents on primary ores, including the carbonates, sulphates and basic silicates, of which the most important are malachite, azurite, chalcanthite and chrysocolla, with a vast number of rarer minerals, including phosphates, arsenates, vanadates and a host of others. Lastly in the cycle of chemical change come the minerals of secondary enrichment, precipitated usually at the lower margin of the oxidation zone by reduction of descending oxidized solutions, including a great variety of complex sulphantimonides and sulpharsenides, with cuprite and native copper. The most important of these secondary sulphides are tetrahedrite, tennantite, enargite, famatinite and bournonite, with chalcocite, covellite and chalcopyrite. It will be noticed that several of these minerals also occur in the primary zone; they often show some variation of physical characters under the two sets of conditions.

Types of Copper Deposits. On the basis of their genetic relationships and mineralogical composition the copper deposits may be divided into nine fairly well-defined groups, as follows:

(*a*) The sulphides of the primary veins, extending indefinitely downwards in depth and often of very low grade: these are formed by solutions of magmatic origin. In some cases they pass downwards into tin ores and upwards into a zone of lead-zinc ores.

(*b*) Segregations of cupriferous pyrite and pyrrhotite, often rich in nickel, formed by two-phase differentiation before crystallization of basic magmas.

(*c*) Bodies of cupriferous pyrite in schistose rocks, formed by metasomatic replacement of unstable rocks by magmatic sulphide solutions: very closely allied to the next group, but not usually seen in actual contact with igneous rocks.

(*d*) Replacement deposits of cupriferous pyrite and pyrrhotite, often referred to as contact-metamorphic deposits. They are formed at the margins of intrusions, some unstable rock, usually limestone, being more or less completely replaced by sulphides carried by magmatic solutions.

(*e*) Native copper in the vesicles of amygdaloidal lavas and forming the cement of conglomerates; these are apparently of direct magmatic origin.

(*f*) Dissemination ores, found in porphyries and other igneous rocks which have been mineralized by their own residual solutions.

(*g*) The oxidized deposits of the superficial layers of the earth's crust, composed of deposits of any of the types mentioned above that have undergone oxidation to carbonates, silicates and occasionally to sulphates and chlorides, with a great variety of complex minerals. This zone is usually limited downwards by the level of ground-water.

(*h*) The rich deposits of native copper, oxides and complex sulphides of the zone of secondary enrichment, often sharply limited above and grading downwards into primary ore.

(*i*) Deposits of copper ores in sedimentary strata, formed at normal temperature and pressure.

Secondary Alterations in Copper Veins. In order to make clear the chemical, physical and mineralogical changes that take place in copper deposits under normal conditions, let us take a simple ideal case of a vertical deep-seated vein consisting of chalcopyrite in a gangue of quartz with small quantities of other elements, as sulphides, either in mechanical intermixture or in solid solution in the chalcopyrite. As denudation gradually removes the overlying rock the time will eventually come when weathering agents, especially oxidation and solution, aided by carbon dioxide, begin to attack the uppermost portions of the vein: chalcopyrite being a rather unstable mineral is quite easily oxidized: the iron molecule forms first ferrous sulphate, which soon oxidizes further to insoluble hydrated iron oxide, and this remains in a solid form as limonite, giving rise to the spongy mass of cellular iron hydroxides generally known as *gossan*. The copper molecule is also oxidized, first to copper sulphate, which either travels downwards in the percolating meteoric water, or may be precipitated on the spot or below in a solid form as a carbonate, malachite or azurite, or as a silicate, chrysocolla. If the meteoric water contains chlorides, as in certain dry regions, atacamite or some other chloride mineral may be formed. The zone of oxidized and reprecipitated minerals generally extends down about as far as ground-water level; that is, to the permanent water-table. Within this zone oxygen and carbon dioxide are dominant and in general the movement of dissolved metallic salts is downwards. This zone is seldom more than one or two hundred feet in depth, often only a few feet. In the copper mines of Chile however, where the climate is arid and ground-water level very low, it seems to extend downwards for many hundreds of feet. The variety of copper minerals actually found in the oxidized zone in different regions is very great, and the ores are often very rich, especially in dry regions. The necessary conditions here are excess of oxygen, without too much leaching action by water.

As the waters containing dissolved copper salts percolate downwards they eventually reach a region where oxygen is deficient, and where in point of fact the conditions favour reduction rather than oxidation. Hence the copper and other metals in solution tend to enter into entirely new combinations and to be precipitated in a solid form, as insoluble compounds. This precipitation is largely brought about by contact with the sulphides and other unoxidized ore-minerals of the primary zone, and commonly results in the formation of secondary sulphides, together with sulpharsenides, sulphantimonides and bismuth-bearing minerals, together with cuprous oxide and metallic copper. In the case of the two last mentioned the reducing action is obvious. The cupric salts of the oxidation zone are reduced to cuprous oxide, cuprite, Cu_2O, or to the native metal, these two being often found in close association. Native gold, native silver and various silver compounds also appear in a concentrated form at this horizon. Perhaps the commonest copper mineral in this zone of precipitation by reduction is chalcocite, Cu_2S, which often undergoes a slight later superficial alteration to covellite, CuS. Bornite, Cu_3FeS_3, is also very common and secondary chalcopyrite is also found. Among the complex sulphides the most important are tetrahedrite, tennantite, famatinite, enargite and sometimes bournonite. Some of these are sometimes apparently also of primary origin, as described elsewhere.

This zone of secondary sulphide enrichment is usually located at or about ground-water level, and is generally of very limited thickness, sometimes only a few feet, but it is commonly the richest part of a mine.

The chemical processes that accomplish the secondary enrichment of copper deposits are obviously very complex, and hitherto little understood. Many attempts have been made to reproduce them experimentally and to reduce them to definite equations, but there is commonly much doubt as to what actually occurs in any particular case. From the physical standpoint it is quite clear that ground-water level is the decisive factor, as this controls the equilibrium between oxidation and reduction.

The British Isles. In the earlier half of the nineteenth century the British production of copper exercised a controlling influence on the world's market: now it has sunk to the most trifling proportions. Nevertheless in its prime British copper mining, as carried on in Cornwall and Anglesey, was a most flourishing industry and yielded great profits. The geological relations of the British deposits also showed some interesting features and are worthy of description. Deposits carrying copper ores of various kinds are very widely distributed among the older rocks of the British Isles and belong to several distinct types. The two chief producing areas were in North Wales and Cornwall. In the former, the famous Parys mine in Anglesey was the most important. The copper-mining area of Cornwall is so closely associated with tin mining that one geological description will serve for both, hence no details of it will here be given, except to remark that some of the older and larger mines were wonderfully productive. Thus it is stated in a circular issued to the shareholders in 1920 that the total copper production of Dolcoath mine has amounted to a value of £2,328,435, in addition to £6,785,998 in tin concentrates.

Copper ores, mainly chalcopyrite, often associated with lead and zinc, occur in a very large number of veins, often of considerable size, lying in the Lower Palaeozoic rocks of North Wales and the Lake District. Copper ores are also found to some extent in some of the veins in the Carboniferous rocks of the Pennine chain, from Derbyshire northwards, and in the Carboniferous limestone of North Wales. Although now of little or no commercial value the latter type of occurrence is of some interest in its bearing on the origin of the ore-deposits of lead and zinc in the Carboniferous. This question is discussed in a general way in Chapter X.

Perhaps the most interesting of the British copper deposits is the famous mine at Parys Mountain, Anglesey. This is of considerable age: from 1783 to 1785 the annual output was about 3000 tons of metal and the produce is said to have been worth over half a million pounds a year for many years. When in its prime this was the world's biggest producer of copper and controlled the markets. An excellent account of its geology and history has recently been published by Dr Greenly[1].

Parys Mountain is a hill about 500 feet high near Amlwch, in

---

[1] Greenly, "Geology of Anglesey," *Mem. Geol. Survey*, 1919, pp. 823–43 with further references.

the north of Anglesey. It consists of Ordovician and Silurian graptolitic shales, with a sill of felsite several hundred feet thick intruded at and near the junction of the two systems. The general structure is a deep isoclinal fold bounded and traversed by at least three thrust-planes. The rocks are very highly altered and the so-called lodes are essentially belts of maximum alteration, especially impregnation by pyrite and chalcopyrite. The mineralization took place just after the great post-Silurian earth-movements. Twelve principal "lodes" are recognized, of which the chief are the Great Lode in the Silurian and the Careg-y-doll and North Discovery lodes in the Ordovician: they run along the strike and dip about N. by W. at high angles, usually 45° to 60°. The lodes are more or less well defined, but with one possible exception they have no true walls. The gangue is mainly quartz and the ores are chalcopyrite with some chalcocite, pyrite, blende and galena. The last two occur as an intimate mixture called bluestone. There is some evidence of primary zoning: thus in the Great Lode pyrite, chalcopyrite and bluestone occur in order from north to south and from old descriptions there was probably once an upper zone of very argentiferous galena with some arsenic mineral. Over the southern lodes some true gossan appears to have existed.

The Great Lode was worked in two contiguous open cast pits over a total length of 620 yards and covering in all about 12 acres. The pits were from 110 to 140 feet deep and the lode, which carried from 3 to 5 per cent. of metal, wedged out downwards. The North Discovery lode has been entirely worked out, but the old descriptions of it read more like a true vein; it dipped N. at about 60°, was from 8 to 10 feet wide, and had well-defined hanging wall and footwall. It carried 20 to 25 per cent. of metal and is said to have produced at least a million sterling. The richest specimens seen in collections came from this lode, and form a typical quartz-chalcopyrite ore. Underground work has long ceased, but the mine is still worked by a leaching process.

It is unnecessary to describe in detail the numerous copper-bearing veins that are found in the Lower Palaeozoic rocks of North Wales and the Lake District. They may be summed up in general terms as fissure veins or brecciated lodes, carrying chalcopyrite and pyrite with usually more or less blende and galena: the gangue is usually quartz, or fragments of crushed country rock, with sometimes calcite and barytes. They frequently occupy fault-fissures or shatter-belts. Their geological relations are similar to those of the lodes carrying mainly lead and zinc, described in Chapter XIV.

The Ecton copper mine, on the borders of Staffordshire and Derbyshire, is of some historic interest as well as of theoretical significance. It was here that drilling and blasting were first used in England, by German miners brought over by Prince Rupert in 1636, and by the year 1778 the principal deposit had been worked to a depth of 200 fathoms, a great depth for the days before winding engines were introduced. The published descriptions of this ore-body are not very clear, but it seems to have consisted of a number of veins, some running E.–W., others N.–S., with many smaller branches, the whole occupying a sort of pipe in the Carboniferous limestone, which is here much disturbed. The upper portions carry mainly galena and blende, while in depth chalcopyrite and erubescite are dominant. The gangue consists of yellow calcite and fluorspar, with barytes and chalybite. The existence of ore-zones is here very obvious, copper underlying zinc and lead. The galena is said to be poor in silver and the usual upper zone of richly argentiferous lead ore seems to be missing, probably owing to denudation, since this deposit appears to represent a deeper zone of the usual type of Derbyshire lead-zinc deposit.

Moonta and Wallaroo, South Australia. An interesting example of copper lodes of distinctively pegmatitic type is afforded by the rich deposits of the Moonta and Wallaroo district in South Australia, an area which up to the end of 1916 had yielded copper to the value of £19,000,000. These lodes are in genetic connection with a pre-Cambrian granite intrusive into felspar-porphyry and schists, and they do not pass up into the overlying nearly horizontal Cambrian. The lodes form three fairly well-defined series, striking nearly N. and S. The most typical are the pegmatites of Moonta, which are composed chiefly of quartz, microcline, biotite, haematite, tourmaline, fluor-spar and apatite; molybdenite is present in small quantity in some lodes. These minerals vary much in relative abundance and the lodes pass laterally into more or less pure quartz-copper veins. The chief metallic minerals are chalcopyrite, pyrite and pyrrhotite, with small amounts of galena, blende and scheelite. The lead and zinc minerals are the latest to form and sometimes enclose calcite. In the upper parts of some of the lodes bornite is abundant, but it is not clear whether this is primary or secondary.

At the surface is an oxidation zone from 100 to 150 feet deep carrying carbonates and atacamite, with a sub-zone of metallic copper and cuprite at its base; and below this in many localities were large enriched masses of chalcocite, clearly of secondary origin, with a little covellite in places: these are now worked out. The alterations of the primary ore in the upper zones are perfectly normal to a dry climate, where the ground-water is rich in chlorides, forming atacamite and various carbonates, including malachite and azurite.

This deposit clearly belongs to the pegmatitic type of copper-deposit with tourmaline and fluorine minerals, and shows strong affinities to the upper parts of the Cornish lodes, and to the tourmaline-copper veins of Chile.

Copper Deposits of the United States. At the present time the United States is by far the largest copper producer, the output in 1918 being 848,000 tons of metal, or 60·8 per cent. of the world's total of 1,395,000 tons. The copper mines of the United States also illustrate extremely well most of the important types, so that it will be profitable to consider a few selected examples in some detail. In the first place it should be said that most of the ores are of low grade: in 1916 the average copper content of all the ore mined in the United States was only 1·7 per cent., and the success of the industry obviously depends on working by methods specially adapted for treatment of such ores, necessarily on a large scale. The famous copper deposits of Lake Superior, of whose richness so much is heard in more or less popular literature, actually carry about 1 per cent. of metallic copper, nevertheless these are profitably worked at a vertical depth of 5000 feet.

The largest and most important deposits belong in the main to the four following types:

(a) Native copper with zeolites in the vesicles and cavities of lavas, and native copper in the cement of conglomerates, both apparently of direct magmatic origin. Lake Superior.

(b) "Replacement veins" in igneous rocks: metalliferous veins penetrating igneous rocks, usually granite, monzonite or diorite, with alteration and mineralization of the wall-rock. Butte, Montana; Globe, Arizona.

(c) "Porphyry copper," disseminations of copper ores in igneous rocks, extrusive or intrusive, also of magmatic origin. The same name is sometimes used to cover similar disseminations in schists. Bingham, Utah; Ely, Nevada; Miami, Arizona (schists).

(d) Contact-metamorphic deposits, formed by replacement of limestones by magmatic solutions, usually at and near intrusive contacts. Morenci, Arizona; Bisbee, Arizona.

The examples here given in most cases are merely one or two of the most typical of a large class, but the Lake Superior occurrence is unique on a large scale.

The Lake Superior Copper Region. The Keweenaw Peninsula projects from the middle of the southern shore of Lake Superior for a distance of about 70 miles in a north-easterly direction and has a width of about 40 miles. It consists mainly of rocks belonging to the Keweenawan System, the uppermost division of the Algonkian (Upper pre-Cambrian) of this region. The general dip is to the N.W. and the peninsula forms the southern limb of a syncline of which the axis lies under the lake. Along the middle of the peninsula runs a great fault, which brings down nearly horizontal Cambrian sandstone on its southern side. The Copper Range, which forms a belt from 4 to 6 miles wide to the north-west of the fault, consists mainly of a succession of flows of basic lavas, chiefly basalts, invaded by dykes of more acid composition, porphyries and quartz-porphyries. With the lavas are interbedded sandstones and conglomerates, the latter being largely composed of pebbles of the intrusive rocks, but with many basaltic boulders. Many of the lava-flows are highly vesicular and amygdaloidal, especially their upper and

lower portions, and both these and the coarse conglomerates are
richly impregnated with metallic copper. In the conglomerates
the copper is mainly in the cement; in the lavas it is in the
vesicles, along with quartz, calcite and many zeolites, especially
prehnite, laumontite, thomsonite, apophyllite, analcite and
natrolite; epidote is also abundant. The richer deposits, both in
the lavas and conglomerates, form great tabular ore-shoots,
striking and dipping with the beds and often of wide lateral
extension. There are also veins cutting the ore-bodies, but these
are not of much commercial importance, though often rich at
the intersections.

The greater part of the production now comes from the
amygdaloids. Some of the flows are as much as 100 feet thick

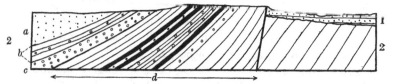

Fig. 35. Geological section through Keweenaw Point, Lake Superior.
1, Cambrian Sandstones. 2, Keweenawan Series; a, Sandstone; b, Trap
and Conglomerate; c, Great Conglomerate; d, Trap and Conglomerate,
with three ore beds (of altered trap and conglomerate) shown in black.
(After Van Hise and Leith, Monograph LII, U.S. Geol. Survey, 1911,
Fig. 75, p. 574.)

and the mineralized vesicular portions of the flows may be as
much as 30 or 40 feet thick. The average copper-content is only
about 1 per cent. Individual mines working this type of deposit
are too numerous to mention.

The Calumet conglomerate, which is from 12 to 25 feet thick,
dips N.W. at 39° at the surface and has been worked down the
dip for 8000 feet, or nearly a mile in vertical depth. Near the
surface it carried about 4 per cent. of metal, but at 5000 feet
only about 1 per cent. The Allouez conglomerate has also been
worked, but the other conglomerates are unpayable.

There has been a good deal of discussion as to the origin of
these deposits. In the first place it is clear that there is a close
connection between the zeolites and the copper and furthermore
that both were formed very soon after the eruption of the lavas,
since blocks of amygdaloid in the conglomerates are in exactly

the same condition as the undisturbed rocks. It is now generally accepted that the mineralization was brought about by magmatic solutions connected with the eruptive activity that produced the lavas and intrusions. Nevertheless it is not clear why the copper is in the metallic state: perhaps the most striking feature is the absence of sulphides, and the presence of zeolites indicates a comparatively low temperature, perhaps not much over 100° C. The mineralization therefore belongs to the solfataric type and must have been carried on very near the surface. It is also evident that some strong reducing agent must have been at work to precipitate the copper as metal.

The Copper Deposits of Butte, Montana. In this area the connection between orogenic movements, igneous activity and mineralization is brought out with special clearness. The sequence of events was as follows[1]. The main folding of the Rocky Mountains, in Middle Cretaceous times, was followed by andesite eruptions in the Upper Cretaceous, and this again by extensive thrust-faulting. The intrusion of the main Boulder bathylith took place in the Eocene, while normal faulting, accompanied by rhyolite flows, extended from the Oligocene to the Pliocene. The intrusion of the bathylith was divided into four phases, indicating progressive differentiation: (1) basic. diorite and pyroxenite, (2) quartz-monzonite (the main mass), (3) aplites and pegmatites with tourmaline and other pneumatolytic minerals, (4) quartz-porphyry with metalliferous quartz-veins. The last three phases all carry ores, chiefly copper, lead and zinc, with silver, gold, bismuth and manganese. The earlier andesite flows gave rise to native copper, magnetite and galena, with gold and silver, while the later rhyolite flows are accompanied by gold veins.

The mineralization of the monzonite phase is chiefly associated with outlying bosses or cupolas rather than with the main mass: it consists mainly of auriferous pyrrhotite, pyrite and magnetite: fissure veins contain copper, lead and zinc. The veins of the aplite-pegmatite phase carry galena, blende, chalcopyrite, pyrite, arsenopyrite and rhodochrosite. The final quartz-porphyry phase

[1] Billingsley and Grimes, *Trans. Amer. Inst. Min. Eng.* vol. LVIII, 1918, pp. 284–361. Summary in *Geol. Mag.* vol. LVI, 1919, p. 280.

gave rise to the richest copper deposits, some of the most important in the world: they lie rather deep within the bathylith and consist of enargite, tennantite, tetrahedrite, bornite and chalcocite. The special points of interest here are the occurrence of chalcocite as a primary mineral and the abundance of

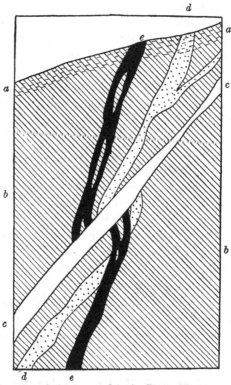

Fig. 36. Ideal section of the Anaconda vein, Butte, Montana. *a, a,* oxidized zone; *b, b,* granite; *c, c,* shatter belt; *d, d,* quartz-porphyry; *e, e,* vein. (After Emmons, Bull. 625, *U.S. Geol. Survey,* 1917, plate VI simplified.)

enargite: this shows that here arsenides to a large extent take the place of sulphides in a primary ore-deposit. It is unnecessary to refer here to the processes of secondary enrichment in this area, which appear to be of the usual type. An important feature, however, is that the copper has been almost completely leached out from the upper portions of the lodes, which were originally worked for silver only to a depth of from 100 to

500 feet. Copper mining was a later development in depth in this field.

The workable copper deposits occur in definite fissure veins in the bathylith. There are two chief sets, one striking E.–W., of which the most important is the Anaconda vein, the other striking N.W.–S.E. Altogether six different series of fissures are recognized. When all the veins are considered together a certain regularity of arrangement can be made out. There is firstly a main or central copper zone without zinc or manganese; secondly, an intermediate zone with much copper accompanied by blende, rhodonite and rhodochrosite; thirdly, an outer zone, mainly carrying rhodonite, rhodochrosite, blende and pyrite, with silver. Here then copper and arsenic belong to a deeper zone than zinc, lead and manganese, and the whole is clearly due to differentiation of a monzonitic magma.

"Porphyry Copper" Deposits. According to Emmons[1] the process of formation of a primary disseminated copper deposit of this type is usually as follows:

Stage 1. Intrusion of an acid or intermediate magma on a large scale, forming a boss or laccolith of diorite, monzonite or porphyritic granite.

Stage 2. Fracturing of great masses of the igneous rock, probably mainly by earth-stresses, in part perhaps by contraction on cooling.

Stage 3. Alteration and mineralization of the fractured rock by magmatic solutions derived from the same or a related intrusion. The chief processes are sericitization, silicification and deposition of primary sulphides in the fissures.

This results in the formation of a low-grade primary ore, rarely carrying as much as 1 per cent. of copper, and often only 0·5 per cent. However, when denudation brings such bodies near the surface the usual processes of oxidation and secondary enrichment get to work and produce oxidized ores and secondary sulphides. Of recent years it has been found profitable to mine and concentrate even some low-grade primary porphyry coppers on a large scale and deposits of this type are now among the chief producers in the United States.

[1] Emmons, *The Principles of Economic Geology*, New York, 1918, p. 350.

As a typical example of a porphyry copper deposit we may take part of the Bingham, Utah, copper area. At Upper Bingham is a great intrusion of monzonite-porphyry, much shattered by a vast number of small irregular fracture planes and intensely bleached, silicified and sericitized. The oxidized zone is very poor owing to excessive leaching: below it is a zone of chalcocite, covellite and secondary chalcopyrite: below this again is the primary zone of very low grade. The zone of secondary enrichment is unusually thick: after about 100 feet of leached capping has been stripped, the workable part extends down usually for some 400 feet, occasionally as much as 800 feet. This carries about 1·5 per cent. of copper and from 20 to 30 cents. a ton in gold and silver. The unaltered primary ore is of too low grade to work at present. In 1915 the production was about 74,000 short tons of metallic copper and more than a million dollars in gold and silver.

At Ely, White Pine County, Nevada, a mass of highly-mineralized monzonite-porphyry about 9 miles long is intruded into Palaeozoic limestones and shales. It shows, especially towards its centre, the usual type of fracturing, sericitization and deposition of sulphides, with an oxidized capping about 100 feet thick and a secondary chalcocite zone about 200 feet thick, locally extending to 600 feet and averaging about 1·7 per cent. of copper, as against 0·4 per cent. in the primary ore. There are also lodes containing lead, zinc and gold.

The Globe and Miami district in Arizona produces a large amount of copper from disseminations in the Pinal schist, of pre-Cambrian age, near the Schultze granite intrusion. The schist is shattered, altered and mineralized in the usual fashion, the primary ore being chalcopyrite formed by magmatic solutions derived from the granite. The altered rock is a grey schist composed of quartz, sericitic mica and some felspar, with biotite, hornblende, tourmaline, magnetite and chlorite, originally an impure felspathic sandstone. An oxidized capping at the surface carries more copper than at Bingham, mainly as carbonates and chrysocolla. The secondary ore-body, carrying chiefly chalcocite, forms a belt 9 miles long and a quarter of a mile wide, and some hundreds of feet thick, with 2 to 2·5 per cent. of metallic copper.

The whole might be regarded as a contact deposit, but its origin is obviously essentially the same as that of the true porphyry copper deposits.

Limestone Replacement Deposits. The deposits of this type really belong strictly to the category of contact metamorphism, since they are accompanied by many of the characteristic features and minerals of the thermal alteration of calcareous

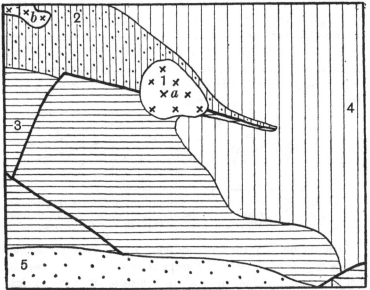

Fig. 37. Geological map of the Bisbee district, Arizona. 1 *a*, granite-porphyry; 1 *b*, granite; 2, pre-Palaeozoic; 3, Palaeozoic; 4, Cretaceous; 5, Quaternary. (After Ransome, Prof. Paper 21, *U.S. Geol. Survey*, 1904, map 2.)

rocks. One of the best known examples is the Bisbee district in Arizona, near the Mexican border[1]. The oldest rock here is again the Pinal schist, which was invaded in pre-Cambrian times by a granite and then overlain unconformably by Cambrian quartzite followed by limestones, ranging from Cambrian to Carboniferous, with a total thickness of some 4000 feet. In Mesozoic times these were again invaded by masses of granite-porphyry before the pre-Cretaceous denudation. The mineralization was the result of the Mesozoic intrusive phase. Around the

[1] Ransome, Prof. Paper 21, *U.S. Geol. Survey*, 1904.

porphyry masses the limestones contain garnet, tremolite, diopside and other silicates, with lenticular masses carrying pyrite,
chalcopyrite, bornite and magnetite. These ore-bodies are very
irregularly distributed, but obviously related to the intrusions.
The chief ore-bodies lie in a syncline of the limestones with
gentle dips, between two faults. A notable feature here is a very
deep oxidation zone which is independent of the present water-
level and is related to the old pre-Cretaceous land surface: it is
now again exposed by removal of the Cretaceous and perhaps
later rocks. The earlier workings were mainly in this oxidized
zone, which is notable for very fine specimens of carbonate ores
(azurite, etc., from the Copper Queen mine). Now, however, the
chief production comes from enriched chalcocite ores and even
from the unaltered primary zone.

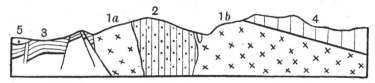

Fig. 38. Geological structure of the Bisbee district, Arizona. 1 a, granite-
porphyry; 1 b, granite; 2, pre-Palaeozoic; 3, Palaeozoic; 4, Cretaceous;
5, Quaternary. (After Ransome, Prof. Paper 21, *U.S. Geol. Survey*, 1904.)

Cupriferous Pyrite Ore-bodies. A very large proportion
of the copper production of Europe is obtained from masses of
copper-bearing pyrite enclosed in schistose or more or less
highly altered sedimentary rocks. Such are largely developed
for example in the Huelva district of Spain, Norway and the
Urals. They naturally show a good deal of variation in detail,
but they possess certain general characteristics in common. The
ore-bodies are usually more or less definitely lenticular in shape,
steeply inclined or vertical and often arranged *en echelon*,
parallel to the schistosity of the enclosing rocks: consequently
the outcrops usually show a more or less oval form, often tailing
out rather indefinitely at the pointed ends. They are often
closely associated with, though not as a rule in actual contact
with, igneous intrusions of a somewhat similar form, generally
of rather basic composition, such as gabbro or diabase. The ore
is generally very pure and massive, consisting mainly of pyrites,

and containing 1 or 2 per cent. of copper, apparently in most
cases as a fine mechanical admixture of chalcopyrite. Gold and
silver are often present in small quantities. Phenomena of
oxidation, especially gossan formation, and secondary enrichment are conspicuous.

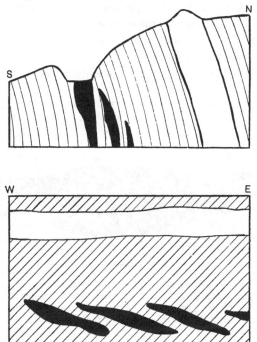

Fig. 39. Cross section and plan of the San Platon Mine, Huelva, Spain.
Fine lines, slates; white, porphyry dyke; black, ore-bodies. The richest
copper is found on the side nearest to the dyke in each case. (After
Collins, *Bull. Inst. Min. Met.* No. 206, 1921, simplified.)

Rio Tinto District, Spain. This copper-bearing region lies
mostly in the province of Huelva in the south-west corner of
Spain and extends just over the border into Portugal. It forms
the southernmost portion of the ancient plateau known as the
Iberian Meseta, a region strongly folded by the Armorican
earth-movements of late Carboniferous and Permian times. The
country rock consists of slates of Carboniferous and possibly of
late Devonian age thrown into strong east and west folds and

invaded by masses of granite, quartz-porphyry and other more
basic igneous rocks. The forms of the intrusive masses are closely
related to the folding and the ore-bodies are great lenticular
masses of sulphides associated with the porphyries and aligned
in a general east and west direction. They are always steeply
inclined, often nearly vertical, in accordance with the dip of the
enclosing rocks. Some of these masses, which are called lodes,
are of very great size and afford typical examples of the forms
mentioned in the last section.

The ore-bodies occupy a belt extending from Aznalcollar in

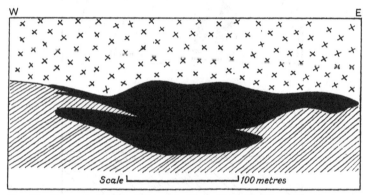

Fig. 40. Cueva de la Mora Mine, Rio Tinto district. Plan of La Corta ore-
    body on the fifth level. Porphyry marked with crosses, slate with fine
    lines, ore black. The richest copper ore is found at the eastern ex-
    tremity. (After H. F. Collins, *Bull. Inst. Min. Met.* No. 206, November,
    1921.)

Seville, through Huelva to Aljustrel in Alemtejo (Portugal).
They are lenticular both in plan and section and both ends
generally feather out along the strike. The longest masses include
the San Dionisio and South lodes of Rio Tinto, which have a
combined length of over 2 kilometres. The great majority, however, vary from 300 to 700 metres. The San Dionisio lode is 250
metres wide and the rest are usually in proportion, maintaining
a fairly constant ratio of length to breadth of about 5 or 6 to 1.
The general depth also seems to be fairly proportional to the
length; the San Dionisio lode is known to be over 500 metres in
depth, but few others exceed 300 metres. They are in fact to
be regarded as flat circular cakes of ore, set up on end and

denuded to different depths: some are known underground but
do not reach the surface; of others only the lowest portion has
escaped denudation.

The primary ore consists essentially of massive pyrite, with
very little impurity at any rate in the central parts of the ore-
bodies. When entirely unaltered it contains from 0·8 to 2 per
cent. of copper as finely disseminated chalcopyrite, and small
quantities of arsenic, lead, zinc, gold, silver, with traces of bismuth, selenium and tellurium. The upper part of the lodes to
an average depth of about 30 metres consists of a very well-
developed gossan, a massive iron ore containing over 50 per
cent. of metallic iron. This is sharply limited downwards by the
level of ground-water: and the line of contact between gossan
and sulphide sometimes shows a band of earthy material carrying
considerable values in gold and silver. Below this the zone of
enriched sulphides, assaying from 3 to 12 per cent. of copper,
largely as chalcocite, extends to an average depth of about
50 or 60 metres below the gossan. Below this comes the lean
primary ore.

Most of the lodes are worked by enormous open cast pits, and
the treatment of the ore is essentially as follows: first it is
leached with water, by which the copper is removed in solution,
to be afterwards reprecipitated. The pyrite is then roasted for
the manufacture of sulphuric acid, and the final residue is sold
as an iron ore (purple ore or Blue Billy), being exported to
England or Germany. Thus all the constituents of the ore are
utilized in one way or another. The best-known mines are Rio
Tinto and Tharsis in Spain, and Mason and Barry in Portugal.
The industry has till lately been very profitable and the output
very large, but industrial troubles and labour disputes have of
late caused a great falling off in yield and in profits.

There has been much discussion as to the genesis of these
ore-bodies; at one time they were considered to be of sedimentary
origin, but this view is not now seriously held by any one, and
it is generally agreed that they are genetically connected with
the porphyry intrusions. These are here acid rocks, instead of
basic, as in most other similar instances, but the general principle
is the same. The ore-bodies may be considered to be due to the

independent intrusion of a sulphidic partial magma separated
by two-phase differentiation from the porphyry magma according to the law of limited miscibility of fused silicates and
sulphides, as discussed theoretically in Chapter II (pp. 22–25).

Copper-bearing Sulphides in Norway. In various parts
of Norway masses of cupriferous sulphides have long been
known and worked. The geology of these ore-bodies has attracted
much attention from many writers, and they are noteworthy
from the fact that it was here that the magmatic origin of
sulphide ores was first placed beyond doubt by the important
researches of Vogt in the early nineties of the last century[1],
following on the work of Kjerulf[2], and in opposition to the
German school, who regarded such masses as of sedimentary
origin.

There are in Norway four chief areas where such occurrences
are found, besides numerous smaller scattered ones: attention
will here be confined to two such fields, the Trondhjem field in
62°–64° N. lat., and the Sulitjelma district north of the Arctic
Circle.

These ore-bodies are always found in Lower Palaeozoic slates,
in close association with, and sometimes actually enclosed by,
intrusions of basic igneous rocks, originally gabbros, now often
more or less completely converted to hornblende-schists (amphibolites). This association is so universal that it cannot be merely
accidental. The ore-bodies are always lenticular in form, and
broadly concordant to the cleavage of the slates, though occasionally slightly transgressive in detail. Many of them show
great lateral extent in comparison with their thickness, and are
to all intents and purposes sills, while others are phakoidal in
form. Some of them clearly lie at or just below the base of a
gabbro laccolith. The dominant mineral in most cases is pyrite,
but in some cases, as at the Mug mine at Röros it is almost
entirely replaced by pyrrhotite. The copper occurs as chalcopyrite, in much the same manner as at Rio Tinto. There is also
as a rule a little blende, with still less galena, while tetrahedrite
and arsenopyrite are rare. The chief non-metallic minerals are

[1] Vogt, *Zeits. f. prakt. Geol.* Jahrgang 1894, pp. 41, 117, 173.

[2] Kjerulf, *Udsigt over det sydlige Norges Geologi*, 1879, pp. 231–58.

quartz, often in lenses which appear to be secondary and independent, mica, hornblende and actinolite, garnet, felspar, epidote, sphene and chlorite: tourmaline, spinel, kyanite, barytes and fluorspar are rare. This assemblage of gangue-minerals alone, when reasonably considered, is sufficient to prove an

Fig. 41. Section across the Röros pyrite deposit, Kongen Grube. (After Vogt, *Zeits. prakt. Geol.* 1894, p. 123.)

igneous origin for the ores. A banded structure is common and in some cases quite conspicuous.

The following analyses, (1) and (2), are given by Vogt as typical of bulk-samples of the Röros ores: for comparison we give a (not quite complete) analysis of ore from Rio Tinto (3).

	(1)	(2)	(3)
Sulphur	49·00	45·00	47·76
Iron	42·35	37·90	43·99
Copper	2·48	2·07	3·69
Lead	0·24	0·38	0·01
Zinc	2·01	6·42	0·24
Arsenic	—	—	0·83
Manganese	—	0·07	—
Silica	1·85	4·60	1·99
Alumina	1·25	2·01	—
Lime	0·23	0·37	0·23
Magnesia	0·22	0·70	0·07

To judge from the percentage of copper the ore from Rio Tinto must come from the upper enriched portion. Between (1) and (3) the similarity is striking; the most notable difference is the presence of 0·83 per cent. of arsenic in (3): this element is very scarce in the Norwegian deposits.

Such pyrite masses are found of all sizes, from small lenticles a few feet long, up to the great ore-bodies of Röros and Sulitjelma, measurable by hundreds of yards. Thus, for example, the ore-body of the Mug mine at Röros is about 1200 yards long, from 100 to 150 yards wide and about 10 feet thick. The Storvarts mine in the same district works another body about 1500 yards long, 200 yards wide and also about 10 feet in maximum thickness. Many of the smaller masses however do not show such a wide lateral extension in proportion to their thickness and are perhaps more correctly described as "stocks." They all belong more or less clearly to the phakoidal type.

The Copper Deposits of Namaqualand. For a good many

Fig. 42. Plan of the Ookiep ore-body, Namaqualand, at crossing of quartz-diorite dyke and fault. Wavy lines, Archaean gneiss; black, ore. (After Kuntz, *Trans. Geol. Soc. S. Africa,* vol. VII, 1904, plate XX, fig. 1.)

years a large output of copper was obtained from certain mines in Namaqualand, in the extreme north-western corner of Cape Colony: these deposits present some unusual and instructive features[1]. The country rock of this extremely arid and barren area consists of gneisses and schists of various kinds belonging undoubtedly to the fundamental complex and probably equivalent to the Swaziland Series of S.E. Africa. In the copper belt, a region about 60 miles inland, these are traversed by several very large dykes, mostly norites or quartz-norites, striking east and west, and cut at intervals by fissures running N.W.–S.E. The ore-bodies are found at the intersection of the dykes and

[1] Kuntz, *Trans. Geol. Soc. S. Africa,* vol. VII, 1905, p. 70. Ronaldson, *ibid.* vol. VIII, 1906, p. 158.

fissures and consist mainly of bornite and chalcopyrite, with secondary copper minerals, mixed up with the normal minerals of the norite dykes and with blocks of the gneissose country rock. The Ookiep ore-body is 1000 feet long by 200 feet wide on the surface and 300 feet deep, while the Nababeep ore-body is of the same order of magnitude. The genesis of these ores is rather doubtful, but the field relations certainly suggest that

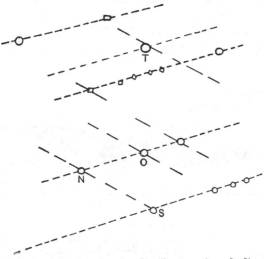

Fig. 43. Diagrammatic plan showing distribution of ore-bodies in relation to faults and dykes, Ookiep-Concordia Copper district, Little Nama-qualand. N, Nababeep Mine; O, Ookiep Mine; S, Springbokfontein; T, Tweefontein. Circles indicate actual mines, squares indicate pro-spects and abandoned mines. (After Kuntz, *Trans. Geol. Soc. S. Africa*, vol. VII, 1904, plate XIX, simplified.)

they were formed by precipitation in the norites of copper from solutions that arrived by way of the barren N.W.–S.E. fissures. It is suggested that the presence of certain ferruginous bands in the gneiss had some influence on the precipitation. Fig. 43 shows in a general way the lie of the ore-bodies.

Some of these ores were remarkably rich, running up to 20 per cent. of metallic copper, and were exploited on a large scale for many years by the Cape Copper Company and others. So far as can be judged the bornite ores here appear to be primary, and these occurrences therefore constitute an unusual type.

Unfortunately it is not clearly demonstrated that they are of direct igneous origin: they do not, from the field-evidence, appear to have been formed by differentiation from the norite magma: at least there is no positive proof of this.

Copper in Sedimentary Rocks. A quite considerable proportion of the copper production of the world is derived from deposits of a type very different from those hitherto considered, and indeed forming an exceptional class of ore-deposits: these are the copper ores worked in apparently unaltered sedimentary strata. Of these the copper-shales of Mansfeld, Saxony, worked on a large scale since the twelfth century, are the classical

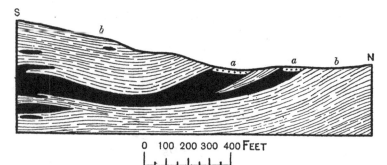

Fig. 44. Longitudinal section of Skorovas Mine, Norway. *a, a*, gossan; *b, b*, schist; ore, black. (After H. H. Smith, *Bull. Inst. Min. Met.* No. 218, October, 1922, by permission of the author and the Council of the Inst. Min. Met.)

example. These are not however the most typical cupriferous sediments, since if we take the world over, it will probably be found that more copper comes from sandstones, and especially from red sandstones, than from shales. Workable concentrations of copper ore in sediments are to be regarded as characteristic of one facies of sedimentation, that is, the "red beds," or deposits formed under arid conditions, and no doubt the prevalence in such strata, both at the time of their formation and also later, of saline solutions is an important genetic factor in the case.

The origin of such copper concentrations is obscure, and perhaps not referable to any general law. In some instances the copper may be "primary," having existed in the sediments since their deposition; in other instances it may be "secondary," having been introduced at a later stage. But in all cases the process of deposition probably took place at the normal temperature and pressure. In the "primary" deposits it must have been so, from the nature of the case; in the "secondary" deposits there is little or no evidence of any igneous or metamorphic action, and indeed igneous rocks are as a rule conspicuously absent. The only visible secondary changes are those due to simple oxidation and hydration in the superficial zone. But it is clear in many instances that the ores were laid down under reducing conditions. Another noteworthy point is that the copper is often associated with lead, and what is of more interest, with vanadium, uranium and radium in a variety of curious minerals described in the sections dealing with those elements (see Chapter XVIII).

Before discussing further the genesis of these peculiar ores it will be well to describe one or two typical examples.

Copper in the Trias of England. Copper ores are found to a certain extent at several localities in the Triassic area of Cheshire, Staffordshire and Shropshire. Most of these occurrences are unimportant, though of some interest, but one of them, at Alderley Edge, has been worked on a fairly extensive scale, and shows some peculiar mineralogical features. Alderley Edge is a conspicuous escarpment of Keuper sandstones with a steep dip to the west and sharply cut off on the north by a cross-fault. At the base is a conglomerate and above this a thick series of white and yellow sandstones. The copper occurs as blue and green carbonates and other hydrated compounds in the cement of the conglomerate and disseminated through various beds of the sandstone in variable quantity. Some samples are said to contain as much as 12 per cent. of metallic copper. There is also a certain amount of black cobalt ore, and along a small fault-plane masses of a very remarkable rock, a sandstone with a cement of crystalline galena. At Mottram St Andrews, about a mile north-east of the Alderley Edge mine vanadium ores have been worked in some quantity. The same association of copper and cobalt in Keuper sandstone is also found at Bickerton in the Peckforton Hills. In these occurrences the lie of the ore-bodies suggests that the mineralizing solutions came up along fault-planes, and their horizontal extension appears to be very limited.

The Copper Deposits of Katanga. On a very much larger scale than the foregoing, but apparently of similar general character, are the copper deposits of the Katanga district in the south-eastern corner of the Belgian Congo, near the borders of Rhodesia. The copper-bearing strata are sediments of somewhat

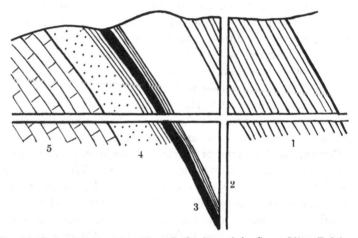

Fig. 45. General section of main ore-body, Star of the Congo Mine, Belgian Congo. 1, schist; 2, cellular quartzose rock; 3, richest ore bed, with walls of schist; 4, sandstone; 5, dolomite, used as flux. (After Ball and Shaler, *Econ. Geol.* 1904, vol. IX, fig. 137, p. 624.)

uncertain age, and perhaps belonging to more than one formation. Some of them have been correlated with the Transvaal System of South Africa, others with the Karroo System, but the evidence does not seem very satisfactory. Most of the rich copper deposits lie in the highly-inclined strata of the older series, consisting of slates, sandstones and silicified dolomite. The ore-shoots are lenticular in form, commonly more or less parallel to the bedding, but sometimes apparently determined by the direction of cleavage. They are characteristically associated with sharp folding, faulting and brecciation. They vary much in size: one of the Kambove ore-bodies is said to be 3000 feet long and from 240 to 400 feet wide. The most abundant copper minerals are malachite and chrysocolla, azurite and melaconite being less abundant. Cuprite, native copper and various rare minerals are also recorded, including a copper-cobalt phosphate[1]. The presence of a considerable amount of cobalt is noteworthy and recalls the association in the British Trias. Some samples of these ores are very rich, running from 6 to 14 per cent. of copper and the reserves are large: the copper-belt is said to be about 100 miles long and from 30 to 60 miles wide. Several mines, including the Star of the Congo and others, are worked on a large scale by modern methods and a smelting plant has been erected with a capacity of 100 tons of copper per day. In 1916 the total output was over 21,000 tons, and it is expected that this will be greatly increased by an electrolytic leaching process for the treatment of low-grade ores. It is thus evident that the copper resources of this area are very large, and likely to be of much importance. The genesis of this type of ore-deposit is still very obscure: the ores visible at the surface are all of an oxidized character, but some borings to a depth of 200 feet or so appear to have struck chalcopyrite and other sulphides, which are probably the primary ores. There is little or no evidence of any igneous rocks in the neighbourhood and further development must be awaited before any useful conclusions can be drawn.

Copper in Red Sandstones. Disseminations of copper ore in sandstones have been found in many parts of the world, especially in Russia and in the south-western United States. It is a significant feature of most of these deposits that they occur chiefly in sandstones belonging to the arid facies of sedimentation, associated with beds of gypsum, rock-salt and other saline compounds. This association is too common to be accidental and it has been suggested that the deposition is due to precipitation in strong salt solutions (salt lakes) of copper salts derived from the denudation of older copper ores. Along with the copper deposits are often found lead, vanadium and uranium in very peculiar mineral combinations, described in Chapter XVIII. The copper ores of Cheshire previously described may be referred to

[1] Ball and Shaler, *Econ. Geol.* vol. IX, 1914, pp. 617–30; Buttgenbach, *Ann. Soc. Géol. Belge*, vol. XLI, 1914, pp. 11–51.

this type, since the mineral association is the same, although in this case the ores certainly appear to be of later date than the sandstones. Most of the deposits of this class belong to the Permian or Trias, in America sometimes to the Jurassic, in which the same facies of sedimentation continued. Those in the Kirghiz Steppes between the Urals and the Altai mountains are in sandstones supposed to be of Palaeozoic age. The ores, which are often associated with plant remains, consist of chalcocite, bornite or chalcopyrite, usually oxidized to malachite, azurite and chrysocolla near the surface.

The Copper Shale of Mansfeld, Saxony. The best known of all occurrences of copper-bearing sediments is the famous Kupferschiefer of Mansfeld, Saxony, which has been worked since the year 1199 and still yields large quantities of copper: it is believed that it was extensively exploited during the war. The Kupferschiefer is the middle division of the German Permian lying between the Rotliegende below and the Zechstein, or Magnesian Limestone, above: it thus corresponds to the Marl Slate of N.E. England. Although extending over a very large area the copper-bearing bed is less than 2 feet thick and consists of black shale with abundant plant and fish remains and finely-disseminated sulphides, chiefly chalcopyrite, with bornite, chalcocite, pyrite and rarely tetrahedrite and galena. The ore also contains zinc, vanadium, cobalt and nickel in small quantities. Its average copper content is about 2·75 per cent. There are a few small veins of barytes and calcite. This deposit certainly appears to have been formed by reduction and precipitation of copper compounds in water of peculiar composition, probably very rich in decaying organic matter. The copper and other metals must have been leached out of the surrounding rocks by meteoric waters, and concentrated in this lake or shallow sea.

Origin of Copper in Sediments. A comparison of the different occurrences of copper ores in sedimentary rocks shows quite clearly that they cannot all be explained in the same way. In the case of the Mansfeld shale, for example, it certainly seems probable that the copper was deposited at the time of formation of the shale, by precipitation from solutions and reduction to sulphide. In the sandstone deposits of the British Trias and of

Katanga the field relations suggest a subsequent origin by infiltration of solutions that may have come upwards or downwards;
probably the former, since connection with faulting is manifest.
Some of the sandstone ores of Russia and the United States
may, on the other hand, be contemporaneous or nearly so,
though the frequent replacement of plant remains by copper
minerals suggests a slightly later date. At any rate the association of copper ores with "red beds" is striking and evidently
of genetic significance, although it seems probable that two or
more different groups of phenomena, of varying origin, are here
included.

CHAPTER XIII

TIN

THE use of the metal tin has been known from a very early stage of civilization as a constituent of bronze, the forerunner of iron as a material for cutting tools; bronze was and still is largely employed also for coins and for a variety of castings, both ornamental and useful. In later times tin has found employment in a great number of metallic alloys for special purposes and at present a large amount is needed for coating iron or steel plates, the "tins" of ordinary colloquial language ("cans" in America). Hence tin-mining is a very ancient industry and is still of great commercial importance. At the present time tin commands the highest price of any of the metals produced on a large scale, and the amount of capital invested in tin-mining is very large. So far as is known the world's supply of ore is limited and prices seem likely to maintain a high level. The geology of the tin deposits is pretty well known and presents some features of great general interest, which throw considerable light on other and more obscure problems of ore-genesis.

The earliest evidences for the use of tin, in bronze, have been found in the eastern Mediterranean region, but it is not known whence the ore was obtained; some of the current statements of trade in tin in very early times by the Phoenicians in western Europe are not very well authenticated and the whole subject is still involved in much obscurity. It is clear, however, that the art of making bronze has long been known also in China, but whether this was an independent discovery is uncertain.

Tin Minerals. The mineralogy of the tin ores is unusually simple: only two minerals occur in workable quantities; namely, cassiterite and stannite. Cassiterite has the composition SnO_2, and is by far the more common ore. Stannite is a complex sulphide with tin, copper, iron and zinc. Both of these minerals are of primary origin. Wood-tin is a peculiar amorphous, fibrous or radiating mineral, consisting of SnO_2 and therefore similar in composition to cassiterite, but believed to be of secondary origin, resulting from the decomposition and oxidation of stannite. Tin is also a constituent of a number of other minerals, none of which are worked commercially, most of them being merely mineralogical curiosities.

Ores of tin are nearly always associated with those of other metals in workable quantities, and it is usually possible to recognize in each region a definite mineral paragenesis, as will appear in succeeding sections. Furthermore, tin ores are more clearly restricted than those of most other metals to a particular type of igneous intrusion, in this instance, the more acid end of the series, usually deep-seated plutonic granites, less commonly the hypabyssal or extrusive forms of the same magma, and in a great many cases tin ores accompany the class of changes generally classed as "pneumatolytic": the common association of tinstone and tourmaline is well known.

Since cassiterite is a very resistant mineral it tends to accumulate to a great extent in superficial deposits and a large proportion of the world's supply of tin is won at the present time from "alluvial deposits" of various kinds, some of which are not strictly alluvial in the true scientific sense of the word, but rather of the nature of residual deposits. It is probable, however, that in the not distant future lode-mining for tin will increase in relative importance.

The distribution of deposits of tin ore in nature is curiously limited and shows some features of interest. Such deposits are in nearly all cases associated with the intrusion of large masses of granite into highly folded sedimentary rocks in the deeper-lying portions of a great mountain chain, and it is remarkable that so many of the chief tin deposits of the world are associated with earth-movements and intrusions of Permo-Carboniferous age. The reason underlying this fact, if indeed there is a reason, is at present unknown. The chief exception is in the case of the Tertiary tin-silver deposits of Bolivia, which are peculiar in other ways, as will appear later.

In general terms it may be said that tinstone occurs as an original mineral disseminated in granites, as a constituent of pegmatites, dykes and quartz veins of magmatic origin, and in various types of lode-fillings and disseminations along fissures, very frequently in brecciated lodes, as hereafter described in detail. Stannite, which is much less abundant, is usually found as a lode-mineral, along with other sulphides. Both of these are minerals of the primary unaltered deep zones, whereas wood-tin is a mineral of the zone of secondary alteration. The general characteristics of the tin-bearing deposits may be more profitably summed up after a description has been given of certain typical examples.

Cornwall and Devon. The geology of the tin-bearing district of the west of England, though extraordinarily complicated in detail, is simple in its main features. As seen on the present surface, five large granite bosses and several smaller ones are intrusive into a series of sedimentary rocks, ranging in age from the Carboniferous down to an unknown antiquity. Essentially the mining region consists of the southern limb of a large syncline, with the newer rocks on the north and the oldest in the extreme south. If an east and west line be drawn approximately through Truro, all the rocks to the south of this, except the granites, are Lower Palaeozoic or older; the rest of Cornwall is Devonian, as well as the south of Devon, while the middle part of that county is occupied by a broad belt of Carboniferous strata, of a special local type to which the name Culm is generally applied. Besides the general synclinal structure, all the sediments show the most intense local folding, overthrusting, compression and cleavage, as well as conspicuous metamorphic effects from the granite intrusions. The gneissose rocks of the Lizard peninsula and the schists of the Start district are undoubtedly pre-Cambrian in age, while the rest of the slates, grits and greenstones of western Cornwall may be either pre-Cambrian or Lower Palaeozoic. Since they are mostly unfossiliferous this point cannot be decided; in one division, the Veryan beds, Ordovician fossils have been found[1].

The map in Fig. 46 shows the salient features of the geology of the mining region: as will be seen there are five principal granite masses, named as follows from west to east: Land's End,

[1] Some geologists still consider that the whole of the "killas" of western Cornwall is Devonian, and it is so indicated on the majority of geological maps.

Carn Menellis, St Austell, Bodmin Moor and Dartmoor. There
are also smaller masses, such as St Michael's Mount and Godolphin, near Penzance; Carn Brea and Carn Marth to the
north-west and north of the Carn Menellis mass; St Agnes, near
the north coast, and others too small to be marked on a small-
scale map, but of some importance from the mining point of
view. All these granites are very similar in their general characters and relations and there can be no doubt that they are
all local domes and cupolas on the surface of a great bathylith
that underlies the whole district at no very great depth. This is

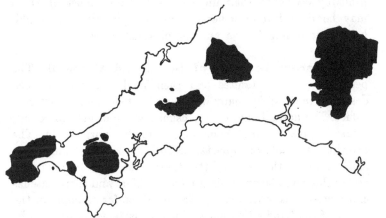

Fig. 46. Map of the granite masses (black) of Cornwall and Devon.

a point of much importance in mining geology, as will appear
later. All the granites are more or less heavily mineralized,
especially round their margins, and the mineralization extends
for a considerable distance into the slaty rocks. The number of
different metallic compounds found in these areas is extraordinary, and in point of fact Cornwall is one of the most
richly mineralized regions of the whole world, both for abundance and variety of minerals. The principal metals which
are or have been obtained, in addition to tin, are copper,
tungsten, arsenic, lead, zinc and iron, with smaller quantities of
antimony, bismuth, molybdenum, cobalt, nickel, silver, gold and
uranium.

The pre-granitic rocks of the mining areas are for the most

part sediments of fairly or very fine texture, which have been compressed and folded into structures of almost inconceivable complexity, as well as strongly cleaved: one result of this is that the *killas*, as it is called locally, is a remarkably sound rock, so that little timbering is required, even in deep mines. Penetrating the killas and apparently, as a rule, folded with it are a number of basic sills, called collectively greenstones. That these are older than the granites is shown by the fact that they have undergone thermal metamorphism by the granite, often to a striking degree. As a rule the greenstones are not mineralized, probably owing to their impervious texture. Most of them may be described in general terms as epidiorites or amphibolites, according to the amount of crushing that they have undergone.

Tin-bearing Granites of Devon and Cornwall. The petrology of the granites and their modifications has been described in detail by many writers: a brief summary must here suffice. The normal granites are usually white or grey in colour, coarse in texture and generally markedly porphyritic. The principal minerals are quartz, orthoclase or albite, muscovite and biotite, with often a little tourmaline even in unaltered rocks: this must be an original mineral. Here and there patches and masses with finer texture are found, as for example in the centre of the Land's End mass: this is considered by some to be a separate, later intrusion. The foregoing description applies to the fresh unaltered varieties from non-mineralized areas. In the neighbourhood of the ore-deposits the granite is usually much decomposed and stained red or yellow, and in the actual mineralized portions it has undergone very far-reaching changes, to be described hereafter.

Nearly everywhere pegmatites and aplites are to be found, with much the same minerals as in the normal granite, but differing in texture. The most striking feature of all these altered granites, however, is the abundance of tourmaline. The formation of this mineral is in most cases a secondary process and every stage can be seen from a granite with a mere trace of tourmaline to a rock consisting exclusively of that mineral and quartz (schorl-rock).

This change comes under the general heading of pneumatolysis, of which three types can be seen in the Cornish granites[1]:

(1) Tourmalinization.
(2) Greisening.
(3) Kaolinization.

The first two of these are intimately connected with the formation of ores of tin and other metals; the third is of great economic importance as the origin of china-clay, but requires no further consideration here.

Pneumatolytic Changes in the Granites. The process of tourmalinization is remarkably well displayed in many parts of Cornwall and Devon. In many districts the apparently unaltered granite contains more or less tourmaline, which often appears to be an original constituent. More commonly, however, the tourmaline is plainly of secondary origin, having been formed by the action of vapours or solutions during the later stages of the cooling of the granite intrusions. The common order of events is that the biotite is first attacked and replaced by crystals of tourmaline, often of fairly large size. In the next stage the changes affect the felspar, which is converted into a mass of tourmaline and quartz, the former mineral often showing a remarkable radial arrangement of fine needle-like crystals, generally of a bluish colour under the microscope. An excellent example of this is the well-known rock called luxulyanite, now found only as occasional scattered boulders of disintegration at Luxulyan, in the parish of Gready, near St Austell[2]. Very similar types are also found near Carbis Bay, St Ives, and at the Levant mine. These rocks are particularly striking from the large size of the pink orthoclase crystals. Another variety found at Trevalgan, west of St Ives, has large white or creamy felspar crystals in a black matrix of quartz and tourmaline. In extreme cases the whole rock is converted into a mass of quartz and tourmaline only: this varies in appearance from a very fine-grained type to one of very coarse texture: the general name

---

[1] "The Geology of the Land's End District," *Mem. Geol. Survey*, 1907, pp. 53–60.
[2] Bonney, *Min. Mag.* 1877, p. 215.

of schorl-rock is applied to these. A well-known example of the finer type is the Roche rock near St Dennis; while the schorl-rock with large prisms of tourmaline is locally quite common, especially in various parts of the Land's End granite.

Tourmalinization also extends into the quartz-porphyry (elvan) dykes and other local modifications of the granites, while there is commonly an extensive development of the same mineral in the metamorphosed sediments near the contacts, giving rise to many types of tourmaline-schist and other analogous rocks.

As would naturally be expected, tourmalinization is strongly marked in the tin- and copper-bearing lodes, many of the distinctive types of vein-stuff largely consisting of this mineral, which is often accompanied by chlorite, giving rise to many varieties of capel and peach (see p. 98). It will thus be seen that tourmalinization is a very widespread and distinctive feature in connection with tin deposits.

The second type of pneumatolytic alteration, the formation of greisen, is much more restricted in its distribution. Greisen is a rock consisting of quartz and mica, with or without topaz, but typically carrying little or no tourmaline. Its formation appears to be due to the action of magmatic vapours or solutions rich in fluorine, but without boron in appreciable quantity. Greisen is most commonly found forming narrow veins, or thin layers on the walls of open joints which have clearly served as channels for the passage of the active chemical agents set free from the crystallizing magma. The mica is very often a variety rich in lithium, and greisens often carry a large variety of metallic and other minerals, including ores of tin, tungsten, molybdenum and uranium, with apatite, zircon, etc. This process is not confined to the granite, but is also seen to a certain extent in the elvans and in the sediments. Famous localities for greisen are St Michael's Mount, near Penzance, and Cligga Head on the north coast, but it is fairly common in many mining districts, though never in large amount. In very many cases, as at Cligga Head and St Michael's Mount, the centre of each greisen vein is occupied by a more or less well-defined band of coarsely crystalline quartz, carrying tinstone, wolfram or other ore-minerals.

From this the manner of formation of the veins is clear: the greisen layers are altered granite and the quartz is the actual filling of the fissure, solidified after the passage of the pneumatolytic solutions.

The Tin Ores of Cornwall. The distribution of workable deposits of tin ore in Cornwall is for the most part limited to lodes and veins. Some of the latter are clean-cut fissures filled with ore and gangue minerals and possessing definite walls, but the more important lodes are generally somewhat indeterminate, many of them shading off gradually into normal rock. The lodes traverse both granite and killas and in any one district they show a certain regularity of arrangement, being often clearly divisible into two sets, at right angles or nearly so, known as lodes and cross courses. Some of them are clearly faults and one set is sometimes shifted by the other. In accordance with this origin, as faults, it is often found that the lodes are really zones of brecciated material, "shatter belts," much altered by pneumatolytic processes and cemented by deposition of quartz, tourmaline, chlorite, fluorspar and other secondary minerals, as well as by ores. Owing to the abundance of tourmaline and chlorite the lode stuff is often blue or green in colour, even when derived from white granite. To this material the names capel and peach are often applied, in a somewhat vague way: as a working classification it is as well to call the blue tourmaline-rock capel, while the green chlorite-rock can be called peach, but some miners use these terms in the opposite senses or indifferently. Fine crystals of cassiterite are occasionally found in vugs in the veins, but the greater part of the tin exists in a finely-disseminated condition in the lode stuff and in the country rock immediately adjoining.

Besides tin, some of the lodes are rich in copper, wolfram and mispickel, while others contain mainly galena, blende and iron minerals, and these minerals are found to show a certain regularity of distribution, which is doubtless of genetic significance. Since about 95 per cent. of the total production of tin now comes from the Camborne-Redruth area it will be well to describe this somewhat fully.

The Camborne-Redruth mining district lies on the north-west

side of the Carn Brea granite mass, which itself is situated to the
north-west of the larger Carn Menellis boss: the Carn Marth
boss lies immediately to the north of Redruth and all three are
known to be connected underground. The principal lodes of the
district strike about E. 20° N. and W. 20° S.; some underlie to
the north, some to the south, while others are vertical. This lode
series is cut and in some instances shifted, by another set of
lodes known as cross courses, more or less at right angles to the
first set. The lodes penetrate both the granite and the killas, and
the most productive portions have a definite relation to the
granite-killas contact. The upper parts of the lodes, and especially
the upper parts of those lying in the killas at the present surface,
were very rich in copper, which was found to give place gradually
to tin in depth. All over the district are many old shallow mines,

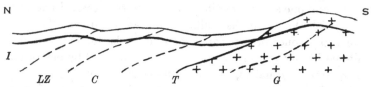

Fig. 47. Generalized section of the metalliferous zones in connection with
the granite of Carn Brea, Cornwall. *G*, granite; *T*, tin zone; *C*, copper
zone; *LZ*, lead-zinc zone; *I*, iron zone (after Davison).

which were worked mainly for copper and abandoned when the
lower limit of profitable copper-mining was reached. Below the
copper come rich deposits of tin-ore, with frequently wolfram
and arsenical pyrites in addition. It has been found that the
tin-content of the lodes in its turn falls off seriously at an average
depth of about 200 fathoms below the granite-killas contact, or
about 360 fathoms below the surface on the line of the principal
lodes: even the rich Dolcoath ore-shoot, the deepest known, is
unpayable below about 500 fathoms. The cross courses, and
especially those portions of them further from the granite,
contain chiefly ores of lead and zinc, while still further away
veins of iron ore are known. From these data it is possible to
construct a generalized diagrammatic representation of the
zonary distribution of the principal ores, as shown in Fig. 47.

In this district as a rule the richer lodes underlie to the north;
those with a southerly dip are on the whole less productive,

though there are exceptions, while the vertical lodes are of still
less importance. The lodes are as a rule steeply inclined, rarely
departing more than about 30° from the vertical: in the deepest
levels of Dolcoath mine, however, over 300 fathoms below the
top of the granite, the dip is 45° south. A notable exception to
the general rule is the Great Flat Lode along the southern
margin of the Carn Brea granite, which in one place is only 10°
from the horizontal. Lodes at about 30° from the horizontal
are also known at West Wheal Seton, north of Camborne, and

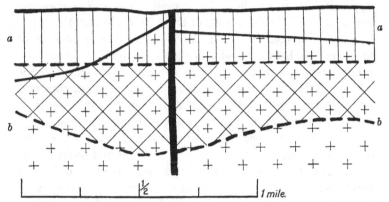

Fig. 48. Longitudinal section to true scale of the Great Lode in Dolcoath
Mine and its eastward continuation, showing the depth of dominant
copper and tin ores. *a, a*, killas; *b, b*, granite. The copper zone is
indicated by vertical lines, the tin zone by diagonal lines. Founded
on data given by Hill and MacAlister, "Geology of Falmouth and
Truro," *Mem. Geol. Survey*, 1906.

at Killifreth, in the Gwennap district. It is notable that in
many instances the underlie of the principal lodes is the same
as that of the elvans in the same district. Some of the lodes
have been traced continuously for several miles along the strike,
but their maximum extent is quite uncertain since they die out
gradually and become unpayable and are consequently not
worked to their actual terminations.

The workable width of the lodes is naturally very variable;
some of them are only a foot or so wide while in other cases
there may be as much as 40 or 50 feet of payable stuff. In the
larger brecciated lodes the boundaries are extremely vague and
the width cannot be stated with any degree of accuracy. The

average of a large number of measurements made many years ago gave a mean width of a little under 4 feet, but this figure is not of much value. Where the tin-content falls off gradually from the centre outwards the payable width naturally depends on a combination of economic factors, especially the efficiency of the dressing plant, the cost of labour and the current price of tin concentrates. Hence frequent assays are obviously essential as a control. It is not easy to make any definite statement as to the tin-content of the lodes, which varies widely, but as a rough approximation it may be said that most of the stuff now worked contains from 1 to 2 per cent. of black tin (cassiterite).

When examined in detail the lodes show a good deal of variation. As before stated, some are clean infillings of fissures, others are zones of alteration in country rock along fissures, while a third type consists of alteration and cementation of a crush-breccia. In the first type the ordinary comby structure is sometimes seen, with more or less regular successive layers of various ore and gangue minerals: in some cases a definite order of deposition can be made out. Tin, wolfram and quartz often occur against the walls, while sulphides, fluorspar and other veinstones occupy the middle parts. Sometimes the country rock for some distance outside the fissure itself is altered to capel, with or without tin. This forms a transition to the next type, where the actual fissure vein is very narrow, merely a crack, and the lode consists of capel and peach derived from the country rock impregnated with tin ore. Occasionally a lode is made up of a number of parallel stringers of this kind enclosed in a broad band of altered country rock. Brecciated lodes of the third type are very common and important in the Camborne district. They may consist of an interlacing network of stanniferous and quartz veins with fragments of country rock, and in some instances it can easily be proved that the lodes have been sealed up and again brecciated several times[1]. The great lodes of Dolcoath, East Pool and South Crofty are of the brecciated type: many of them are very rich in wolfram and arsenic. Some of these lodes, especially in depth, show a very considerable

[1] "The Geology of Falmouth and Truro," *Mem. Geol. Survey*, 1906, pp. 131–8.

resemblance to pegmatites, many of the crystals of wolfram
being very large. In many localities cavities or vugs, often of
considerable size, have been observed in the lodes, but they are
not very common: in the old days many fine mineral specimens
were obtained from these.

As typical and representative examples of the disposition of

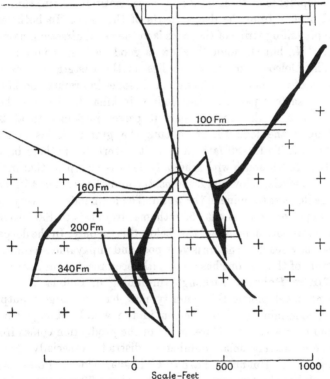

Fig. 49. Section of the principal lodes in East Pool Mine, Cornwall (by
permission of East Pool and Agar, Ltd). Crosses, granite; white, killas;
black, lodes.

the lodes in this district we may select two famous mines, Dolcoath and East Pool. Both these mines date from the days
when copper ore was the principal product of the district, so
far as lode-mining was concerned. While the workings were
shallow, lying in the killas, the output was mainly copper, but
in depth the copper died out and in the granite the yield was

mainly tin: it is many years since any appreciable quantity of copper ore was produced. Wolfram and arsenic ores are valuable by-products. It is estimated that Dolcoath produced altogether about 230,000 tons of copper ore containing 16,000 tons of copper, while East Pool and Agar, now one company, yielded 88,000 tons of ore with 5000 tons of copper. But the modern developments at these mines are based entirely on tin, tungsten and arsenic from the deeper parts of the lodes. In both setts the prevailing strike of the lodes is in the usual direction, namely E. 20° N., but the underlie shows a good deal of variation[1].

The Dolcoath mine, which lies at the eastern end of the parish of Camborne, works numerous lodes in granite and killas. In the eastern part the main lode is in killas to the 80 fathom level, but in the western part it passes into granite at 230 fathoms: the lode therefore cuts the granite killas contact obliquely. To the 80 fathom level it is vertical: it there begins to turn to the south with gradually decreasing dip, so that in the deepest levels, below 500 fathoms, the underlie is about 45°. This is the deepest tin mine in the world. The main lode was very rich in copper down to about 150 fathoms: from this to 190 fathoms there was both copper and tin; below this tin only. In the deepest levels now reached the ore is very poor and unpayable, so that the bottom of the tin zone has apparently been reached, or nearly so.

Tin in Bolivia. Although tin-mining in Bolivia is but of recent development this country now has the largest output from lode-mining of any region in the world. Though tin is found over a very wide area most of the production comes from a comparatively small number of districts, especially Oruro, Morococala, Huanuni, Pazña, Llallagua, Uncia, Potosí and Chorolque: the foregoing account for about 70 per cent. of the total: Uncia alone contributes about one quarter of the Bolivian output, and Llallagua yields nearly as much. Nearly all these localities and many others are in the department of Potosí[2].

The Bolivian tin district is a remarkable one, in that it constitutes a very definite metallogenetic province with a peculiar mineral association, namely tin, tungsten, bismuth and

[1] Maclaren, *Mining Mag.* vol. XVIII, 1917, p. 249.
[2] Davy, *Econ. Geol.*, vol XV, 1920, pp. 463–496.

silver. The deposits are found in the Cordillera Real throughout the greater part of the length of the country, but do not extend to any appreciable degree into either Peru or Argentina. These veins were described long ago by Stelzner as the Potosí type, but some of the distinctive characteristics noted by him have not stood the test of more extended investigations and the differences between these veins and those of Cornwall and Saxony are not so marked as was formerly thought.

An examination of the whole region undertaken by Armas[1] in 1911 showed that there are two zones of tin deposits in Bolivia with a south-east trend and lying about 35 miles apart. The south-west zone, in which lie Oruro, Potosí, Chorolque and others is mainly associated with Tertiary igneous rocks, while the north-east zone, including Araca, consists mainly of slates and quartzites. Generally speaking, in both zones there is a gradual relative decrease in tin and increase in silver from north to south, with some notable exceptions.

The associated igneous rocks range from rhyolite to dacite, or, as they are called by other authors, quartz-porphyry and porphyrite. They seem to include both extrusive and intrusive types. As to the age of these rocks most writers appear to consider them Tertiary, though some authors have placed them in the Cretaceous. With these are found sediments mostly of lower and middle Palaeozoic age, consisting of shales, sandstones and greywackes, usually of a dark grey colour: it appears, however, that some rocks of very similar appearance have been found to contain Tertiary fossils. Veins are found in both the intrusive rocks and in the sediments, in the latter mainly in the sandstones and quartzites. Those in the igneous rocks, which are usually the more important, are as a rule well-defined fissure veins usually from 2 to 4 or 5 feet wide: some of the wider ones are brecciated lodes with a cement of secondary minerals. The mineral composition of the lodes is very variable: as a rule pyrites is most abundant, while cassiterite usually comes next in amount: there are also found tetrahedrite, jamesonite, bismuth, bismuthinite, antimonite, galena, blende, chalcopyrite, wolfram and mispickel: many of these contain silver. The chief

---

[1] Armas, *Ann. des Mines*, Ser. 10, vol. xx, 1911, pp. 149–230.

gangue minerals are quartz, tourmaline, calcite, barytes, chalyb-ite and occasionally fluorspar. There are also a number of remarkable and rare minerals containing germanium. In the upper part of the veins there is much secondary silver enrich-ment, with ruby silver, stephanite and native silver. Much of the tin is obtained from the ferruginous gossan of the oxidized layer, where it is naturally concentrated by solution of other minerals. The richest ores may contain as much as 8 per cent. of tin. The ordinary concentrate, barilla, usually runs from 55 to 65 per cent. tin.

As a typical example of a Bolivian tin district we may take the Uncia-Llallagua district, the largest producer of all. The Cerro de Llallagua is a mountain consisting of a central mass of quartz-porphyry penetrating a thick series of dark or red shales and shaly sandstones; there are also some basic intrusions. The Uncia ore-bodies on the south-east slope of the mountain are associated with brecciated shatter-belts; sometimes the veins follow these belts, in other places they cut across them. The veins are mostly parallel, striking N. 30° E. and dipping steeply either way: the average width is about a foot. Some of them are true fissure veins filled with fine-grained cassiterite of a very compact nature, with some bismuthinite, wolfram and scheelite, native copper, chalcopyrite and pyrite. Wolfram is now worked extensively as a by-product. The gangue minerals are quartz, aragonite and chalybite. The Uncia mines in 1915 produced 8416 tons of tin concentrates.

The Llallagua mine on the same mountain, about 3 miles north of Uncia, works part of the same group of veins and its output in 1915 was 5812 tons of concentrates. In this property the veins cut the slates, as well as the quartz-porphyry, but they do not carry much tin in the sediments and tend to pinch out. Otherwise the relations are almost precisely similar.

The Pazña tin district now produces about 1000 tons of con-centrates annually from veins in quartzites and slates, varying in width from 3 to 30 feet and associated with dykes and masses of quartz-porphyry. The primary ore consists of cassiterite, pyrite and quartz with a little tourmaline and a few narrow stringers with wolfram at a higher level. The ore is said to

average 5 per cent. tin. The mines of the Huanuni district lie on
the south side of Posoquoni, a mountain 14,700 feet high, composed of grey or pink quartzites with grey shales, intruded by
dykes and masses of quartz-porphyry. The veins are numerous
and vary in width up to 30 feet: they are shatter-belts cemented
by cassiterite and sulphides, with some quartz, chalybite and
tetrahedrite.

Tin in the Malay Peninsula. At the present time the most
important tin-producing area in the world is in the long stretch
of country extending from Lower Burma through Siamese
territory into the Malay Peninsula and continued in the Dutch
islands of Banka and Billiton off the north-east coast of Sumatra.
The following table shows the total output of tin from this part
of the world in recent years, in long tons.

	1914	1915	1916	1917	1918
Malaya	49042	46766	43870	39833	37370
Banka	13973	13773	14548	13246	12378
Billiton	4000	5750	5000	5500	4500
Siam	6800	7800	7800	8765	9153
Burma	300	500	900	1200	1000

In some instances these figures are estimates only, but probably not far from the truth. By far the greater part of this
output is from alluvial and other superficial deposits, but lode-
mining is of increasing importance.

This tin-bearing region forms one of the zones of the great
folded Burmese-Malayan arc[1]. It may be described in the most
generalized terms as an axis of crystalline schists and ancient
sediments, invaded by vast masses of granite which carried the
tin. The age of the older rock-series is somewhat uncertain; the
crystalline schists are probably pre-Cambrian, while the less-
altered sediments are perhaps Lower Palaeozoic; the Raub
limestone of the Malay States may be Carboniferous, while other
stratified rocks are claimed as Permo-Carboniferous. The age of
the folding is not known with certainty, but is probably more
or less contemporaneous with the intrusion of the granites,
which may themselves be Permo-Carboniferous or Mesozoic in
age. The intrusions were accompanied by conspicuous pneu-

---

[1] Suess, *Das Antlitz der Erde*, vol. I, 1885, p. 579.

matolysis and there is a high degree of mineralization. In the southern part of the zone, tin is dominant, in the north tungsten.

The islands of Banka and Billiton consist of sedimentary rocks of unknown age, possibly Palaeozoic, including quartzites, sandstones, slates and micaceous schistose rocks, intruded by granites. The strike and dip of the bedded rocks are both very variable; in Billiton the dips are usually from 60° to 90°, while in Banka the inclination may be as low as 20°. The granite, which is considered by Verbeek to be probably Permo-Carboniferous, carries hornblende and mica and is accompanied by aplite and quartz-porphyry. Primary tin ores are found both in the granite and in the sediments, largely in quartz-veins, also in lodes and irregular stockworks. The greater part of the output however is alluvial[1].

The alluvial tin deposits of Banka are of the usual type, namely, a bed of tin-gravel, mainly composed of pebbles of quartz with cassiterite, resting on the bed-rock, which is commonly granite, with an overburden of 20 or 30 feet of fine gravel, sand, and clay. These deposits form as a rule alluvial flats in valleys and occasionally spread out into plains of some width, which are often swampy, and plenty of water is available for working the tin.

The details of the geology of the primary tin deposits of the Malay Straits region have been, and still are, the subject of much controversy; furthermore the production nearly all comes from alluvials, hence no good purpose would be served by a minute discussion of this difficult subject[2]. One point must, however, be mentioned with some emphasis, namely, that a number of the tin propositions worked successfully on a large scale by the methods usually applicable to alluvial deposits, are as a matter of fact nothing but rock in place, of various kinds, including granite, phyllite and quartzite, that have undergone an unusually high degree of weathering under the influence of the tropical climate. Also it seems very probable that some at any rate of the so-called "boulder clays" attributed to ice-action

[1] Verbeek, *Jaarb. v. h. Mijnwesen in Ned. Oost-Indie*, vol. xxvi, 1897.

[2] See Scrivenor, several papers in the *Quart. Journ. Geol. Soc.* 1909–13, *Mining Mag.* vol. xix, 1918, p. 254, and elsewhere.

in the Permo-Carboniferous period, are actually igneous or other rocks weathered in place, and containing cores of still undecomposed rock, forming the boulders. Another very peculiar type of tin occurrence found in the Federated Malay States takes the form of primary deposits of cassiterite in limestone. In parts of the area pneumatolytic phenomena are remarkably well developed, the resulting rock-types showing a strong resemblance to those of Cornwall. Towards the north, however, and especially in Tavoy, where wolfram is abundant, tourmaline is rare or absent. This has been explained on the supposition that the lodes of Malaya belong to a deeper zone of mineralization than those of Tavoy, and it has hence been argued that there should be a tin zone below the wolfram lodes of the last-named area[1].

With regard to the true secondary tin deposits of the Malay Peninsula, in spite of their enormous commercial importance, there is not much of value that can be said from the geological point of view. They are in point of fact perfectly normal river-deposits formed in a region of varied relief, rapid weathering and high rainfall. Consequently they show infinite variation in detail, since their occurrence is purely a matter of topography. In a tin-bearing region any stream-deposit or surface accumulation may carry tin; most of what is said in the chapter on alluvial gold deposits is equally applicable to tin deposits and it would be a waste of space to discuss the subject twice on general principles.

Tin in the Commonwealth of Australia. Reference has already been made to the fact that the Burmese-Malayan arc of folding, with its accompanying granite intrusions, is continued down the eastern side of the Australian continent, and into Tasmania. Throughout the greater part of the stretch of highlands between the north of Queensland and Tasmania the rocks of this chain are well mineralized, showing very markedly the tin-tungsten-molybdenum-bismuth facies of deposit. Besides tin, important quantities of the other metals mentioned are produced, especially tungsten in northern Queensland, molybdenum in southern Queensland and New South Wales, with bismuth in both states. Queensland and New South Wales

[1] Jones, *Trans. Inst. Min. Met.* vol. xxix, 1920, pp. 320–76.

produce a good deal of tin, but Tasmania is the most important in spite of its comparatively small size. Curiously enough there is only a small amount of tin in Victoria. Since the geology of the deposits in Queensland and New South Wales that produce tungsten and molybdenum is described pretty fully in the chapters on those metals, attention will be confined here chiefly to Tasmania, which shows some interesting and special features of its own.

The chief tin producing area of Queensland lies in the far north in the Cape Yorke Peninsula and may be conveniently divided into the three following fields: Herberton, Kangaroo Hills and Cooktown. In the extreme south of the state, not far from Brisbane, is the Stanthorpe field which geographically belongs rather to the New South Wales group. In the last-named state the Vegetable Creek mine in the New England district is the largest. In most of these areas a large proportion of the tin is obtained from alluvial deposits, either modern gravels, or deep leads, like the similarly named gold deposits of Victoria.

The tin deposits of Tasmania are not only of great commercial importance, but one of them at least, that of Mount Bischoff, is of a rather unusual character and of considerable interest.

The tin fields of Tasmania lie chiefly in the north-west and north-east corners of the island. On the north-west side the most important areas are at Mount Bischoff and Heemskirk, in the north-east the Ringarooma district, especially the Blue Tier and Anchor mines. The Blue Tier area consists of a porphyritic two-mica granite intrusive into Silurian slates and sandstones. The tin occurs chiefly in a non-porphyritic variety of granite of rather finer grain than the main mass, usually kaolinized or silicified and of a greenish colour. This variety can often be seen under the normal granite. It generally carries about 1 per cent. of cassiterite, occasionally up to 2 per cent., along with wolfram, scheelite, molybdenite, chalcopyrite, galena and fluorspar. Tourmaline is rare. The tin appears to be mainly a dissemination in the greisenized rock in the neighbourhood of large E.–W. or N.–S. joints.

The Pioneer and Briseis mines on the Ringarooma River work deep leads and each produces in normal times about 500 tons of black tin per annum. The Briseis tin gravel lies under a sheet of olivine-basalt, which is rotten and can be worked by hydrau-licking; the Pioneer gravel is under 40 feet of hard cemented overburden.

In the north-west of the island a well-defined mineral belt runs through Mount Heemskirk, Zeehan and Mount Rosebery. The sedimentary rocks are invaded by granite which has induced much metamorphism. Quartz veins lie in both schist and granite and carry cassiterite, wolfram, chalcopyrite, bismuthinite, mag-netite, rutile, tourmaline, topaz and fluorspar: they are often greisenized. At Heemskirk pipes are found at the crossings of fissures. At Zeehan stannite occurs, with other sulphides and wolfram, in quartz veins with siderite and sometimes fluorspar.

The great tin working of Mount Bischoff is of a very remark-able character, perhaps unique. It is situated on the top and sides of a hill about 2600 feet high, near Waratah, and all the workings are open, the various parts being called *faces*. They are not however all of the same kind. Some of them are residual or eluvial deposits of a scree-like nature, while others are quarries on dykes and lodes. The principal geological feature of the hill is the intrusion of quartz-porphyry dykes into Cambrian slates and sandstones, with accompanying metamorphism and much pneumatolysis of the dykes themselves. The main dyke is from 15 to 150 feet wide and has been traced for over 5000 feet. It is much altered, the felspar being replaced by quartz and topaz with some tourmaline. It is in fact a greisen very rich in topaz. Besides the mineralized dykes there are also true lodes: the Queen lode, a true fissure vein lying in both slate and porphyry, varies from 8 inches to 18 feet in width, averaging 3 feet over a length of over 3000 feet. It runs from 1·5 to 2 per cent. of tin oxide, with quartz, pyrite, fluorspar and iron oxides.

The greater part of the production, however, has come from great masses of residual debris formed by weathering and surface concentration of material from the dykes. The White Face on the south side of the hill is mainly scree material from the main dyke, with some quartz-porphyry in place: the Brown Face on the east of the summit is now a great horse-shoe shaped excavation with the great dyke on three sides of it; it lies mainly in residual material.

Northern Nigeria. Tin mining in the northern provinces of Nigeria is quite a recent development, having only actually begun in 1909, but it soon reached very considerable proportions; the output for the four years 1916–19 averaged slightly over 8000 tons of concentrates per annum. The whole of the production has hitherto come from alluvial or other detrital deposits, but lodes are known to exist in many places, though none have yet been worked at a profit. The most important deposits are found on the Bauchi plateau at a height of 2000 to 4000 feet above the sea. Here an ancient complex of gneisses and crystalline schists has been invaded by a younger granite bathylith, of which the highest portions now appear as bosses, cupolas and dyke-like masses, the latter being often best described as quartz-porphyries. In certain places the normal biotite-granite is replaced by an alkaline facies with riebeckite. There are also associated acid and basic dykes of somewhat later date. The granite is considerably shattered and mineralized along fissures, often forming belts of greisen and topaz rock with tin and wolfram. So far as is known, these lodes are not very large or important, and it is possible that the very rich alluvial and residual tin deposits are the detritus of higher portions of the granite, nearer the original surface of the intrusions and now mainly removed by denudation. Extraordinarily rich blocks of tin ore have been found here and there in the gravels, but they have not yet been traced to their source. At some comparatively recent period the older rocks were largely buried under a covering of alluvial and volcanic rocks, called the Fluvio-Volcanic series. These may be of Mesozoic or early Tertiary age. In the middle Tertiary a period of elevation occurred, raising the plateau to the present level and causing much denudation and rearrangement of the older tin-bearing alluvium. Still later large parts of the area were buried under sheets of basalt, which again covered up many of the tin-bearing deposits on the valley floors.

CHAPTER XIV

LEAD AND ZINC

LEAD and zinc are so commonly found in association that to describe them separately would involve an immense amount of repetition, hence in this work they will be treated together. The geology of the lead-zinc deposits is on the whole fairly simple, so far as regards the field-relations of the deposits, but the theoretical explanation of the genesis of some of the commonest types presents unusual difficulties, perhaps more so than in the case of any other common ores. Many lead-zinc ores are unmistakably vein-deposits of high-temperature origin, referable to magmatic processes, but others, and perhaps the majority of the more important occurrences, show special features that are difficult to explain in a satisfactory manner. It was the study of the lead-zinc ores perhaps more than any others, that led to the development of the idea of lateral secretion, and even yet meteoric waters play a very important part in theories of genesis.

Another point to be borne in mind here is that a large proportion of the world's output of silver is obtained as a by-product in the mining and smelting of lead ores, and very frequently it is only the combination of the two metals in one ore that makes its exploitation a workable proposition. In fact it depends on market conditions whether the lead or the silver is to be regarded as the principal product. It is only in recent times that zinc ores have acquired great importance and in the early days of mining many rich lead-zinc deposits were worked for lead (and silver) only, the zinc ore being thrown on the dump. Many such dumps have been profitably worked over by modern methods. Another by-product, obtained from zinc-smelting, is cadmium, which has certain industrial applications.

Lead and Zinc Minerals. Both lead and zinc occur as constituents of a large number of minerals, belonging both to the primary and secondary sulphides and to the group of oxidized ores. By far the most important, however, of the first class are the simple sulphides, galena and blende. From these the more important of the oxidized ores are derived. Many of the complex sulphides are mainly mineralogical curiosities, while some are of value for other constituents. The very peculiar zinc minerals of the Franklin Furnace district in New Jersey are dealt with specially in a separate section and for the sake of simplicity are not included in the following table.

Lead Minerals		Zinc Minerals	
galena	PbS	blende	ZnS
jamesonite	$Pb_2Sb_2S_5$	—	—
cerussite	$PbCO_3$	calamine	$ZnCO_3$
anglesite	$PbSO_4$	—	—
pyromorphite	$3Pb_3P_2O_8 . PbCl_2$	hemimorphite	$Zn_2SiO_4 . H_2O$
mimetite	$3Pb_3As_2O_8 . PbCl_2$	—	—

Zinc sulphate is of course a very soluble salt and does not occur as an ore.

Zinc sulphide also occurs in a dimorphous form, as wurtzite, having the same composition, but crystallizing in a different system (hexagonal). Some of the zinc sulphide ores described in German text-books as Schalenblende, usually a platy or botryoidal growth, are probably wurtzite. In some books of reference hemimorphite is called smithsonite, while in others the carbonate is called smithsonite and the silicate calamine. This has led to much confusion. The lead molybdate, wulfenite, is described under the heading of molybdenum (p. 421).

Galena is a silvery-grey or lead-coloured mineral with strong metallic lustre, crystallizing in the cubic system, usually as cubes or a combination of the cube and octahedron. It is readily distinguished by its very perfect cubic cleavage and high density, about 7·5. Its general appearance is very similar to that of metallic lead.

Jamesonite is a steel-grey metallic mineral, described more fully under antimony, as it is an important ore of this metal.

Cerussite. An orthorhombic mineral, when crystallized often as complex twins, but often massive and compact. When pure it is colourless, sometimes grey, and has an adamantine or pearly lustre. Its density is about 6·5.

Anglesite. Also a colourless or grey orthorhombic mineral

forming tabular or prismatic crystals with a density of about 6·3 and usually more or less transparent. It also commonly occurs in a massive form.

Pyromorphite and *Mimetite* are very similar in appearance, usually forming barrel-shaped hexagonal crystals, yellow, orange or green in colour. There is also a vanadium compound, vanadinite, and the three form a good example of an isomorphous series.

Blende, Zinc-blende or *Sphalerite.* A mineral of very variable appearance, ranging from colourless through yellow and brown transparent forms to a black metallic variety. The crystals belong to the tetrahedral class of the cubic system and there are six very good cleavages. Hardness and density both about 4.

Calamine belongs to the rhombohedral system, like calcite, but is rarely found in good crystals: it is usually botryoidal or stalactitic, sometimes earthy, and white, grey, greenish or bluish in colour.

Hemimorphite. Orthorhombic crystals, often in bunches, or in concretionary and stalactitic lumps, white or pale green in colour, often stained yellowish or pale brown.

The only commercial source of the metal cadmium is as a by-product in the smelting of zinc ores. Occasionally the mineral greenockite, CdS, is recognizable in the ore, but usually the metal only appears as an impurity in the process of refining the zinc.

General Occurrence of Lead and Zinc Ores. Workable deposits of lead and zinc occur under a great variety of geological conditions. The most characteristic and widespread types of occurrence however belong to two fairly well-defined classes:

(*a*) Vein-deposits formed at a high temperature (magmatic veins).

(*b*) Deposits in limestone, of somewhat doubtful origin.

It is a matter of grave doubt, however, whether it is possible to draw any hard and fast line between these two classes. This point can be more profitably discussed at a later stage. Furthermore, in the study of lead-zinc ores the question of ore-zones of primary origin assumes great importance with reference to the possible persistence in depth of such deposits, or their replacement at lower levels by other metals. This of course is totally distinct from the question of the existence of zones of oxidized and primary ores and the possibility of an enriched zone of secondary origin below the oxidized ores. We may say at once that in this group the last-mentioned effect is of little or no importance.

The number of extensive lead-zinc deposits in the world is so great that it is only possible to select a few typical examples to illustrate some of the more important types with special reference to genetic significance. It so happens that the lead and zinc deposits of Great Britain, though now insignificant compared with the world's output, are of considerable geological interest, and no apology is needed for describing them pretty fully.

Lead and Zinc in Great Britain. Lead mining is one of the most ancient industries in the British Isles and was carried on actively by the Romans in several districts. It also flourished under the stimulus and protection of the religious orders in the thirteenth to fifteenth centuries, when so much lead was needed for the magnificent buildings of that age. At the reformation mining suffered a collapse from which it was long in recovering, like all other industries in this country, and it scarcely became active again till the introduction of pumping and winding machinery. In the early days zinc ore was a waste product, as before said, and the utilization of zinc naturally stimulated lead mining as well. The silver production was always also a matter of much importance.

The important lead-zinc deposits of this country may be referred to three geographical and geological groups:

(*a*) Cornwall.

(*b*) The Lower Palaeozoic areas of Wales, the Lake District and the south of Scotland.

(*c*) The Carboniferous Limestone of northern England and North Wales.

The Cornish deposits are referred to incidentally in the chapters on ore zones and on tin, and will not be further considered here: the other two groups must be dealt with in some detail[1].

Veins and deposits of more or less regular form, carrying varying amounts of lead, zinc and copper, are very common in several areas of Ordovician and Silurian rocks in central and north Wales, in Shropshire, in the Lake District, and in the Southern Uplands of Scotland. The proportions in which the different metals, especially copper, are present, vary considerably, as also does the silver content of the lead, and there is some evidence of a tendency to a zonary arrangement, with rich silver-lead at the top, then lead carrying less silver with zinc below, and in the lower zone dominant chalcopyrite and pyrite. However, this rule is not universally established. The gangue minerals also vary, the commonest being quartz, calcite and barytes. The ore-bodies are sometimes clean-cut veins with definite walls, but often brecciated zones or shatter-belts cemented by ores and gangue minerals. These have no definite walls and the ore runs out into veinlets and stringers in the country rock. These lodes often show a striking parallelism over large areas and are clearly of tectonic origin, many of them being actual faults, although the amount of displacement is often difficult to determine, owing to the uniformity of the rocks. In some localities veins are mainly confined to the coarser types of sediment, such as grits and flags, but in other areas they seem to be more abundant in the slaty rocks.

Central Wales. Mineral veins are very numerous in that dreary expanse of Ordovician and Silurian rocks forming the northern part of Cardiganshire and the western part of Montgomeryshire: some of the mines in this region were at one time extraordinarily productive, and the industry was of considerable importance. Even now there are considered to be possibilities of future development. Recent researches have shown that the whole metalliferous area is composed of a series of Bala and Llandovery rocks, probably about 10,000 feet in total thickness

---

[1] Almost the whole of the information here given as to lead and zinc in the British Isles is taken from the *Special Reports on the Mineral Resources of Great Britain*, published by the Geological Survey. In some cases the Director was kind enough to supply advance proofs of volumes then in preparation and since published.

and consisting entirely of grits, mudstones, shales and slates without any visible igneous rocks.

The latest writer on the subject, Prof. O. T. Jones, has for mining purposes subdivided these rather monotonous formations as follows[1]:

Cwmystwyth formation (dark grits and shales).

Frongoch formation (grey and greenish shales, flags and mudstones).

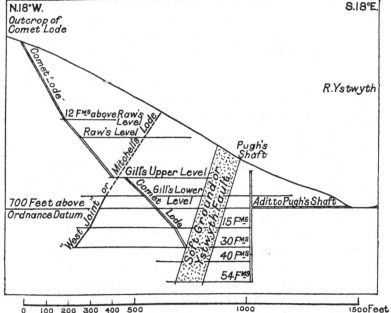

Fig. 50. Transverse section of lodes and Ystwyth fault at Cwmystwyth, Cardiganshire (O. T. Jones, *Spec. Rep. Min. Res.* vol. xx, fig. 4). *By permission of the Controller of H.M. Stationery Office.*

Gwestyn formation (dark pyritous shales).

Van formation (dark mudstones with massive grey grits).

The whole area consists of a closely-packed series of anticlines and synclines, the axes of which range a few degrees east of north and west of south, the axes of the folds usually having a

[1] O. T. Jones, *Spec. Rep. Min. Res. Gt. Brit.* vol. xx, "The Mining District of North Cardiganshire and West Montgomeryshire" (*Mem. Geol. Survey*), 1922, p. 3. Most of what follows is taken from this publication.

decided pitch to north or south. These folds are traversed by several important faults and many minor ones, the faults being the determining factor in the distribution of the ores; in fact most of the lodes are themselves faults. The north and south faults are usually parallel to the axes of the folds: these are usually barren. The ore-bearing faults run approximately east and west, or E.N.E.–W.S.W. Many of them have a large displacement, and they are often to be described as shatter-belts or zones of complex faulting rather than as clean-cut fissures. Some of these shatter belts are as much as 100 feet wide and may include several parallel bands of ore. They often also have a considerable underlie, 25° to 35°, occasionally as much as 45° from the vertical. The fault-filling consists of angular fragments of country rock, cemented by quartz, blende or galena, with sometimes chalcopyrite, pyrite, calcite and dolomite. They are thus typical brecciated lodes. The walls are often strongly slickensided.

As a typical example of the deposits of this district we may take the Van lode in Montgomeryshire, $2\frac{1}{2}$ miles N.N.W. of Llanidloes. The Van mine, situated on this lode, has been one of the most productive in the British Isles and reached its high-water mark in 1876, when it produced 6850 tons of lead ore, worth £103,000, and 2460 tons of blende, worth £9600. The average yield of silver is about 10 oz. per ton of lead. In ten years this mine paid £348,000 in dividends on a paid up capital of £63,750. The principal lode runs about E. 26° N. and dips at about 70° to the south. It is very wide and a large proportion of the ore came from a rich shoot in the plane of the lode, but the most remarkable development occurred in the form of "flats" or expansions of the ore within a group of massive grits, mainly along vertical fractures in the grits, but also extending along horizontal planes and in other irregular directions. These flats follow in a general way the dip and strike of the grit formation, and show some analogy to ore-deposits in fissured limestones[1].

---

[1] Jones, *op. cit.* p. 157. Warington Smyth, "On the Mining District of Cardiganshire and Montgomeryshire," *Mem. Geol. Survey,* vol. ii, pt ii, 1848, p. 655. Le Neve Foster, *Trans. Roy. Geol. Soc. Cornwall,* vol. x, 1879–86, p. 33.

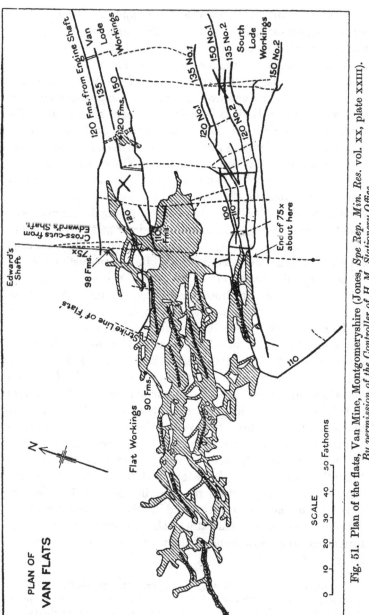

Fig. 51. Plan of the flats, Van Mine, Montgomeryshire (Jones, *Spe. Rep. Min. Res.* vol. xx, plate xxiii). By permission of the Controller of H.M. Stationery Office.

The Van lode has been traced for a total distance of about 9 miles and in its western portion barytes and witherite make their appearance as gangue minerals.

Shropshire. In the Shelve district of Shropshire and Montgomery lead mining dates back to Roman times, as shown by the discovery of pigs of lead stamped with the name of the Emperor Hadrian. The output reached its maximum in 1875, when 7932 tons of lead ore concentrate were produced, chiefly from the Roman Gravels, Tankerville and Snailbeach mines. Of these Snailbeach was by far the most productive.

The metalliferous veins are almost confined to a definite division of the Arenig strata, known as the Mytton Flags: they do not extend down into the Stiperstones quartzite, or pass up into the Hope shales above. The veins chiefly lie in two parallel N.–S. anticlines of Mytton Flags, separated at the surface by a syncline of Hope shales, and may be divided into two chief groups. In those striking N.W.–S.E. the gangue is calcite and galena extends up to the surface, but in the more important E.–W. group there is usually a capping of barytes, with galena and blende below, the last-named mineral increasing in amount in depth. These barytes-bearing veins are now being extensively exploited by Shropshire Mines, Ltd., which company has acquired control of a large area and has already shipped much barytes equal in quality to the best German product. Some veins in the Longmyndian rocks east of the Stiperstones contain a little copper ore, and secondarily-deposited barytes[1]. This is a most interesting area for the student of ore-genesis, since it shows so clearly both zonary arrangement of ores, and the influence of a particular rock type on deposition.

North Wales. Veins carrying lead and zinc, often with copper, have at one time been extensively worked in the older Palaeozoic rocks of Carnarvonshire, as at Bettws-y-Coed, Geirionydd, Llanrwst, Trefriw and Trecastell in the Conway valley[2].

---

[1] Hall, "Lead Ores," *Imp. Inst. Monograph*, 1921, pp. 49–51. *Report on the Shropshire Mining District*, privately printed (by the courtesy of the Managing Director of Shropshire Mines, Ltd). B. Smith, *Spec. Rep. Min. Res. Gt. Brit.* vol. xxiii, Part i. Hall, *Min. Mag.* vol. xxvii, 1922, pp. 201–9.

[2] Smith and Dewey, *Spec. Rep. Min. Res. Gt. Brit.* vol. xiii, part ii.

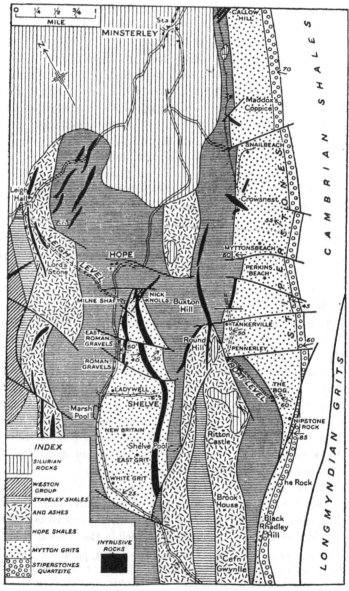

Fig. 52. Geological map of the mining district of West Shropshire (Hall, *Min. Mag.* vol. xxvII, 1922, p. 203, by permission of the Editor).

A complex copper-lead-zinc ore is also found in considerable quantity at Parys Mountain, Anglesey, but was not worked to any extent during the palmy days of that mine, owing to metallurgical difficulties in its treatment.

The Lake District. In the central parts of the Lake District and especially within a radius of about 10 miles around Keswick mineral veins are abundant. In some of these copper is dominant, as is also the case near Coniston, but in others the chief minerals are lead and zinc ores, and some veins are very rich. The greater number of the lead-zinc mines are situated in a belt of country running through Keswick in a general N.E.–S.W. direction, and it is perhaps significant that this extension is parallel to the general strike of the country and coincides roughly with the anticlinal axis that appears to run through Skiddaw. Along this anticline the Skiddaw granite was intruded in Devonian times. The veins in the extreme north-east, in the Caldbeck Fells, lie in the volcanic rocks of the Borrowdale series, but the others are in the Skiddaw slates. Somewhat further south, on the eastern slopes of Helvellyn and about the head of Ullswater another series of metalliferous veins lie in the Borrowdale series.

I. *Caldbeck Fells.* The most important mine here is Roughtengill, which at one time had a large production. There are here two principal veins striking about E. 15° N., cut by a cross course running N. 20° W., or nearly at right angles to the principal veins. Some copper was found here, but the principal production was lead and zinc: these veins were particularly rich in many species of oxidized lead minerals, and some very fine specimens were obtained. Driggeth mine was situated about 1½ miles further east on the same veins. The relations of these veins to the tungsten and molybdenum deposits of Grainsgill are not clear.

II. *The Threlkeld area.* Several productive mines are or were situated below the southern face of Saddleback (Blencathra), near the village of Threlkeld. The old Blencathra mine in the Glenderaterra valley, though long abandoned, is one of the most interesting, since it lies along the line of one of the greatest fractures in the Lake District, namely, that passing from the head of Windermere over Dunmail Raise to St John's Vale,

determining the lowest pass across the main watershed. Moreover, the workings of this mine were in the metamorphic aureole
of the Skiddaw granite, and the veins may very probably run

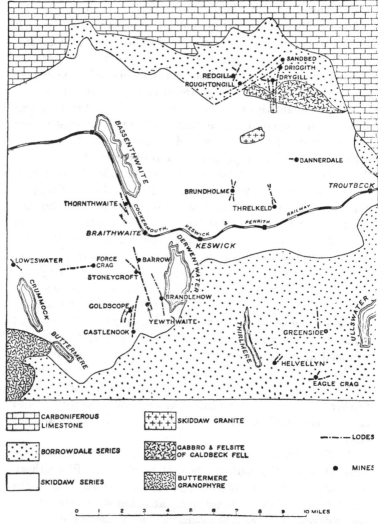

Fig. 53. Geological sketch-map of the Lake District, showing the distribution of the lead and zinc mines (Eastwood, *Spec. Rep. Min. Res.* vol. XXII.
fig. 1). *By permission of the Controller of H.M. Stationery Office.*

into this granite. Some of the old mines of the Threlkeld group have been successfully worked for zinc and lead up to the beginning of the industrial crisis; the chief of these is the Threlkeld mine, formerly known as Woodend. The Woodend and Gategill veins join in its workings, striking more or less north and south and dipping towards the west. There is a rich ore-shoot at the junction, but the Gategill vein is poor in blende.

III. *Mines west of the Derwent.* Many of the mines of this group have been worked since very early times, chiefly for copper. In the reign of Elizabeth there was a boom in copper-mining about Keswick and German miners were brought over. Some of the veins, however, yield principally lead, and a few of them, *e.g.* Thornthwaite, near Braithwaite, have been recently worked. Some of the more important of the mines were: Gold-scope, largely copper but with two good lead veins, 60 fathoms apart, running N.E.–S.W., an unusual direction; Yewthwaite, north-west of Catbells, mainly lead; the same vein is also worked in the Barrow mine on the other side of the Newlands valley; Brandlehow, on the south-west side of Derwentwater, has produced large quantities of blende, galena and cerussite; the vein runs nearly north and south and dips east—its maximum width is about 6 feet. This mine used to produce about 300 tons of lead per annum. The Thornthwaite mines, on the west side of Bassenthwaite Lake, work seven large veins, all striking nearly north and south, except two which are N.E.–S.W. cross courses. The ores are galena, cerussite and blende. The Force Crag mine, west of Braithwaite, differs from the others in producing barytes and manganese in addition to lead and zinc. Stolzite, $PbWO_4$, is said to have been found here.

IV. *The Ullswater district.* Of late years a good deal of lead and zinc has been produced from mines situated on the eastern slopes of Helvellyn, near the boundary between Cumberland and Westmorland; the most important producer is the Greenside lode, which strikes nearly north and dips to the east. The foot-wall is sharp but the upper boundary of the lode is very vague with many stringers of ore and in places a width of 40 feet is workable. The ore is mostly galena and there is some barytes.

Other lodes in this same area and at Wythburn have also been worked in past times.

To sum up: the lodes of the northern part of the Lake District carry varying proportions of lead, zinc and copper; as a general rule the lead-zinc lodes strike north and south or N.N.W.–S.S.E., while the copper lodes run east and west, but there are exceptions. Many of the lodes are brecciated, and the commonest, almost the universal, gangue mineral is quartz, while barytes is sometimes found in quantity. Fluorspar has apparently never been observed in the veins in the Lower Palaeozoic rocks, except at Grainsgill, which is an aberrant occurrence. Most of the lead ore is rich in silver, in both the granular and coarsely crystalline varieties of galena. Oxidized minerals, in great variety, are common in the upper parts of the lodes. Cerussite and pyromorphite are specially noteworthy.

The Leadhills District, South Scotland. The tract of country on the borders of Dumfries and Lanark, commonly called the Leadhills, has long been known as a producer of metals. The country rock consists of Ordovician greywackes with occasional bands of slate, the whole forming part of the usual structure of the Southern Uplands. Veins have been worked over an area of about 2 miles from north to south, and 3 miles from east to west, on both sides of the Clyde-Nith watershed, around the villages of Leadhills and Wanlockhead, and a large number of veins have been proved, about 70 in number, mostly striking a few degrees west of north and dipping steeply to the north-east. Some of these veins are very large and very rich. About a hundred years ago a mass of solid ore 14 feet wide is said to have been worked out. The veins are less productive, or disappear altogether in the slates, but are often found again on the other side of a slate band. The usual gangue is quartz or calcite. Gold is found in the streams of the neighbourhood, but is not known anywhere *in situ*. These mines have been worked from a very early date, perhaps by the Romans[1].

---

[1] Mitchell, *Min. Mag.* vol. xxi, 1919, pp. 11–20. Wilson, "The Lead, Zinc, Copper and Nickel Ores of Scotland," *Mem. Geol. Survey*, 1921, pp. 9–43.

Lead and Zinc Ores in the Carboniferous Rocks of North Wales. An important mining district is situated in the counties of Flint and Denbigh, where the Carboniferous limestone occurs on both sides of the Vale of Clwyd, its outcrop being repeated by faulting. Fig. 54 gives a generalized map of this region. There is also a smaller occurrence on both sides of the Bwlchgwyn gap, extending from Hope Mountain nearly to Llangollen. On the western side of the Vale of Clwyd the ores are unimportant, but in the eastern range are the well-known mining areas of Holywell and Halkyn, while the famous Minera mine is just south of Bwlchgwyn, which is the line of the great Bala fault.

In the area north of the Bala fault the lodes run in two well-defined sets, one mainly east and west, called *veins*, the other running north and south, called *cross-courses*. Both sets are usually fault-fissures: the first and earlier east and west veins shift the strata considerably; the cross-courses again shift the veins to a small extent. In the Minera district the chief veins trend N.W.–S.E. The chief veins of both areas contain blende and galena, the latter carrying silver up to about 15 oz. per ton. The cross-courses have no blende and are poor in silver, but often show a little copper. In the upper levels the ores are often oxidized to cerussite and calamine. The commonest gangue minerals are quartz or amorphous silica and calcite: fluorspar and barytes are rather rare. The blende of the Halkyn district has a peculiar and characteristic red colour. Some of the veins are very large and productive, but water presents a great difficulty and schemes are on hand for extensive drainage tunnels, to unwater the veins down to sea-level, or nearly so: below this depth working will probably always be impossible[1].

The Pennine Region. From the earliest times the most important lead-producing area in the British Isles has always been the Pennine anticline, from Derbyshire to Northumberland. This may for convenience be divided into three parts: Derbyshire; West Yorkshire; and the northern area about the head-

[1] Bernard Smith, *Spec. Rep. Min. Res. Gt. Brit.* vol. XIX, "Lead and Zinc Ores in the Carboniferous Rocks of North Wales," 1921.

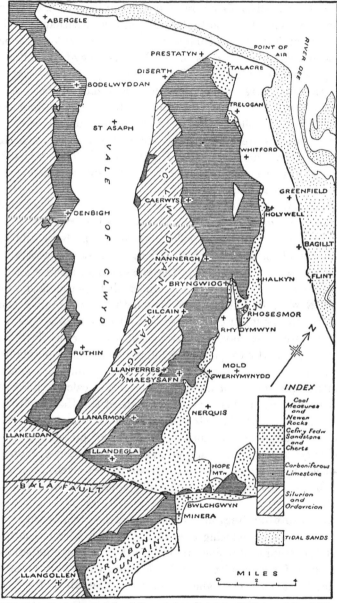

Fig. 54. Geological map of the mining district of Flint and Denbigh
(B. Smith, *Spec. Rep. Min. Res.* vol. XIX, fig. 1). *By permission of
the Controller of H.M. Stationery Office.*

waters of the Tyne, Wear and Tees, in Durham, Cumberland and Northumberland. In recent times a good deal of zinc has also been produced in certain places, as well as much fluorspar and barytes. The present production is only small, but the geological relations shown are of some interest, forming a good example of a widespread type of the occurrence of such deposits, namely in limestones. In all these areas by far the commonest form of ore-body is the fissure-vein, which is usually also a fault. It so happens, however, that the principal active producers are in nearly all cases in some way abnormal for the district. This point must be strongly emphasized.

Derbyshire. Lead-bearing veins are exceedingly numerous, both in the main anticline and in the small inliers of Ashover and Crich: they all carry mainly galena, with calcite, barytes and fluorspar in varying proportions. The richest deposits lie in the highest part of the limestone, up against the base of the Yoredale shales, on the north and east sides of the main outcrop, and are known to extend a long way under the shales. Some of the principal rakes (fissure veins) can be traced for several miles. Other common forms of ore-body are pipes, pockets, flats and caverns. Near the outcrops the galena was largely oxidized to cerussite, but most of this is now worked out. Zinc ores are uncommon, the only important occurrence being at Millclose mine, to be described presently. Good examples of rake-veins are to be seen at Castleton, Eyam and Tideswell Moor in the north, in the Winster and Brassington district, and in the Ashover and Crich inliers, both of the last-named areas being remarkable for the abundance of fluorspar[1]. All over the district there are often interbedded with the limestones beds of contemporaneous lava (toadstones), which often form the lower limit of the workable portion of the lodes. At Ashover the so-called toadstone is a tuff of unknown thickness. It is unnecessary to describe the normal types of deposit in any further detail, as most of the mines are now closed down and many of them worked out.

[1] Wedd and Drabble, "The Fluorspar Deposits of Derbyshire," *Trans. Inst. Min. Eng.* vol. xxxv, 1908, p. 501. This paper includes a useful map of the Derbyshire veins.

The only important active producer in Derbyshire is the Millclose mine, which works quite an exceptional type of deposit in the form of infillings of solution cavities on either side, but chiefly on the west, of a large vein striking north and south. These lie in limestone, with shale above and toadstone below, and may take the shape of flats and fissures along the joints of the limestone or actual caverns due to solution and only partly filled by ore and gangue minerals, forming a breccia on the floor and an incrustation on the roof. Zinc ores are here abundant and the order of deposition of the minerals is: blende, fluorspar, galena, barytes, calcite[1]. Some of these solution cavities extend as much as 250 yards from the main vein, with which they are connected as a rule only by thin strings and leaders. The limestone is here rather cherty, but there is no metasomatic alteration, as in the *flats* of Weardale and Alston Moor (see p. 300).

Yorkshire. In the north of Derbyshire the limestone disappears under the Yoredale shales and Millstone Grit, but emerges again in Wharfedale, where there are some productive veins: curiously enough these appear to have been most productive where they run up into the overlying Millstone Grit, being rather poor in the limestone. An interesting mining area, now actively producing, lies in an anticline of limestone at Greenhow, near Pateley Bridge. Very fortunately the Bradford Corporation has lately driven an aqueduct tunnel through this ancient mining field, some 400 to 500 feet underground, showing that some of the veins are still productive at that depth. The veins of this area mostly run east and west, and there are some flats and typical pipes, containing galena with calcite, fluorspar and barytes. The rather curious relations shown by some small cross-veins prove that the mineralization is later than all the rather complicated faulting and fissuring of this district. At one time there was an important lead-mining industry in Swaledale, especially on the northern slopes of the valley between Keld and Marske, where rather shallow veins strike E.–W. or S.E.–N.W., carrying galena with calcite, quartz and barytes, but mining practically ceased some time ago. It is a remarkable fact that the great limestones of the Ingleborough district are

[1] Parsons, *Trans. Fed. Inst. Min. Eng.* vol. XII, 1896–7, p. 115.

practically barren of ore, although all the conditions seem most favourable.

The Northern Area. Deposits of lead and zinc ores are remarkably abundant in that wild and high-lying tract of country on the borders of Northumberland, Cumberland, Durham and

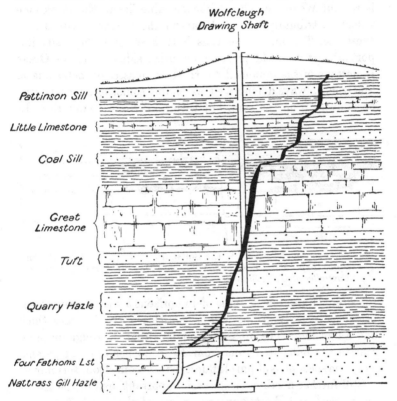

Fig. 55. Section of Wolfcleugh Mine, Weardale (S. Smith, *Spec. Rep. Min. Res.* vol. XXVI, fig. 1). *By permission of the Controller of H.M. Stationery Office.*

Yorkshire, around the head-waters of the Tyne, Wear and Tees, to the east of the summit of the Cross Fell escarpment. The best-known areas are upper Weardale (Boltsburn), the Allendales and Alston Moor. Here again the dominant type of deposit is the fissure-vein, but the chief producers are flats, which differ in important respects from the flats of Derbyshire. In Weardale the Yoredale series comprises some 700 feet of sandstones, shales and limestones, with a dolerite intrusion, the Whin Sill, intercalated at varying horizons. Most of the important fissure-veins are chiefly productive in the Great Limestone, and contain galena and fluorspar, with chalybite in the upper levels. Blende is rare in Weardale. The Boltsburn mine lies in the Rookhope valley, a tributary of the Wear on the northern side above Westgate. The main vein runs N.E.–S.W., but the really important feature is the presence on either side of it in the Great Limestone of the famous flats: these are masses of metasomatic limestone, always dolomitized, extending sometimes for 50 yards or so away from the vein, but with no definite boundaries. In

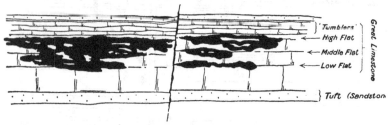

Fig. 56. Section of "Flats" in the Boltsburn Mine, Weardale (S. Smith, *Spec. Rep. Min. Res.* vol. XXVI, fig. 2). *By permission of the Controller of H.M. Stationery Office.*

the altered limestone at three main horizons are scattered masses of galena, usually in bunches or lenticles up to 3 inches thick and a foot long. The ore is so abundant that the rough material sent to the mill often runs from 15 to 25 per cent. of galena. Very fine specimens of purple fluorspar also come from this mine[1].

Still further north comes a very important area in the Allendales, East and West, and the valleys of the South Tyne and Derwent. Although all of these have been large producers it is impossible to describe them in detail. A brief account may, however, be given of some unusual deposits found in the area conveniently known as Alston Moor, which is mainly in Cumberland; here the Vieille Montagne Company, a Belgian concern, has of late years produced great quantities of zinc, for the most

[1] For a very good description see H. Louis, "Lead Mines in Weardale," *Min. Mag.* vol. XVI, 1917, p. 24.

part from flats in the Nenthead area. These flats lie in the Great Limestone, usually at three different horizons and are very irregular in extent, sometimes extending for hundreds of feet from the parent vein. They consist of lenticles of ore and spar alternating with dolomitized limestone. Between 1872 and 1918 the production of zinc ore from Nenthead and Nentsberry was about 175,000 tons. However, this type of deposit, like the similar ones at Boltsburn, is quite exceptional and the main production of lead all over this northern area has come in past times from veins of the usual type, which are extremely abundant. It is noticeable, however, that they are as a rule productive only on high ground: the modern valley system has obviously cut right through the mineralized zone, which is only some 600 or 700 feet thick, and there is no sign of passage into a zone of any other ore in depth[1].

Origin of the Lead–Zinc Ores of the British Isles. As before stated, the genesis of some types of the British lead-zinc deposits presents a problem of great difficulty, to which no satisfactory solution can yet be given. In the case of the Cornish lodes, there is no reasonable doubt that they are to be referred to the same cause as the tin, wolfram and copper lodes, that is, to the intrusion of the post-Carboniferous granites. But with the Welsh, Pennine, Lake District and Scotch deposits the matter is much more complicated. These lodes are not in obvious genetic connection with a definite suite of intrusions, or with any other set of significant phenomena: furthermore, it is a matter of extreme difficulty to assign any precise geological age to the mineralization. Those in the Carboniferous limestone are at any rate post-Carboniferous in date: they may in great part be referable to the Armorican-Pennine crust-disturbances but some at any rate appear to extend up into the Triassic rocks, as in the Mendips and Cheshire. Again, the age of the veins in the Lower Palaeozoic rocks of central Wales, the Lake District and southern Scotland is quite uncertain. They may represent a deeper zone of the Armorican mineralization, or they may belong to an earlier Caledonian phase of igneous activity.

[1] Smith and Carruthers, *Spec. Rep. Min. Res., Geol. Survey*, vol. XXV, 1923. Carruthers and Strahan, *id.* vol. XXVI, 1923.

No instance has apparently yet been discovered in which a mineral vein in the Lower Palaeozoic rocks is definitely traceable upwards into the Carboniferous or later series.

Leaving Cornwall out of account, it is possible to establish two great groups of lead-zinc veins in Great Britain, each possessing characteristic features as regards their manner of occurrence and mineralogical composition. The first group includes the deposits in the Lower Palaeozoic rocks of central Wales, the Lake District and southern Scotland. In these copper often occurs in considerable amount, while the commonest gangue mineral is quartz, with subordinate calcite, dolomite and barytes, while fluorspar is notably rare, or altogether absent. In the Carboniferous group, on the other hand, copper is rare or quite absent, while the principal non-metallic minerals are calcite, dolomite, barytes and abundant fluorspar. In some veins siderite is common, thus resembling the highest lodes of Cornwall.

Another notable difference is that the lodes in the Lower Palaeozoic rocks commonly belong to the brecciated type, being often merely shatter-belts cemented by ore and gangue minerals, with flucan, clay belts and the usual concomitants of this type, while the Carboniferous veins are usually obvious cavity-fillings and metasomatic replacements. This difference is perhaps mainly attributable to the original character of the rocks themselves, rather than to the mineralization-process. In both cases many, if not most of them, are fault planes, and the two classes of rocks, Palaeozoic grits and slates on the one hand, and Carboniferous limestone on the other, behave very differently under pressure, while in the latter solution-cavities and open joints are of the greatest importance.

Turning to mineralogical characteristics, it is generally agreed that vein-quartz accompanies deep-seated lodes, e.g. in Cornwall[1], while dolomite and calcite belong to higher levels. The prevalence of fluorspar in the Carboniferous veins and its general absence from the others at first sight presents a difficulty, since this mineral is usually regarded as belonging to the deeper zones. However, Wedd and Drabble[2] have clearly shown that

---

[1] Cronshaw, *Bull. Inst. Min. Met.* September, 1921.

[2] See footnote, p. 297.

in Derbyshire this mineral occurs in quantity only in the upper-most zones of the Carboniferous limestone, which is there a homogeneous formation, highly pervious to solutions and vapours, and overlain by the much less pervious Yoredale shales and sandstones. This feature is accounted for by the high degree of volatility possessed by fluorine compounds, which tend to rise through the much-fissured limestone till stopped by an im-pervious roof. The abundance of fluorspar in itself suggests a deep-seated magmatic origin for the vein-stuff, since we know that fluorine is a characteristic constituent of late-magmatic vapours or solutions, as shown by the common phenomena of pneumatolysis. Against this idea must be set the statement so frequently made that the veins cease to be productive when followed down to the volcanic rocks, toadstones or dolerites, that occur at various horizons in Derbyshire, or to the Whin Sill in the northern area. It is uncertain to what extent this objection is really valid, since it is doubtful whether the rocks below the volcanic horizons have ever been properly prospected. It is obviously difficult and costly to follow a vein down through a great thickness of such unproductive rock, and the objection cannot be regarded as demonstrated. Since, however, it is commonly believed that the veins are confined to the upper part of the limestone and die out downwards, it has been argued that the vein fillings must have come from above, or at least sideways; hence the theory of lateral secretion was evolved to meet similar cases. The greatest objection to both theories, percolation from above, or extraction from the surrounding rocks, lies in the explanation of the source of the metals and of the fluorine. Till this explanation is forthcoming it seems most satisfactory to conclude that the source of the vein material was below.

However, it has to be admitted that there are no visible intrusives from which it could be derived. Both the igneous rocks of Derbyshire and the Whin Sill appear inadequate from this point of view, and fluorine is characteristic of acid and not of basic rocks. It must be remembered, however, that the Pennine Chain is an anticlinal structure, which must be somehow supported from below, and this support may be afforded by the intrusion of a granitic bathylith along the axis of the anticlinorium.

The veins in the Lower Palaeozoic rocks also occur in areas which are known to have been highly disturbed at many periods in the earth's history and as such are specially liable to intrusion of igneous material. One such period occurred during the Devonian (Caledonian) movements and the mineralization may have taken place then, although there is no direct evidence of this. On the other hand it is possible to regard these veins as the deeper portions of veins that originally ran up into overlying Carboniferous or later rocks, the upper part having been removed by denudation. In this case they must of necessity be assigned to a later date. Unfortunately we possess no definite evidence on this point[1]. Although these areas were probably land during the early part of the Carboniferous, there can be little doubt that they were covered by a thick blanket of Upper Carboniferous sediments which were later removed by Permo Carboniferous denudation following the Armorican-Pennine uplift. During this period of disturbance the conditions must have been favourable for igneous activity and mineralization and it is suggested that the veins in question were formed at this time, those now seen in the Lower Palaeozoic rocks being the deeper portions of veins originally extending upwards to the Carboniferous, and those now visible in the limestone being at or near the upper limit of penetration of the ore-bearing solutions, and therefore characterized by the presence of the more volatile constituents, especially fluorine. In accordance with the usual zonary arrangement, the copper is found in the roots of the veins in the lower rocks with quartz, while the lead and zinc rise to the higher levels, along with the calcite and fluorspar.

Although the mineralization thus probably began before the Permian period, it may have been, and probably was, a long-continued process, since copper ores and galena are found in many places in Triassic rocks, in Cheshire, Staffordshire and Shropshire, and the lead veins of the Mendips are said to extend into the Dolomitic Conglomerate. There is always, however, the possibility that these deposits may be due to later solution and re-deposition of the earlier ores by meteoric waters.

[1] On the age of this mineralization see especially two valuable papers by Finlayson, *Quart. Journ. Geol. Soc.* vol. LXVI, 1910, pp. 281–327.

The Zinc Ores of Franklin Furnace, New Jersey. This well-known and important occurrence of zinc ores constitutes an altogether exceptional or perhaps even unique type, having regard to the peculiar character of the mineral assemblage there found. It lies in a region of highly metamorphic, mainly gneissose, rocks, and the ore-body itself is essentially a sheet of highly crystalline coarse-grained marble folded into a pitching syncline with limbs of very different sizes, so that a cross-section has rather the form of a fish-hook. The very peculiar zinc and manganese-bearing minerals are embedded in the marble. The ore-body outcrops at the surface for about half a mile and has been followed some 2000 feet in depth. It lies close to but not actually at the contact of limestone and intrusive granitic gneiss. The principal zinc minerals are franklinite, a zinc-manganese spinel, zincite, ZnO, and willemite, $ZnSiO_4$; the gangue minerals are calcite and sometimes pink manganese carbonate, rhodochrosite; rhodonite, $MnSiO_3$, is also present.

The genesis of this peculiar mineral assemblage has given rise to much discussion: it has been suggested that the metals were introduced by the intrusion of the granite, but the most probable explanation is that the ore-body was originally an impure siliceous limestone containing oxidized replacement-ores of zinc and manganese, probably carbonates: this was metamorphosed by the intrusion of the gneissic granite and a series of reactions occurred analogous to those characteristic of dedolomitization, thus:

$$ZnCO_3 \rightarrow ZnO + CO_2$$
$$\underset{\text{calamine}}{} \quad \underset{\text{zincite}}{}$$

$$ZnCO_3 + Mn_2O_3 \rightarrow ZnO.Mn_2O_3 + CO_2$$
$$\underset{\text{franklinite}}{}$$

$$2ZnCO_3 + SiO_2 \rightarrow Zn_2SiO_4 + 2CO_2$$
$$\underset{\text{willemite}}{}$$

These reactions are precisely similar to those forming periclase, spinel and forsterite from magnesian limestones[1], and this is very suggestive of a like origin.

The Zinc Ores of Missouri and Neighbouring States. One of the most important zinc-producing areas of the world,

[1] Hatch and Rastall, *Petrology of the Sedimentary Rocks*, 2nd edition, London, 1922, p. 262.

which also yields a good deal of lead, is situated in Missouri and in the neighbouring states of Arkansas and Oklahoma. The dominant geological feature of this region is the Ozark uplift, a low dome of very wide extent, with a comparatively steep dip only on the southern side. This consists of a core of crystalline rocks, overlain successively by Cambrian, Ordovician, Silurian, Devonian and Lower Carboniferous strata; a very large part of this succession consists of limestones and dolomites. After the deposition of the Lower Carboniferous (Mississippian) limestone, there was a local uplift, accompanied by extensive denudation of its surface, leading to the formation of a karst topography, with swallow holes, forming a very uneven surface. This was then overlain by the Upper Carboniferous (Pennsylvanian) shales. The major part of the ore-deposits are closely related to this erosion surface: they consist of "runs," and "sheet ground," which are very like the rakes and flats of Derbyshire: these do not need detailed description. The point of interest is the view that now prevails among American mining geologists as to the origin of these ores[1].

In the first place there are no signs of igneous activity in this area: secondly, the ores occur in four principal areas near the outcrop of the impervious Pennsylvanian shales which surround the dome as a ring: thirdly, the older formations and the deep-seated waters derived from them contain appreciable amounts of lead and zinc, especially the Cambrian and Ordovician lime-stones. It is suggested that when the impervious Pennsylvanian shales were removed by denudation from the central part of the dome an artesian circulation was set up, with an intake in the central area and flow down the dip with an ascending discharge at the inner margin of the shale, leading to deposit of ore in the immediately underlying limestone, which at the same time was largely dolomitized and silicified. Ore-deposit took place in this way in several scattered areas, of which the Joplin district on the south-west is the most important, the location of these areas being mainly determined by the removal of the impervious underlying Chattanooga shale, of Devonian age; where this is unbroken there is no ore, except along fault-lines. It is believed

[1] Siebenthal, Bulletin 606, *U.S. Geol. Survey*, 1916.

that in the ore-bearing areas there has been considerable residual secondary enrichment by carbonates and other oxidic ores.

This explanation, though sufficing to explain the present concentration of the ores in the Carboniferous rocks, does not in reality touch the question of the primary genesis of the metals. It merely indicates a possible method for the redistribution and concentration of ores already in existence. The early phases of the story are doubtless connected with the observed occurrence of ores with a fairly definite arrangement in the crystalline nucleus of the dome, where deposits of nickel, cobalt, copper, iron occur as well as lead and zinc, apparently with a normal zonary arrangement. The full sequence of events was probably as follows: normal mineralization of the pre-Cambrian rocks from deep-seated sources, with a zonary arrangement: denudation and distribution of such ores into the Lower Palaeozoic sediments: local concentration as above outlined by artesian circulation into Carboniferous limestone: local enrichment at and near the surface by meteoric waters. Hence it appears that this occurrence has little or no analogy with the lead-zinc mineralization of the British Carboniferous limestone, since it is an instance of secondary and not of primary deposition.

Broken Hill, New South Wales. This well-known occurrence takes a very high rank among the lead-silver-zinc mines of the world, its total production of all minerals to date having amounted to over £111,000,000 in value. It is also of very great geological interest and its genesis presents some special features. The Broken Hill district is situated in the Barrier Ranges, a very arid and remote area in New South Wales, near the boundary of South Australia. Its natural outlet is in point of fact through Adelaide, with which city it is connected by rail: transport through New South Wales is very difficult, owing to the nature of the country. The literature of the subject is now extensive and an authoritative and very complete memoir has just appeared[1]. From this most of the following is taken.

The mineralized area consists of an inlier of ancient rocks,

---

[1] Andrews, "The Geology of the Broken Hill District," Memoir 8, *Geological Survey of New South Wales*, 432 pp., 1922.

probably of Archaean age, called by Mawson the Willyama series[1]. These consist of a variety of sediments, mainly originally mudstones and shales, with subordinate fine sandstones, invaded by sills of igneous rock, both acid and basic, with intense thermal metamorphism, converting the sediments to schists and gneisses, sillimanite-gneiss being a specially noteworthy type. This was followed by further crushing and metamorphism, probably by gaseous emanations from igneous magmas, giving rise to such minerals as magnetite, apatite and garnet. After this came the special lead-zinc mineralization along channels which are essentially zones of folding, shearing and crushing, as described later. There are also in the neighbourhood lodes containing tin and wolfram, as well as platinum, gold, copper, cobalt and nickel. These will not here be described further.

Lying unconformably on the Willyama rocks is another series of sediments, much less altered, called the Toorowangee series. This comprises quartzites, limestones, and what is believed to be a glacial conglomerate (boulder clay). This series has been assigned by different authors to a late pre-Cambrian or Cambrian age. Above this again is the Mootwingee series (perhaps Devonian) and outliers of Cretaceous rocks.

The chief types of ore-body found in the neighbourhood of Broken Hill may be conveniently summarized as follows[2]:

Silver-lead and zinc: the great Broken Hill lode (see description later).

Silver-lead: galena and quartz, with or without fluorspar. Mount Robe group.

Silver-lead without fluorspar: galena, siderite, quartz. Great Western Basin.

Tin-wolfram: cassiterite and wolfram with tourmaline and greisen (pegmatitic). Northern Basin.

Platinum: platinoid minerals with cobalt, nickel, copper, haematite and quartz, in sills and tongues of serpentine.

There are also unproductive lodes with magnetite, apatite, hornblende, haematite and quartz.

[1] Mawson, "Geological Investigations in the Broken Hill area," *Mem. Roy. Soc. S. Australia*, 1912.

[2] Andrews, *loc. cit.* p. 135.

The Broken Hill lode proper outcrops at the surface over a length of rather more than 3 miles: in plan it forms two irregular arcs, striking nearly north and south, with their convexities to the north-west. In longitudinal section also the top of the workable lode forms more or less of an arch, sinking towards both ends. Its form in cross section is complicated, but in many places it appears to form a rather unsymmetrical saddle. There has been much discussion as to its true form. It is at any rate clear that the rich ore-body occupies parts of what is in reality, when reduced to its simplest terms, a great zone of shearing, complicated to a high degree by local folding, faulting and thrusting in a vertical plane. The width of the visible outcrop varies from 10 to 100 feet. The whole shear-zone dips mainly to the west at a high angle, but is occasionally vertical, or dips steeply eastward. Owing to the prevailing westward dip it has become customary to speak of the hanging-wall and foot-wall portions of the lode. On the foot-wall (eastern) side the country rock is mainly gneiss, granulite, aplite, amphibolite and vein-quartz with belts of altered sediment. The hanging-wall consists mainly of altered sediments, with sills of augen-gneiss, amphibolite, granulite and pegmatite. In the ore-zone itself all these rocks are much mixed up and contorted, as well as having undergone replacement by new minerals.

The largest and most continuous ore-body lies near the foot-wall side, which is a zone of very strong rock-flow and dislocation: bedding is often still traceable in the lode-stuff, except when this consists of solid masses of sulphides. The hanging-wall ore-bodies, the supposed shorter limb of the saddle, appear to be in reality largely bulges and projections from the foot-wall lode, of irregular form, but often clearly related to corrugations in the folding. In some sections two or more of these bulges can be seen, one above the other. The generalized form, however, as before stated, is essentially that of an asymmetric saddle, the eastern side being by far the deepest, almost everywhere.

In general terms the lode material may be said to consist of galena, blende, a peculiar pegmatite with green felspar, garnet, rhodonite, zinc-spinel, fluorspar, calcite, quartz and biotite. Other sulphides present include chalcopyrite, pyrite, pyrrhotite and various silver compounds. Curiously enough the primary silver sulphides, as distinguished from the silver-lead, appear to be in close association with the chalcopyrite. If the lode be considered in longitudinal section, then the central part may in the main be regarded as a mass of lead and zinc sulphides in a gangue of rhodonite, garnet, green felspar, pegmatite, apatite, fluorspar, quartz and silicified country rock, whereas at the north and south ends rhodonite and garnet are quite subordinate: that is, there is much less manganese present.

In the upper levels there occurs, as might be expected, an oxidized zone: the chief minerals of this are kaolin, manganese dioxide, iron oxides, secondary silica, with cerussite and other oxidized lead minerals and silver chloride, bromide and iodide. The noticeable feature here is the almost complete leaching out of the zinc, owing to the high solubility of its sulphate.

The Bawdwin Mines, Burma. These mines were worked by the Chinese from a very early date, probably about A.D. 1400, but development by modern methods only began about 1909. The mines are situated in Tawngpeng, one of the smaller Northern Shan States, about 600 miles north of Rangoon. The ores occur in a kind of dome of rhyolite and rhyolite-tuff of pre-Cambrian age, that pushes up through the Pangyun (Cambrian and Ordovician) series: on the west is an overthrust belt of Nam-Hsim sandstone (Upper Silurian) and beyond this comes the pre-Cambrian Chaung-Magyi series with the intrusive Tawngpeng granites.

The rhyolite series forms a narrow strip running at the surface for about 3 miles from north-west to south-east: it consists of rhyolite lavas and tuffs, with sandstones and felspathic grits, the whole overlain by the basal conglomerate of the Pangyun series. The rhyolites are pink, brown or chocolate, or of a speckled pinkish grey colour, the tuffs being chiefly white or grey.

The ore-bodies, except the aberrant Goldhole deposit, are confined to a well-marked ore-channel, which is a line of faults and crushed shear zones and plunges south at a very low angle, dipping under the Pangyun series; it is traceable along its strike for 8000 feet and is from 400 to 500 feet wide: on the north it is cut off abruptly by the Goldhole fault. In this channel are three principal lode-systems: (1) the West Burman and Maingtha lodes with the Chinaman ore-body; (2) the central Shan Palaung lode, parallel to the Burman and 150 feet east of it; (3) the East lode.

The Chinaman ore-body is by far the most important. It is an enormous replacement of zinc-lead-silver ore on the hanging-wall side of the ore-channel; the axis of the thickest part strikes N. 25° W. and dips 70–80° W. The hanging-wall is regular, the foot-wall ill-defined, showing a gradual passage into the country rock. The ore-body is cut by shear-zones. Near the top is much anglesite and cerussite, but little calamine. Zinc appears to increase in amount in depth.

The Burman lode is rather thin, but persistent: it lies largely in much faulted country and consists chiefly of a series of small parallel stringers. At the 300 foot level it averages 31 inches of ore with 33 per cent. lead, 16 per cent. zinc and 37 oz. silver per ton. The Shan lode, which is 2 to 3 feet wide, carries about the same percentage of lead and rather more zinc. The Palaung lode is of lower grade but carries about 2·7 per cent. of copper. The Dormouse lode is wide, up to 5 feet at a depth of 327 feet: it is poor in lead but has 37 per cent. of zinc. In all cases silver varies approximately with the lead content, as might be expected.

The chief primary ores are galena and blende, usually intimately mixed and of very fine grain, but there is some coarse galena. Pyrite is common in scattered grains and there is some chalco-pyrite, chalcocite and covellite. In the oxidation zone anglesite is common, with cerussite, pyromorphite and traces of cobalt bloom. Smears of blue and green copper ores are common on shear planes. The order of deposition of the metals appears to be zinc, copper, iron, lead, the lead extending much further into the country rock where the boundary is not sharply defined.

The Goldhole ore-body does not lie in the Bawdwin channel but is an independent thin flat body striking N.N.W.–S.S.E. and dipping west at a low angle. It consists mainly of pyrite with some chalcopyrite and chalcocite.

With regard to the primary origin of the ores, they are believed to be connected with the intrusion of the Tawngpeng granite, by penetration of magmatic material along shattered fault zones, replacing congenial rocks. The granite is not seen in contact with the rhyolitic rocks, but is visible at the surface not far away and probably underlies the whole area at no great depth[1].

[1] The above account is mainly derived from Coggin Brown, *Records Geol. Survey India*, vol. XLVIII, 1917, p. 121. See also Loveman, *Trans. Amer. Inst. Min. Eng.* Bull. 120, 1916, p. 2119.

CHAPTER XV

IRON

In a modern civilized community iron is by far the most important metal from the technical and industrial point of view and the aggregate value of the world's annual output of iron and steel in their many forms, both manufactured and unmanufactured, far exceeds even that of the gold produced in the same period, while their commercial importance is also infinitely greater so far as concerns the employment of capital and labour. One of the most difficult problems that had to be dealt with during the late war was the maintenance of sufficient supplies of iron and steel for munitions and for the innumerable other purposes connected with naval and military operations, in addition to their normal applications in time of peace. This urgent necessity led to the exploitation and development of many hitherto unworked sources, to the resuscitation of extinct industries and to great development of activity in many existing undertakings; all this in spite of acute labour difficulties and enormously increased working costs for supplies and transport. One of the natural consequences of this state of affairs was an increase in the employment of metallurgical processes by which low-grade ores of home production could be substituted for higher-grade imported materials in order to economize tonnage, and a consequent increase in the production of such ores at home by means of labour-saving devices of various kinds.

As a consequence of this abnormal state of affairs much attention has lately been paid to the iron ore resources of the world and of individual countries, and a voluminous literature has lately appeared on this subject, involving much repetition and overlap.

Sources of Iron. Although native iron is known to exist in meteorites, in small quantities in basaltic rocks, and in certain masses in Greenland weighing a few tons, these sources are of negligible importance, and commercial iron is obtained exclusively from ores. These include a considerable number of varieties, some of which are definite minerals, while others are ferruginous rocks and other deposits containing a greater or less proportion of iron in various forms of chemical combination. The principal iron ores are as follows:

(A) Iron-bearing minerals[1]

		Iron content when pure per cent.
magnetite, Fe_3O_4 ...	...	72·4
haematite, Fe_2O_3 ...	...	70·0
göthite, $Fe_2O_3.H_2O$	...	62·9
limonite, $Fe_2O_3.xH_2O$	...	variable
chalybite (siderite) spathic iron ore } $FeCO_3$		48·3
pyrite, FeS_2	... ...	46·6

The so-called titaniferous iron ore is sometimes ilmenite, $FeTiO_3$, and sometimes a mechanical mixture of magnetite and ilmenite.

Besides these iron also exists in combination in a very large number of other minerals, but most of them are too refractory or contain too small a percentage of the metal to be profitably worked. Many, however, by their natural decomposition give rise to various ironstones and secondary deposits of the above ores.

(B) Ironstones.

Under this somewhat vague designation are included a number of types of rock, usually of sedimentary origin, which contain sufficient iron to be worked profitably as a commercial source of the metal. They generally contain some of the above-mentioned iron-bearing minerals associated with a varying amount of impurities of all kinds, according to their mode of origin and subsequent alterations. Some of the chief varieties are oolitic ironstone, clay ironstone and black band ironstone among ordinary sediments, as well as remarkable varieties of iron ore associated with the crystalline schists. The characters and origin of the principal types will be considered later in detail.

The workable deposits of iron ore show a wide range of variety both in their character and mode of origin, since they

[1] It is to be regretted that in many technical writings mineralogical names are often used in a very loose and incorrect manner: thus, for example, the "brown haematite" of many authors is as a rule limonite, sometimes göthite.

can be formed by a considerable number of different geological processes. From the genetic point of view they can be conveniently classified according to the following scheme:

(1) Magmatic segregations.
(2) Lodes and veins.
(3) Contact deposits and metasomatic replacements.
(4) Original stratified deposits.

Of these classes (1) and (4) are of primary origin, since they are contemporaneous with the rocks in which they occur: the contact deposits and metasomatic replacements are essentially of metamorphic origin, while lodes and veins are subsequent in age to the surrounding rocks. Hence it is seen that deposits of iron ore may belong to any one of the principal classes into which rocks are divided by the petrologist. In practice it is often very difficult to decide whether certain deposits are more correctly to be referred to the contact deposits or to metasomatic replacements and the two classes grade into one another. In form also as well as in origin a wide range of variation is to be seen, the lodes and veins and the stratified deposits being most regular in this respect, while the contact deposits and replacements may possess almost any shape whatever. The origin of some of the iron-bearing crystalline schists is still a matter of considerable doubt, and at present it is hardly possible to assign them definitely to any one of the above classes: they probably include representatives of both igneous and sedimentary rocks, as will appear in the detailed descriptions given in later sections.

Magmatic Segregations. The iron ore deposits classed under this general heading may be conveniently divided into two groups, namely those associated with acid and basic rocks respectively. The ores of the first group consist mainly of magnetite, often with more or less apatite, while the second group are specially characterized by the presence of compounds of titanium, especially ilmenite and titaniferous magnetite: the latter group is consequently of considerably less commercial value than the former, which comprises some of the most important high grade iron ore deposits of the world.

Magmatic segregations are formed during the differentiation

and cooling of igneous magmas, either before or after intrusion into their present position: the general principles underlying the processes of differentiation are discussed in Chapter II and need not be repeated here. They include two different types, according to whether the differentiation has taken place before intrusion, in a deep-seated magma basin, or after the molten material reached its present position: masses of iron ore belonging to both types are known.

The Magnetite Ores of Northern Sweden. In the province of Norrbotten, the most northerly part of Sweden, are found the largest known masses of magnetite, which are both of high scientific interest and of great and increasing commercial importance. The chief deposits are at Kiirunavaara, Luossavaara and Gellivaro, all of which are worked on a large scale and there are also many smaller occurrences.

The rocks of the Kiruna[1] district in the immediate neighbourhood of the ore-bodies consist mainly of highly-inclined masses of greenstone, syenite-porphyry and quartz-porphyry, all very rich in soda, striking nearly north and south and dipping at 50° or 60° to the east. Eastwards these are followed, probably unconformably, by the schists and sediments of the Hauki complex, the whole being of pre-Cambrian age, perhaps Algonkian. The ore-body of Kiirunavaara and Luossavaara follows the strike and dip of the porphyries: the foot-wall is syenite-porphyry (keratophyre) and the hanging-wall quartz-porphyry (quartz-keratophyre). There are other less important ore-bodies in the neighbourhood showing very similar relations. The Kiruna ore-body is a little over 3 miles long with a maximum width of about 300 feet perpendicular to the walls. It has been followed by diamond drilling to a total depth of nearly 1700 feet below the highest point of the exposure, and from magnetic surveys it is estimated to extend down to 7500 feet. The Luossavaara ore-body is about three-quarters of a mile long and 160 feet wide, and also appears to extend to a great depth. The boundaries of the ore are fairly sharp, but veins and small dykes run out into the porphyries and fragments of the latter are found embedded in the ore.

[1] The name Kiirunavaara is commonly for convenience abbreviated to Kiruna.

The ore is mainly magnetite with a little haematite and varying amounts of apatite. Other minerals present in small quantity are augite, hornblende, biotite, sphene and zircon, and occasionally tourmaline: there can thus be no doubt, on mineralogical grounds alone, of the magmatic origin of the ore. This, however, receives the strongest confirmation from the structural relations of the rock and the behaviour of the apatite. This mineral is either evenly disseminated or occurs in patches and streaks, giving a most conspicuous example of fluxion-banding in a heterogeneous magma: this is particularly striking owing

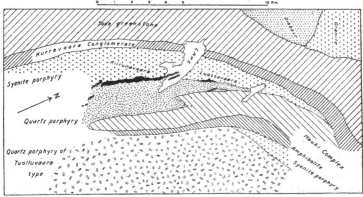

Fig. 57. Geological map of the Kiruna district. From Prof. R. A. Daly, *Origin of the Iron Ores at Kiruna* (by permission of Trafikaktiebolaget Grängesberg-Oxelösund, Stockholm).

to the strong contrast between the black magnetite and the white or pinkish apatite. Some of the dykes especially show alternating streaks of nearly pure magnetite and apatite.

Although the exact mode of origin of the porphyritic rocks is still rather uncertain, it seems probable that they also are intrusive, and that the ore-body, which is the latest, was derived by differentiation from the same magma as the porphyries[1].

[1] Stutzer, *Journ. Iron and Steel Inst.* 1907, p. 105. Sjögren, *Trans. Am. Inst. Min. Eng.* vol. xxxviii, 1907, p. 766. Lundbohm, *Geol. Fören. i Stockholm Förh.* xxxii (2), 1910, p. 751. Geijer, *Econ. Geol.* vol. v, 1910, p. 699; "Igneous Rocks and Iron Ores of Kiirunavaara, etc.," *Geology of the Kiruna District*, Pt ii, Stockholm, 1910. Daly, "Origin of the Iron Ores at Kiruna," *ibid.* Pt v, Stockholm, 1915.

The Kiruna ore consists mainly of magnetite with a variable quantity of apatite, and a small proportion of silicates of iron, magnesia and lime: its general composition averages about 96 per cent. iron ore and apatite taken together, with 4 per cent. silica, titanium dioxide, manganese, magnesia and lime (after deducting the lime contained in the apatite). The ratio of magnetite to apatite, however, varies a good deal and in some samples the phosphorus-content rises as high as 3·50 per cent.: the best grades contain only 0·018 to 0·020 per cent. phosphorus. Formerly the high-phosphorus ores were rejected as useless, but are now largely employed for the manufacture of basic steel. The ore was originally classified into eight grades, from A to H, according to its phosphorus percentage, but now only four grades are recognized, namely, A, C, D and G, on the following basis:

	phosphorus	iron
A	0·018–0·05	69–70
C	0·05 –0·60	67–63
D	1·25 –2·00	63–60
G	over 2 per cent.	60–57·5

The output of grade A is small, not more than 50,000 tons per annum: the export of this grade is forbidden by an agreement between the company and the Swedish Government, and the whole of it is sent by rail, via Luleå, to ironworks in southern Sweden to be used for acid steel. Nearly all the rest is shipped from the Norwegian port of Narvik to southern Sweden, England, Germany and occasionally to America. In the last years before the war the export of this ore from Narvik was on the average a little over 3,000,000 tons per annum[1].

The origin of these ore-bodies has given rise to much discussion. They have been attributed to various causes, including sedimentary deposition with subsequent metamorphism. However, their direct genetic connection with igneous rocks is so obvious that in reality it only remains to consider the details of the process by which they were formed. The first point to be taken into account is that the ore lies on the foot-wall side of an inclined sheet of an acid porphyritic igneous rock, which is

[1] Vogt, "Jernmalm og Jernverk." *Norges Geol. Undersök.* No. 85, 1918, pp. 112–19.

generally believed to be of an intrusive nature. In a case like this two modes of origin are possible; on the one hand injection of the ore as a separate and independent intrusion, on the other hand segregation from the acid rock after its intrusion. Thus the question resolves itself into an alternative between the two chief types of differentiation possible in such cases, deep-magmatic and laccolithic, to use Brögger's terminology. In order to discriminate between these it is necessary to know the field-relations of the rock-types involved, especially the nature of the contact between them, and the presence or absence of inclusions of one in the other. In this case the occurrence of veins and dykes of ore penetrating the quartz-porphyry has been described, thus favouring the idea of later injection of the ore, but on the other hand a most important observation by Daly seems quite conclusive in favour of differentiation in place. This geologist has laid stress on the occurrence in the acid rock of masses and lumps of iron ore, gradually decreasing in number and size away from the lower margin. This indicates that the ore separated from solution during cooling of the magma, aggregated itself into lumps of gradually increasing size and sank under the influence of gravity. This process went on till a thick and nearly solid layer of ore was accumulated at the bottom, until eventually the viscosity of the silicate melt became so great that sinking was no longer possible and part of the ore remained suspended in its present position. The mass of quartz-porphyry is of enormous thickness and it is possible that there was a certain amount of re-fusion of ore in the lower layers due to a higher temperature there prevailing: the banded character of the ore lends some support to this idea, suggesting convection-currents in a viscous semi-fluid mass. It is known, especially from Vogt's researches, that the solubility of magnetite and apatite in silicates is extremely low: they would therefore begin to separate from the magma at an early stage while the viscosity was small and the earlier-formed crystals would thus sink readily to the bottom.

This explanation is good as far as it goes, but it does not help us to understand why an acid intrusion such as this, in other ways quite normal, should have originally contained such an unusually high proportion of iron and phosphorus. This is evidently something much more fundamental, which cannot as yet be explained. The whole case is in many ways exceptional. It has already been laid down as a general law that the ore-deposits connected with the acid rocks separate at a late, often a very late, stage in the cooling, and this is certainly true of the most characteristic types, tin, wolfram, molybdenum, copper, lead, zinc, etc. But in this case the separation was early and took place in a manner showing much more analogy to that usually prevailing in basic rocks. There is, in fact, a very close analogy to the formation of the magmatic apatite-ilmenite deposits of Norway, connected with gabbros.

In a recent publication, however, Geijer controverts some of the conclusions arrived at by Daly and others[1]. He concludes that the porphyries are effusive lavas, and that the ore-body was intruded between them as a sill. This conclusion is based on the observation that contact zones with hornblende have now been observed on the hanging-wall side of the ore-body as well as on the foot-wall, and that the ore occurs in places as a network of veins in the hanging-wall porphyry, such as could not have been formed by sinking of blocks of ore in the lava, which furthermore appears to have been of a highly viscous character. At all events, whatever may have been the exact mechanism, the magmatic origin of the ore is certain, and it affords a very striking example of the formation of an ore-body by differentiation.

Gellivare. Not far from the Kiruna district is the iron-ore mountain of Gellivare, which presents many similar features. The ore-deposits take the form of lenticular masses lying in general conformity with the strike of the other rocks of the district, which consist mainly of syenites with some granites, all more or less converted into gneisses by pressure-metamorphism. These rocks are themselves rich in magnetite, apatite and sphene, and some of the pegmatite dykes cutting them are specially rich in apatite. The lenticles of iron-ore are often separated from the syenites by layers of a peculiar rock known as "skarn," which

[1] Geijer, "Recent Developments at Kiruna," *Sveriges Geol. Undersök.* Årsbok 12, No. 5, 1918.

consists of hornblende, augite, mica, calcite, scapolite, epidote, fluorspar, magnetite and apatite: the origin of this rock-type in this particular case is rather obscure[1], but the presence of fluorspar and scapolite suggests some form of pneumatolysis. The ores are distinguished by a high content of iron and phosphorus, with low silica. There are two types, magnetite and haematite ores, which may occur separately or mixed. They usually possess a granular structure and consequently fall to powder readily when handled: this is somewhat of a drawback, but facilitates magnetic separation, which is used to reduce the proportion of apatite: in some samples this mineral may form 20 per cent. of the whole. There are several ore-bodies, the largest being 4 kilometres long; all have a steep dip to the south-west and have been proved to 500 metres in depth. The reserves were estimated at 270,000,000 tons in 1910.

Grängesberg. The most important occurrence of iron ores in central Sweden is at Grängesberg, near Örebro, about 100 miles west of Stockholm. Here lenticles of magnetite and haematite ores, some with much apatite, lie in gneisses and skarn rocks, cut by pegmatites. They include a wide variety of types, mostly with high phosphorus. The most important type is the apatite-bearing ore of Grängesberg, which is worked in two big open cuts, about 1000 yards long and up to 100 yards wide and 300 feet deep. The magnetite ores run from 59 to 64 per cent. Fe, with 0·8 to 1·5 per cent. phosphorus, while the haematite ore has 60 to 63 per cent. Fe, with phosphorus from 1·25 to 2·0 per cent.[2] Some exceptional varieties, rich in apatite, run much higher in phosphorus.

Fig. 58 shows sections of the Grängesberg mines before any of the ore and rock had been removed by the open-cut workings[3].

The well-known deposit of titaniferous iron ore at Taberg in southern Sweden affords an excellent example of differentiation

---

[1] The occurrence of skarn rocks at the contact of granite and impure limestones is well known and easily explicable as a contact metamorphic effect, with diffusion.

[2] Johansson, *Geol. Fören. Förh.* vol. xxxii, 1910, p. 324.

[3] Högbom, *ibid.* pp. 561–600.

in situ. It is a large intrusion of norite and towards its centre the ilmenite and magnetite are highly concentrated, forming a mass of ore some 1200 yards long by 500 yards wide: this

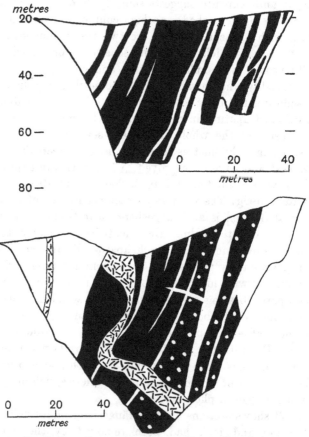

Fig. 58. Vertical sections of open workings, Grängesberg, Sweden (after Johansson, *Geol. För. Förh.* vol. XXXII, 1910, plate XI). Black, apatitic magnetite ore; black with white spots, apatitic haematite ore; white, granulite with amphibolite. In lower figure are two pegmatite dykes (stippled).

contains some olivine, biotite and felspar, showing that it is essentially a basic facies of the norite. The ore contains about 6 per cent. of titanium and on the average only 30 per cent. of iron: hence it is of very low grade.

Very similar in their general mode of origin, though differing in detail, are the great masses of titaniferous iron ore of the Ekersund-Soggendal district on the south coast of Norway and of Routivare in northern Sweden. In both these localities the ore occurs as large streaks and lenticles in basic igneous rocks. The Routivare ore-mountain is about a mile long and 200 to 300 yards wide and about 500 to 600 feet high. The ore is very largely ilmenite and the proportion of titanium dioxide may

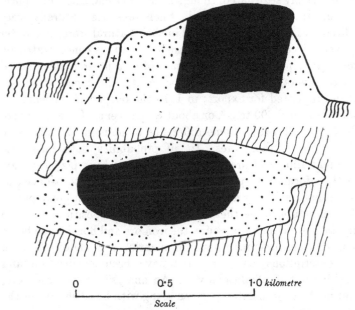

0   0·5   1·0 *kilometre*
Scale

Fig. 59. Plan and section of the Taberg ore-body, Sweden. Wavy lines, gneiss; dotted, norite; crosses, pegmatite dyke; black, ore. (From *Geol. För. Förh.* vol. XXXII, 1910, p. 1042, fig. 25, and p. 1044, fig. 26.)

rise as high as 18 or 20 per cent., while the iron content is low, averaging about 50 per cent.

These highly titaniferous ores are not at present of much industrial importance, since they present considerable metallurgical difficulties; it is possible, however, that in the future they may be successfully concentrated by magnetic separation, and of late years titanium in the form of ferrotitanium has been employed in America in the manufacture of special steel: hence

it is not unlikely that a demand may spring up for ores of this
type, as a cleanser and deoxidizer in steel manufacture[1].

The Banded Ores of Sydvaranger. The extensive iron ore
deposits of the Sydvaranger district in the extreme north of
Norway are an interesting example of a low grade type that can
easily be concentrated to a good workable ore. The district
consists largely of banded gneissose rocks made up mainly of
alternating layers of quartz, hornblende and magnetite, often
quite sharply segregated, though sometimes magnetite and hornblende are intimately mixed. Such ores are naturally very
siliceous and cannot be smelted in their natural state, but water
power is locally abundant and the crude ore is concentrated by
magnetic separation, the percentage of metallic iron being raised
on the average from 36 to 66 per cent.: the concentrates are
then briquetted for export; in 1915 the output of concentrates
was about 600,000 tons[2], or about 80 per cent. of the total production of Norway. Some difference of opinion prevails as to
the origin of these ores. In all probability both magmatic
differentiation and pressure metamorphism have played a part
in their formation: primarily they are at any rate believed to
be of magmatic origin and the banding may be in great part a
flow-structure, while the present mineral composition is due to
crystallization under pressure, the iron ore and hornblende being
derived from basic intrusions: no mention is made of felspar in
the descriptions, so there may have been silicification and
leaching out of alkalies as well. An analysis of the crude ore,
and of the concentrates, is given along with Swedish ores in the
table opposite.

The following table shows typical analyses of the magmatic
iron ores of northern and central Sweden, *i.e.* Kiruna, Gellivare,
Taberg and Grängesberg, and of Sydvaranger in Norway, all of
which appear to be due directly to the differentiation of a magma
rich in iron and phosphorus; this origin is clearly proved by
recent careful studies of the field relations and petrographical
characters of the ores and rocks: it is quite clear that the ores
and the intrusions are genetically connected and it appears

[1] Singewald, Bulletin 64, *United States Bureau of Mines*, 1913.
[2] Vogt, "Jernmalm og Jernverk." *Norges Geol. Undersök*, No. 85, 1918, p. 3.

probable that in many cases the differentiation was brought about by the gravitative sinking of the heavy iron minerals during cooling, the present inclined position of the ore-masses being mainly due to tilting during later earth-movements.

	Kiruna				Gellivare		
	A	C_1	D_1	F	A	C_1	D
Iron	69·50	67·69	61·80	58·68	68·76	66·35	61·54
MnO	0·46	0·12	—	0·11	0·15	0·13	0·11
SiO_2	1·69	2·07	1·94	1·90	1·90	3·72	4·28
Al_2O_3	0·32	0·61	0·38	0·40	1·05	0·95	0·93
CaO	0·23	0·88	6·75	8·41	0·60	1·67	4·68
MgO	0·45	0·46	0·15	0 24	0·73	1·14	1·36
TiO_2	0·08	0·18	0·18	0·12	0·82	0·18	0·31
P	0·02	0·26	2·06	2·76	0·02	0·22	1·39

	Grängesberg			Sydvaranger	Sydvaranger
	magnetite	haematite	Taberg	crude	concentrates
Iron	62·12	62·15	31·46	36·17	66·25
MnO	0·19	0·09	0·40	0·19	0·20
SiO_2	3·26	3·91	21·25	44·93	6·30
Al_2O_3	0·74	0·04	5·55	0·60	0·44
CaO	4·58	5·00	1·65	2·20	0·60
MgO	1·06	0·11	18·30	2·14	0·44
TiO_2	—	—	6·30	—	—
P	1·28	1·42	0·03	0·09	0·01

The Lake Superior Iron Region. For many years past this has been by far the most important iron field in the world: in 1917 the output of ore was no less than 63,481,321 tons, more than four times as great as that of the British Isles, and furthermore the grade of the ore is higher than the British average, so that the total of metallic iron is much larger still.

The ore-fields occupy large, fairly well-defined areas, commonly called *ranges*, in the States of Michigan, Wisconsin and Minnesota, to the south-west and north-west of Lake Superior, with an extension north-eastwards across the lake into the Michipicoten district of Canada. The whole iron-bearing area consists of pre-Cambrian rocks, forming part of the great Canadian shield, in many parts deeply covered by drift. The Cuyuna range, for example, is so deeply drift-covered that it was only discovered by a study of abnormal magnetic distribution. The old rocks are of enormous thickness and deposits of iron ore occur at many horizons, in rocks that must represent an immense lapse of time.

The following table shows in a rather condensed form the principal horizons at which iron ores occur: formations not carrying ores are in most cases omitted.

Algonkian System.

    Keweenawan Series: mainly basic igneous rocks (Duluth gabbro, dolerites, etc.) and granites.

    Upper Huronian Series (Animikie).

        Iron ores of Mesabi, Penokee-Gogebic, Menominee, Calumet, Vermilion, Marquette, Cuyuna and other ranges.

    Middle Huronian Series.

        Negaunee formation of Marquette range.

———   ———   ———   ———   ———

Archaean System.

    Keewatin Series.

        Soudan formation of Vermilion range, Helen formation of Michipicoten, Ontario.

The Lower Huronian and Laurentian, represented by the gap, do not contain workable iron ores. It will be observed that in some districts ore occurs at more than one horizon. To avoid confusion most of the local names of these subdivisions are omitted.

The ore-bodies themselves are masses of haematite, or in some places magnetite, of very varying form and size: they are always associated with types of rock also containing a certain amount of disseminated iron in various forms. When seen in depth and least affected by weathering agents these primary rocks comprise banded cherts, green iron-silicate rocks and iron-carbonate rocks. According to the generally accepted view these were originally sediments of a peculiar kind, since it is believed that a great part of the iron and silica was of magmatic origin, having been poured out on the floor of the sea as a result of igneous activity at the time of formation of the original sediments. According to Leith the further steps in the formation of the ores may be explained as follows: the sediments with the disseminated iron were uplifted within reach of weathering agents and underwent normal denudation, the soluble constituents, even the silica, being leached out by meteoric waters, leaving the least soluble constituent, iron oxide, as a residual deposit. The distribution of these residual concentrations was controlled by rock-structures and the lie of impervious portions, which were sometimes specially compact sedimentary beds, and occasionally igneous intrusions. At a still later date, the whole underwent a high degree of folding, faulting and metamorphism, but it is to be noted that all these changes took place before the deposition of the overlying Cambrian strata. It is inferred that the concentration of the ore took place during a period of arid climate, since the changes produced by weathering extended to a depth of at least 2500 feet below the *present* surface, and it is not known how great a thickness has been removed from above this surface. It is only in comparatively dry climates that weathering and oxidation can extend to so great a depth: with the present moist climate the level of stagnant water is only about 100 feet below the surface. The main point insisted on so strongly by Leith is, that the ore-bodies are

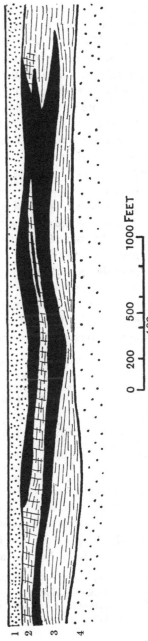

Fig. 60. Section N.–S. through the iron-bearing Biwabik formation, Mesabi district Minnesota (after Van Hise, Monograph 52, *U.S. Geol. Survey*, 1911, plate x). 1, drift; 2, decomposed taconite; 3, taconite; 4, quartzite. Ore-body, black.

residual, formed by removal of other constituents, and not, as frequently stated, due to transport and precipitation of the iron by solutions. The principal proof of this theorem is the fact that ore-bodies never occur in formations that were not originally iron-bearing[1]. As would naturally be expected every gradation can be traced between the original ferruginous cherts (taconite), green iron silicate (greenalite) rocks and massive iron carbonates on the one hand, and nearly pure haematite rocks on the other.

The ores are divided by Van Hise and Leith[2] into the four following categories: (1) soft, brown or red, slaty, hydrated haematites; (2) soft limonite; (3) hard massive and specular haematites; (4) magnetites. Various gradations exist between all these classes. Of the total tonnage shipped in 1906 the first two classes comprised 93 per cent., with 7 per cent. of hard haematite. The magnetite amounted to less than 1 per cent. and is included with the hard ores.

The following analyses show the average composition of all the ore shipped in the years named. There has been for some years past a slow but steady falling off in the percentage of iron, together with a slight rise of phosphorus, as the known highest grade deposits become exhausted.

	Mesabi	Vermilion	Gogebic	Marquette	Menominee
Iron ...	58·83	63·79	59·62	58·60	52·23
Manganese	0·82	0·11	0·77	0·71	0·19
Silica ...	6·80	4·90	8·16	10·20	16·77
Alumina...	2·23	2·93	1·92	1·05	1·41
Lime ...	0·32	0·23	0·37	1·15	1·31
Magnesia	0·32	0·05	0·28	0·46	2·70
Sulphur ...	0·07	—	0·03	0·01	0·01
Phosphorus	0·06	0·05	0·06	0·20	0·07
Loss ...	4·72	0·85	2·82	1·25	2·52

Since the geological (structural) relations of the different deposits are so variable it is difficult to give a generalized account of their forms and extent. Since, however, the forms of the ore-bodies depend on the depth to which oxidation and leaching could extend they are largely controlled by the position

[1] Van Hise and Leith, "The Geology of the Lake Superior Region," Monograph LII, *U.S. Geol. Survey* (a monumental work of 641 quarto pages). Leith, *Economic Aspects of Geology*, New York, 1921, pp. 167–70.

[2] *Loc. cit.* p. 479.

of impervious rock-masses. These may be particular strata in the original sediments, or masses of igneous rock intrusive as sills or dykes, or barriers to free water circulation may be formed by faulting when an impervious rock is brought against a pervious one. Hence many ore-bodies are of synclinal form, or entirely shapeless. Rarely, as seen on a large scale in the Mesabi range, the ore-body is flat and of large horizontal extent.

Banded Ironstones in general. Rocks very similar to those occurring in the Lake Superior region are found among the pre-Cambrian formations of many parts of the world, and have not infrequently given rise to workable ore-bodies. In few localities, however, are the conditions for local concentration so favourable and the ores are often much too siliceous to be worked under present conditions. Good examples of such rocks are to be found in South Africa, where some varieties with a pronounced colour-banding are called "calico-rock." Such are found at many points in the Barberton district and in the Griquatown and Transvaal systems. They consist mainly of alternating bands of haematite or magnetite and red or white chert. Some varieties may well be called haematite schist.

There is also an enormous development of haematite (or magnetite) schists (itabirite) in Brazil. These are most largely developed in the State of Minas Geraes. The thickness of the iron formation varies from 50 to 4000 feet and in this the rich ore occurs in lenses; where weathered at the surface there is formed a conglomerate (Canga) consisting of ore-fragments cemented by limonite, with 50 to 65 per cent. iron and 0·1 to 0·3 per cent. phosphorus, while the highest grade unweathered ore carries 69 to 70 per cent. iron and 0·003 to 0·02 per cent. phosphorus[1]. Some day these will probably be of great economic importance, but at present their development is hampered by transport difficulties, largely due to absence of coal in Brazil. There is undoubtedly a great future for the iron ores of the crystalline schists of the more remote regions when the readily available supplies of western Europe and North America are exhausted.

[1] Harder, *Econ. Geol.* vol. IX, 1914, p. 101, and *Trans. Amer. Inst. Min. Eng.* vol. L, 1915, p. 143.

Metasomatic Iron Ores. Many important iron ores have been formed by metasomatic processes of replacement, especially by the replacement of limestones. The ores produced in this manner belong to several different groups of widely varying petrographical type. This is only natural, since the term metasomatism, as at present employed, covers processes taking place under a wide range of physical conditions. Some forms of metasomatism can only occur at a high temperature and probably under high pressure; in other words, they are processes of contact metamorphism, as the term is generally understood. Among many examples of this kind mention may be made of the famous deposits of Dannemora in Sweden, where the magnetite is believed to be due to chemical replacement of masses of limestone.

However, the ore-bodies that it is proposed to discuss in the present section are those that have been formed under approximately normal conditions of temperature and pressure, by the replacement and transformation of calcareous rocks belonging to unmetamorphosed and sometimes comparatively young geological formations. Two undoubted and highly important members of this class are the haematites of Cumberland and Lancashire and the ores of the Bilbao district of northern Spain. But here we are up against a very large and difficult problem. There is a large class of iron ores, including some very important deposits, which are interstratified with perfectly normal fossiliferous sediments of various ages, and themselves show many of the characters of such normal sediments. Such are, for example, the Jurassic ores of England and Lorraine, the Clinton ores of the eastern United States and the Wabana ores of Newfoundland. With regard to all these the interesting and still unsolved problem is, whether the iron is primary or secondary, whether they were originally deposited as we see them now, or whether the iron was introduced at a later date. As to this there is still room for much difference of opinion, and we can but state the facts, leaving the final decision to the future.

Haematite of Cumberland and Lancashire. The iron ores of this region, generally known in the trade as West Coast

Haematite, have for long been of first-class importance in the iron and steel industry of this country. Probably they were worked by the Romans, or even earlier, but the first definite record is the grant of an iron mine in Egremont to the abbey of Holme Cultram in 1179 A.D. The iron of the district was also worked extensively by the monks of Furness abbey (founded in 1127) and from that day till the middle of the nineteenth century iron was worked in charcoal bloomeries and forges. At the dissolution of the monasteries the industry collapsed, like most others, but slowly revived again, and modern coal-smelting began about 1711. During the second half of the nineteenth century the industry was very active and great development took place. The maximum output of approximately 2,400,000 tons was reached in 1882, but the introduction of the basic process led to a falling off. In 1917 the output was 1,586,429 tons, or 10·5 per cent. of the total British output[1], and in 1920 only 1,257,388 tons.

The most salient character of these ores is the extremely low proportion of phosphorus, which causes them to be admirably adapted to the acid processes of steel-making. Hence they are generally classed as Bessemer ores.

The haematite ores of West Cumberland and Furness occur as masses of irregular form in the limestone members of the Lower Carboniferous system. The variation in their size and form is infinite, and it is almost impossible to describe them in general terms. However, they can usually be referred to some definite geological feature, such as the presence of impervious non-calcareous strata, or faults and fractures. Furthermore all the evidence tends to refer them to replacement of the limestone by material derived from solutions working downwards. In this area the Carboniferous rocks are much faulted: they rest on an impervious floor of Lower Palaeozoic rocks, ranging from Cambrian to Silurian, and are overlain unconformably by highly ferruginous strata of the Permian and Triassic systems. These have been partly removed by denudation, but once extended over the whole district.

[1] Hatch, *Average Analyses of British Iron-Ores, Ministry of Munitions,* 1918, p. 2.

Besides the ore-bodies of irregular form in the limestone there are also a certain number of haematite veins in the Lower Palaeozoic rocks.

In the latest and best special publication on this subject the ore-bodies are classified as follows[1]:

(1) Lodes; true veins with definite hanging and foot-walls.

(2) Vein-like bodies or veins; bodies occurring along fractures or faults, with one definite wall, the other side passing irregularly into the country rock.

(3) Flats; bodies of much greater area than thickness, related to bedding planes of enclosing rock, surfaces sharp or indefinite.

(4) Irregular bodies; masses of any form and varying thicknesses, related in a general way to flats and veins, or forming pockets.

(5) Sops; circular or oval in plan and conical in section, formed mainly in old swallow-holes.

The above definitions have been slightly condensed from the original.

Lodes are most common in the pre-Carboniferous rocks, as, for example, in the Skiddaw slate at Knockmurton and Kelton Fells, near Ennerdale, or in the Eskdale granite. The second type is sometimes seen where the Carboniferous limestone is faulted down against the older rocks, as at Bigrigg. Both flats and irregular bodies are exceedingly common and most of the important mines work masses belonging to one or other of these types. The minor varieties that can be recognized are innumerable. Sops are chiefly found in the Furness district, as at the old Park mines, Askham-in-Furness. For our present purpose attention may be confined to the second, third and fourth groups.

Whitehaven District. In this area the Lower Carboniferous series consists of six distinct beds of limestone, alternating with shales and sandstones. The limestones are numbered from above downwards and the fourth is by far the thickest, being over 300 feet near Egremont, with a few thin shale partings. The whole is overlain conformably by the Millstone Grit and Coal-measures. The upper surface of the Coal-measures is a

[1] Bernard Smith, "Haematites of West Cumberland, Lancashire and the Lake District," *Spec. Rep. Min. Res.* vol. VIII, *Mem. Geol. Survey*, 1919. p. 8.

plane of erosion on which rests a Permian breccia, called Brockram, of varying thickness, and above this, again unconformably,
the Triassic St Bees sandstone. The whole area is much faulted,
and some of the faults end off at the base of the Permian. The
most important faults strike N.W.–S.E. and have a curious
effect in letting down narrow strips of Brockram into the Coalmeasures. Masses of haematite of large size are specially
numerous in the strip of country about 3 miles long and 1 mile
wide extending from Cleator Moor to Egremont: in this area lie
many famous mines, such as Ullcoats, Ullbank, Bigrigg, Woodend, Parkhouse and Crowgarth. Among these we select only one
or two for special description.

The Ullcoats and Florence mines, situated to the south-east
of Egremont station, work the second largest ore-body known
in Cumberland, which is essentially in the nature of a gigantic
flat. The geological relations are clearly shown in the accompanying figure[1]. The westerly portion consists of a nearly horizontal mass of limestone, overlain directly by Brockram and
Trias sandstone: this limestone is very largely replaced by haematite. To the north of the Ullcoats fault the strata are thrown
up about 300 feet and steeply inclined, so that the limestone and
ore come up against the base of the glacial drift at a sharp angle.
The ore here only occurs at the base of the limestone. From
the section the relation of the replacement of limestone by
haematite to the faulting and to the impervious rocks below is
obvious. The greatest thickness of the flat, near the western
fault, is about 140 feet and the ore is shown by boreholes to
extend over at least 90 acres, giving an available total of probably 30 million tons, at an average depth of approximately
1000 feet.

A section across Lord Leconfield's Bigrigg mine, half a mile
from Woodend station, shows still more clearly the relation of
ore-deposition to faulting. This section really explains itself.
The main ore-body lies in the Third Limestone to the south
of the fault, while north of the fault there are flats in the
First and Second Limestones. This is an old mine and is nearly
exhausted.

[1] B. Smith, *op. cit.* fig. 5, p. 60.

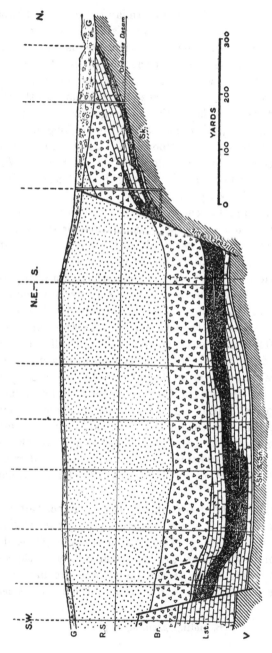

Fig. 61. Diagrammatic section across the Florence and Ullcoats ore-bodies, near Egremont, Cumberland (B. Smith, *Spec. Rep. Min. Res.* vol. VIII, fig. 5). G, glacial beds; R.S. red sandstone; Br. Brockram; Lst. limestone; V. volcanic rocks; Sk. Skiddaw slate. Dark stipple, haematite. *By permission of the Controller of H.M. Stationery Office.*

The third example here chosen is a section of the Montreal mine, near Moor Row station, across the important Cleator Moor fault, which brings Coal-measures against the limestones from the Second to the Fifth. There are numerous shafts on this property. In 1871 the output was 263,000 tons, in 1877 266,000 tons.

A large concealed iron-field is believed to exist in the tract of country south of Egremont, extending nearly as far as High Sellafield[1].

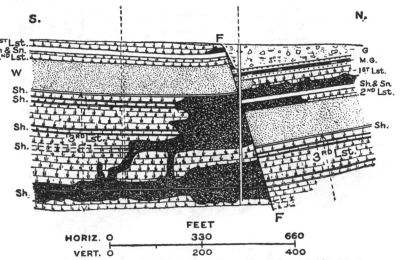

Fig. 62. Section across the Bigrigg Fault at Lord Leconfield's Bigrigg Mine (B. Smith, *Spec. Rep. Min. Res.* vol. VIII, fig. 9). G. glacial beds; M.G. Millstone Grit; Lst. limestone; Sh. shale; Sn. sandstone; W. whirlstone. Dark stipple, haematite. *By permission of the Controller of H.M. Stationery Office.*

Millom District. In the extreme south-west of Cumberland, at the mouth of the river Duddon, lies a small outlier of Carboniferous limestone, containing the Hodbarrow mine, the largest in the north-western district and one of the most important iron mines in the British Isles, with an average annual output in normal times of about 400,000 tons of ore, carrying 56 per cent. of metallic iron and with only a trace of phosphorus (average 0·005 per cent.). The main ore-body here occurs low down in the limestone series, not far above the basal conglomerate,

---

[1] B. Smith, *op. cit.* fig. 2.

which here rests on slates and Coniston limestone. It is essentially
a huge flat, or over part of the area an almost complete replacement of the horizontally bedded limestone. Faults are numerous,
the most important being the Annie Lowther fault, which throws
the limestone down to the west. All the workings are far below
sea-level and the limestone probably extends eastwards beneath
the Duddon Sands, although this is not yet proved. The average

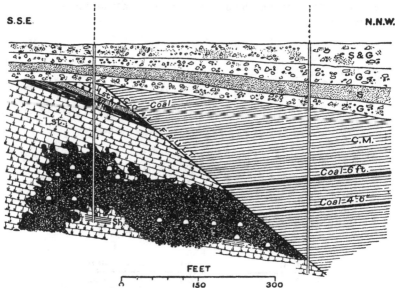

Fig. 63. Section across the Cleator Moor Coal-fault, Montreal Mine,
Cleator Moor, Cumberland (B. Smith, *Spec. Rep. Min. Res.* vol. VIII,
fig. 16). S. sand; G. gravel; C.M. coal measures; Lst. limestone;
Sn. sandstone; Sh. Shale. Dark stipple, haematite. *By permission of
the Controller of H.M. Stationery Office.*

thickness of the ore is about 60 feet and it appears to be bounded
on all sides by faults. The greater part of it has now been worked
out and unless fresh discoveries are made the life of the mine is
estimated at 10 years (from 1919).

Furness District. The ore-bodies occurring in this isolated
portion of Lancashire, which belongs geographically to Cumberland, are on the whole much less regular even than those already
described and partake very largely of the character of pockets
and sops. Nevertheless they are of great importance, and the
total output of some individual mines is large. (In 1913 Lindal

Moor, 90,012 tons; Park, Stank and Yarlside, 120,724 tons; Roanhead and Askham, 179,784 tons.)

As an example of the kind of thing here found we reproduce a section of the Burlington Pit of the Park mines, Askham-in-Furness. Here the limestone rests unconformably against the Skiddaw slates and the ore takes the form of a more or less cylindrical replacement, the centre being filled up by a core of

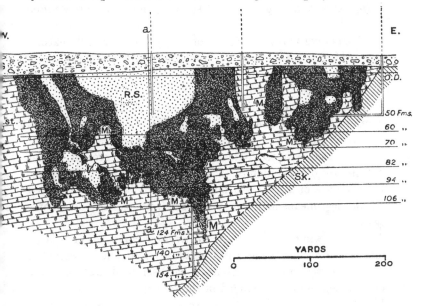

Fig. 64. Longitudinal section of ore-body, Burlington Pit, Park Mines, Furness, Lancs. (B. Smith, *Spec. Rep. Min. Res.* vol. VIII, fig. 23). G. glacial beds; Lst. limestone; R.S. red sand; M. muck and hunger; Sk. Skiddaw slate. Dark stipple, haematite. *By permission of the Controller of H.M. Stationery Office.*

red sand which is a mass of collapsed Trias that fell into a swallow-hole in the limestone. A section taken at right angles to this is very similar. This is a typical sop, and the great majority of the occurrences now worked are essentially similar to this, though varying in detail. In other less common cases the ore forms lenticular masses with their horizontal axes parallel to the bedding planes of the limestone.

Types of Ore. The haematite ores, although all of very similar

composition, nevertheless differ a good deal in appearance and
physical character. Mr Bernard Smith, in the memoir already
quoted at length, recognizes eight different varieties[1]. It is
hardly necessary to give all the characters of these varieties in
detail: they may be summarized briefly as follows:

(a) Compact varieties, purple or bluish-grey masses with no
visible structure, often of metallic appearance.

(b) Kidney and pencil ores, showing botryoidal and radial
structure.

(c) Massive earthy varieties of varying colour and texture,
often soft and incoherent, but sometimes hard and siliceous.

Masses of well-developed crystals (specular ore) are often
found lining cavities and fissures.

Although the ore is always spoken of as haematite, yet it
contains a certain amount of water that is constitutional and
not merely held by absorption. Hence it appears that one or
more of the hydrated oxides of iron must be present, either
göthite, turgite or limonite, in addition to haematite. However,
the presence of a hydrated mineral has not been demonstrated
by physical means and is only inferred from chemical considerations.

The following analyses show the average composition of large
lots of ore derived from different sources: they are "weighted
mean analyses" in which attention has been paid to the amount
of each consignment included in the total[2].

	Cleator Moor Cumberland	Moss Bay Cumberland	Furness Lancs.
Metallic iron ...	52·61	49·70	49·12
Insoluble residue	11·66	14·77	16·67
Calcium carbonate	2·83	1·90	5·73
Phosphorus ...	0·0068	0·0089	0·012
Water	4·62	4·32	7·17

These may be taken as thoroughly representative and show
that the amount of metallic iron is approximately 50 per cent.
or a little more. Alumina and magnesia are always very low
and the insoluble residue is mostly silica (quartz). The ores
therefore are siliceous in character. These analyses were made
during the war period and include a good deal of rather low-

[1] *Op. cit.* p. 26.   [2] *Op. cit.* p. 35.

grade ore. Many individual analyses of high-grade ore are to be found showing a much higher percentage of iron. Some of the early analyses, probably made on specially selected material, show as much as 66 per cent. of iron, but this is now uncommon. The highest figure given by Hatch, out of 31 samples, is 59·32 per cent. (Hodbarrow). Nearly all the others are below 53 per cent. in the dried ore.

With regard to the origin of these ores, all authorities seem to be agreed that they are due to metasomatic replacement of the limestone by iron-bearing solutions. The questions at issue are as to the date of the replacement and the source of the iron. The simplest and most obvious explanation, which is generally accepted, is that the iron was derived from the overlying Permian and Triassic strata, which contain enormous quantities of iron oxide as a cementing material in the breccias and red sandstones. This explanation, however, is disputed by Mr J. D. Kendall, who considers that the change took place before the deposition of the New Red rocks, relying mainly on two lines of evidence: a supposed faulting of the ore-bodies by faults which do not pass up into the New Red, and the presence of pebbles of haematite in the brockram. His theory as to the faulting cannot be regarded as established, and there is no reason why the two separated portions of a limestone should not have been equally replaced after the faulting. In the same way pebbles of limestone may be converted into haematite in the brockrams, since the solutions must necessarily have passed through these beds.

Forest of Dean. Iron ores showing a general similarity to those of the north-west of England also occur to some extent in the Forest of Dean. However, the reserves are small and they are not likely to be of any importance in the future. This area has been well described ·by Professor Sibly[1]. In this area the ores contain a higher proportion of hydrated oxides than in Cumberland and Furness, and some varieties would really be more correctly described as limonite rather than haematite. They are generally known commercially as "brown haematite" and the iron-content of the raw ore generally runs to about 40 per cent. or less.

[1] Sibly, *Spec. Rep. Min. Res. Gt Brit.* vol. x, *Mem. Geol. Survey*, 1919.

Bessemer Ores of Northern Spain. For many years before the war the home supplies of non-phosphoric (Bessemer) ore were inadequate to supply the needs of British steel-makers and large supplies were imported from Bilbao and other places in the provinces of Viscaya and Santander in the north of Spain. This ore is of essentially the same character as the British haematite, being due to metasomatic replacement of limestones of Cretaceous age, roughly equivalent to the top of the Gault and the lowest division of the Chalk (Chalk Marl) of England.

In Viscaya the ore-bodies occur over an area measuring about 16 miles by 4 miles, lying a few miles west of Bilbao, and in Santander some important deposits occur in the districts of Cabarceno and Solares and in Cabarga mountain. One of the most important groups of mines is that of Somorrostro in Viscaya, west of Triano. In this region four types of ore are recognized:

(1) Vena, soft and powdery red or purple haematite.

(2) Campanil, a compact crystalline haematite, often calcareous.

(3) Rubio, limonite, often siliceous and aluminous.

(4) Carbonato, carbonate ore or chalybite.

The first three varieties average about 55 per cent. of metallic iron, and phosphorus generally runs to about 0·010 or 0·015 per cent.

There is also found to a considerable extent a deposit known as *chirta*, consisting of a mass of fragments of ore embedded in red clay. This is a residual or eluvial formation, and the ore is obtained from it by washing. It is essentially limonite.

The Cretaceous strata of this region are considerably folded: the ore-bodies are replacements of the Gault limestone and are often clearly related to synclines or to faults. From a study of the general characters and distribution of the different types of ore it seems quite clear that the calcium carbonate of the limestones was first altered to chalybite (carbonato) and that this, when exposed at and near the surface to oxidation, was afterwards transformed to limonite or haematite. This is therefore an excellent example of a simple type of metasomatism. In this case, however, the original source of the iron is unknown. There are igneous rocks in the neighbourhood, but no connection

has been proved between them and the ores. However, no overlying rocks are known from which the iron could have been
derived, and in some instances a downward extension of the ore
along fissures in the underlying sandstones suggests a deep-
seated origin for the iron-bearing solutions. Hence a derivation
from the known igneous rocks of the district seems the simplest
explanation[1].

Stratified Iron Ores. The ore-deposits included under this
heading are those formed by ordinary processes of sedimentation
among the stratified rocks of various geological periods from the
earliest times to the present. Some of them are of original
formation, being now very much the same in composition and
characters as when they were formed; such are the bog-iron ores
and lake ores of recent date and the clay ironstones and blackband ironstones of the Coal-measures. Others are believed to
be of secondary origin, in that their present iron-content is due
to metasomatic replacement of some of the original constituents
of the rocks; to this group belong the ores of the Jurassic system
of this country and of Lorraine. It should be stated, however,
that some authorities believe that even in these the iron-content
is original and not subsequently introduced: this question will
be discussed later. In the case of the Jurassic ores just mentioned
and similar types found in many other parts of the world, the
really important geological factor bearing on their exploitation
and development is their regularly stratified character: this
greatly facilitates mining and exploration and the estimation
of reserves, in contrast to the extreme irregularity of form and
sporadic occurrence of the haematite masses and similar deposits;
hence it is convenient from the practical point of view to place
the regularly stratified ores in a class by themselves, without
making any dogmatic assertions as to their precise manner of
origin. The one feature that is absolutely clear and certain is
that they are sedimentary rocks, interstratified with other sediments and conforming to all the ordinary rules as to dip, strike,
outcrop, folding and faulting, just as in the case of coal-seams
or beds of limestone, rock-salt and many other non-metalliferous
deposits.

[1] Van der Veen, *Econ. Geol.* vol. xvii, 1922, pp. 602–18.

It will be observed in the detailed descriptions given in the following pages that most of these deposits are "ironstones" in the sense defined at the beginning of this chapter, rather than "ores" in the strict sense of the word; that is to say, they are rocks and not minerals: in fact they commonly consist of mixtures of minerals with substances of rather indefinite composition, such as sand, clay and more or less organic matter. Consequently, as a rule, the only satisfactory way of expressing their composition and characters is by means of a chemical analysis: such names as *blackband, clay-ironstone,* etc., are altogether too vague for practical use.

In the stratified ironstones, as a rule, the percentage of iron is much lower than in the purer types of ore hitherto described: in fact the amount of metallic iron in many ironstones now worked on a very large scale is often below 30 per cent. in the ore as mined. This low grade is, however, compensated for by the ease with which such regular and often thick deposits can be worked: many of them are largely developed in regions of simple geological structure, where the dips are very low and the beds extend for a long distance from the outcrop at a small depth from the surface, so that they can be worked cheaply by quarrying rather than by mining. They also happen to occur largely in accessible regions where transport facilities are easily provided: the occurrence of important beds of ironstone in the Coal-measures was an important factor in the industrial development of some of our coalfields, such as west Yorkshire, Durham and the Black Country, while the Jurassic ironstones of Yorkshire and the Midlands are now worked on a very large scale.

Since the stratified ironstones contain a considerable proportion of material of organic origin they are naturally rather rich in phosphorus and sulphur and the presence of these elements has had an important influence on their development. A high proportion of phosphorus renders an ore unsuited to the older processes of steel-making and for the acid Bessemer process, but the introduction of the basic process gave a great impetus to the production of these ores, since it was found that basic steel made from low-grade ores was well adapted for many purposes for which the more expensive steel made from high-

grade ores free from phosphorus had hitherto been exclusively
employed. This led to a great increase in the production of the
high-phosphorus ores of the midland counties of England and
of Lorraine and Luxemburg. The published statistics indicate
that although the total production of ore in the United Kingdom
fell rapidly from 1913 to 1916 and made a partial recovery in
1917 and 1918, the output of basic steel showed a steady increase
during these years, owing to the urgent need for steel made from
British ore for munitions and other purposes during war-time.
At the present time about 80 per cent. of the total British
output of iron ore is obtained from the stratified ironstones of
the Jurassic system, while in 1912 the output of similar ores
from French and German Lorraine and Luxemburg was no less
than 43,800,000 tons, or nearly three times as great as the whole
British output of iron ore from all sources.

The Coal–Measure Ironstones of Great Britain. In the
early days of the industrial development of this country, up to
the middle of last century, most of the iron ore worked came
from the Coal-measures, the rest being mainly from Cumberland
and Lancashire (Lower Carboniferous). Most of the books used
in schools and colleges still continue to reiterate the statement
that coal and iron come from the same areas, but as a matter
of fact the Coal-measures now provide less than one-twelfth of
the total British output, and the only coalfields where iron is
seriously mined are those of north Staffordshire and central
Scotland. In many other coalfields there are undoubtedly still
great reserves of ironstone, but these can never be worked under
anything like present economic conditions. Mainly owing to the
outrageous demands of labour, it is only in a few places where a
bed of ironstone can be worked simultaneously with an adjoining
coal seam that it can be profitably extracted. These ironstones
are commonly divided into two categories, clayband and blackband, the main difference being that the former require a little
additional fuel for calcination, whereas the latter contain sufficient carbon to calcine themselves when a fire is once started.
The blackband ironstones occur as continuous beds, whereas the
clayband ores are commonly in the form of nodules or small
flattened masses lying in more or less discontinuous layers at

certain horizons in beds of shale: they appear to be usually of concretionary origin and often contain well-preserved fossils. They are in fact very similar to the *doggers* of the Lias formation. Both varieties are essentially aluminous and siliceous iron carbonates, with high phosphorus, and the amount of metallic iron in the raw stone averages about 30 per cent., the blackband ores of north Staffordshire and Scotland being of somewhat better quality, ranging up to 38 per cent., and as above stated, these alone are now worked. The nodular and thinly-bedded clayband ores occur at various horizons in all the coalfields, and have at any rate until quite recently given rise to small local outputs of iron of special quality, as for example at Low Moor, near Bradford, but the excellence of this iron is due to the method of working and not to any peculiar properties of the ore.

The Jurassic Ironstones of the British Isles. As before stated, about 80 per cent. of the present British output of iron ores is derived from the Jurassic system. The whole of these belong to the class of high-phosphorus ores and are on the whole of low grade, averaging possibly 28 per cent. of metallic iron. The proportions of silica, alumina and lime also vary considerably, so that they can be roughly subdivided into siliceous, aluminous and calcareous types.

The British Jurassic ironstones are chiefly found in the Lias and at the very base of the succeeding Inferior Oolite (Bajocian) series: the known ore-deposits of the Corallian series are not at present of importance. The following table shows the production of the different subdivisions for the year 1917.

	Tons
Inferior Oolite	3,169,110
Upper Lias ...	65,985
Middle Lias...	6,182,003
Lower Lias ...	2,699,532
Total	12,116,630

Thus it appears that the Lias produced approximately three-fourths and the Inferior Oolite one-fourth of the total Jurassic output.

The surface outcrop of the Jurassic system extends without a break right across England from Dorset to Yorkshire, but ironstone mining is confined to the northern and middle part of this range, from Oxfordshire to Yorkshire. The Lias ores are chiefly mined in Oxfordshire, Leicestershire, Lincolnshire and Yorkshire and the Inferior Oolite in Northamptonshire and Rutland. Of all these counties the output from the Cleveland district of Yorkshire is by far the largest (4,809,861 tons in 1917), while Northampton and Rutland come next with 3,169,110 tons.

The Frodingham Stone of North Lincolnshire. Ironstone is worked in the Lower Lias only in the north of Lincolnshire, along a strip of country stretching north and south for some 7 miles through Frodingham and Scunthorpe. The strike of the rocks is here north and south, with a very gentle dip, less than 1°, to the east, so that the outcrop is wide and the cover thin for a long distance. When undenuded the full thickness of workable stone is from 25 to 30 feet: it may be described as a ferruginous oolitic limestone: in the upper layers the iron is present as hydrated oxide, but in depth when quite unweathered it is found to be carbonate. The effect of weathering is to oxidize the iron and dissolve out the carbonate, hence the oxidized ore at the surface is richer in iron and silica than the carbonate ore of the deeper layers. However, oxidation is somewhat sporadic and weathered patches are found here and there at a considerable depth. The amount of lime shows a good deal of local variation and certain beds full of large shells (*Cardinia*) are very calcareous. In the unweathered portion a good deal of green silicate is found. Where unaltered the ore is markedly oolitic and the microscopic structure has been figured by Teall[1]. When oxidized, however, the structure is often completely destroyed and the stone takes on a peculiar purple-brown tinge. Where all the carbonate has been removed it is soft and incoherent.

The average composition calculated from analyses of 35 samples is shown in the table on p. 353.

Hence it will be seen that this ore is of distinctly low grade, but it can be worked very cheaply and possesses the additional advantage of being self-fluxing owing to the large proportion of lime present in the carbonate, in fact there is usually rather too

[1] *The Jurassic Rocks of Britain*, vol. IV, pl. II, fig. 7.

much lime, and a certain proportion of siliceous Northampton-shire ore is generally mixed with it. The working is entirely by open-cast, at present under a thin cover, which is stripped by mechanical means and dumped behind the workings, while the ore itself is loaded into trucks by powerful steam excavators after being loosened by blasting; by this means it was found possible before the war to get the stone on rail at about one shilling per ton, exclusive of royalties.

The Middle Lias. The ironstones of the Middle Lias are worked in several counties, which may be divided for con-venience into two groups as follows: (1) Oxfordshire, Leicester-shire and south Lincolnshire; (2) Yorkshire—of these groups the second is the more important. The lithological development of the Middle Lias series differs somewhat in the two areas and this has an important influence on the topography and the winning of the ores, which in the Midlands are mostly worked by quarrying, but in Yorkshire wholly by underground mining.

The Middle Lias Ironstones of the Midlands. Where best developed in the neighbourhood of Banbury in Oxfordshire the Middle Lias has a total thickness of about 150 feet: it includes a very prominent rock-bed, the Marlstone, lying between clays, and therefore tending to form a conspicuous feature. It forms a plateau (in reality a gentle dip-slope) that rises gradually from 500 feet at Banbury to 710 feet at Edge Hill. The Marlstone, which is the iron-bearing stratum, forms the surface of this plateau or lies under but a thin overburden over a large area and can consequently be easily and cheaply worked. It varies in thickness from 12 to 25 feet: the outer parts are always brown from oxidation of the iron, but the centres of large blocks consist of grey or greenish carbonate and silicate. It often contains bunches of fossils, such as *Terebratula punctata* and *Rhynchonella tetrahedra*. The percentage of iron ranges from 17 to 29 in different samples, the silica from 7 to 17 and the lime from 1·5 to 24 per cent., according to the degree of weathering, the lime being more or less dissolved out of the surface layers and the iron and silica correspondingly concentrated. The Marlstone is also worked extensively in Leicestershire, near Eaton, Eastwell

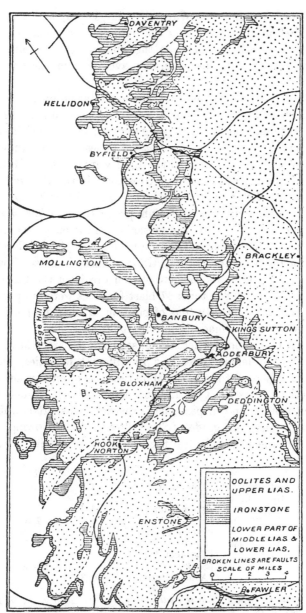

Fig. 65. Sketch-map of ironstone outcrop in the Banbury district (Lamplugh, Wedd and Pringle, *Spec. Rep. Min. Res.* vol. XII, fig. 11). *By permission of the Controller of H.M. Stationery Office.*

and Stathern, Holwell, Wartnaby and Tilton, and also from near Melton Mowbray along the outcrop to a point ten miles north of Grantham. Between this and north Yorkshire there are no workings in this seam. The average composition of the stone is shown in the table on p. 353.

The Cleveland Ironstone. The exploitation of ironstone in the Cleveland district of north Yorkshire began about the year 1815, the first consignments being apparently obtained from the wave-cut coastal platform about Kettleness, locally known as "the scaur," and shipped to the Birtley ironworks in Durham. Regular mining was started at Grosmont, in the Esk valley, in 1839 and at Eston, on the northern outcrop, in 1850[1]. Development was rapid and the maximum production of 6,756,055 tons was reached in 1883. At the present time the output is greater than that of any other British iron-field, although there has been a falling off owing to the competition of cheap imported Spanish ore. In 1917 the total output of iron ore from the Cleveland district was 4,809,861 tons; 47 blast furnaces were in operation and produced 2,095,619 tons of pig iron; a small proportion of this, however, was made from imported ore.

In the Cleveland district several seams of ironstone have been worked at different times and in different localities, but at present only one of these, the Main Seam, is of any importance. The other seams are thin and inconstant, varying much from place to place, and in part of the district transport is difficult owing to the steep grade of the railways. The following table shows in a generalized form the succession of the iron-bearing strata[2].

		ft.	ft.
Inferior Oolite.	Top Seam	4–9	
Upper Lias	Shale		260
	Main Seam	5–12	
	Shale		2–6
	Pecten Seam	$1\frac{1}{2}$–6	
Middle Lias	Shale		3–7
	Two-foot Seam	$1\frac{1}{2}$–$2\frac{1}{2}$	
	Shale		20–30
	Avicula Seam	0–3	

[1] Bewick, *Geological Treatise on the District of Cleveland in North Yorkshire*, 1861, p. 17.

[2] Barrow, "The Geology of North Cleveland," *Mem. Geol. Survey*, 1888.

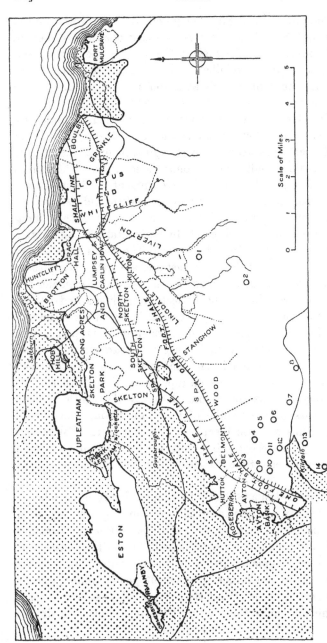

Fig. 66. Sketch-map of the mining area in the Cleveland Main Seam (Lamplugh, Wedd and Pringle, *Spec. Rep. Min. Res.* vol. xii, fig. 2). Beds below the Main Seam dotted. Outcrop of seam shown by thick line. The shale-parting in the Main Seam sets in at "shale line" and is 1 foot thick at line marked "one foot shale." Sites of borings marked by circles with numbers. *By permission of the Controller of H.M. Stationery Office.*

The Top Seam of this table is equivalent to the Northampton ironstone of the midland counties: it was at one time worked on a considerable scale at Rosedale in the south of the district, and was remarkable for containing patches of magnetite. It has also been worked on a small scale at Eston and Upleatham, in the extreme north: here it is of very low grade.

The Cleveland district is an area of high relief, dissected by deep, steep-sided valleys, bounded on the north-west and north by a well-marked escarpment, with high cliffs on the coast. The ironstone seams usually outcrop on the steep sides of the valleys and along the face of the northern escarpment, while the Lower Lias occupies the floors of the valleys, and the ridges between are capped by the harder Bajocian sandstones; hence the mining is exclusively underground, partly by horizontal adits and partly by shafts, sometimes 700 feet in depth. The strata have a very slight dip of 2° or 3° to the south-east and are traversed by some fairly important faults. Along the northern escarpment the Main Seam varies from 9 to 12 feet in thickness and has an average iron-content of 28–30 per cent. in the stone as mined. It is divided into beds by well-marked bedding planes and occasionally shale-bands make their appearance and have to be picked out by hand. In some places the shale between the Main and Pecten seams disappears, so that both can be worked together, as at Eston. Towards the south a shale band appears in the middle of the Main Seam, which itself gradually thins out and disappears, so that in the Esk valley it is no longer recognizable, but the Pecten and Avicula seams were at one time extensively mined and smelted at Grosmont.

When obtained in a perfectly fresh condition from deep levels the Cleveland ironstone is a greenish or bluish-green rock, with conspicuous oolitic structure; nearer the surface it becomes yellowish and brownish owing to oxidation and the ooliths and enclosed shell fragments are often nearly white. One notable feature of the ore is a rather high percentage of alumina, so that it may almost be placed within the category of "clay ironstones." The origin and mineralogical constitution of this and similar ores are discussed elsewhere (see p. 357).

The average composition of the Cleveland stone calculated from 21 complete analyses is given in the table on p. 353. As before remarked, the proportion of alumina is high, ranging from 8 to 12 per cent. in the samples analysed and the silica is also high, so that a considerable quantity of limestone is required as a flux: about 35,000 tons per week are brought into the district, mainly from Durham, for this purpose. Generally speaking, the proportion of iron is highest along the northern outcrop, averaging about 30 per cent. in the stone fresh from the mine, and falling off both to the east and to the south-west, where it may be as low as 25 per cent. The bulk of the ore is used for steel-making, mainly by the basic open-hearth process, a smaller proportion, about 20 per cent., being employed for foundry purposes.

The following are some of the chief mines of the district: Eston, Upleatham, Skelton, Belmont, Lumpsey, Carlinhow, Loftus, Kilton, Spawood, Roseberry, Liverton, Stanghow, Boulby and Grinkle Park. Of these the largest are Eston, Skelton, Loftus, Lumpsey and Carlinhow. In 1917 the Eston, Skelton and Belmont mines, belonging to Messrs Bolckow and Vaughan, produced over 1,400,000 tons.

The Inferior Oolite. The beds immediately succeeding the Lias are perhaps the most variable strata in the whole of the British stratigraphical succession. In the south-west of England they are mostly limestones, with a sand at the base: in Yorkshire they are sandstones and shales of estuarine origin with an occasional bed of low-grade ironstone (but see p. 350). In the Midlands the succession is still more variable and local. It is unnecessary to enter into much detail, but it must suffice to say that over a stretch of country extending from Lincolnshire to Northamptonshire the lowest bed of the series, generally known somewhat unfortunately as the Northampton Sand, includes in places a very valuable deposit of oolitic ironstone, which is now extensively worked at many points and has given rise to a large industry. There are considerable workings in Lincoln and Rutland, but the most typical area is in Northamptonshire, especially round Kettering and Wellingborough. A general description of the development in this region will suffice.

This is a country of gentle relief with shallow valleys floored
by clays of the Upper Lias: the tops of the plateaux between
are often formed by the Cornbrash. Between these come
the local and extremely variable strata of the Inferior and
Great Oolites, consisting of the ironstone at the base, followed
by clays, sands and limestones. One band in particular, the
Great Oolite limestone, is quarried for flux in the local blast-
furnaces, as for example at Islip, near Thrapston.

The ironstone bed varies in thickness from 6 to 15 feet,
averaging about 10 feet. When unweathered it is typically a
green oolitic ore, containing a good deal of ferrous silicate:
locally, as at Islip, near Thrapston, there are beds of blue carbonate ore. The fresh stone is rather poor in lime, differing thus
from the Liassic deposits. When oxidized the stone becomes
bright yellowish-brown and usually shows a well-marked "box-
stone" structure, due to secondary deposition of hydrated iron
oxide along joints, both vertical and horizontal. Fossils are
usually rather scarce, though abundant in a few places. Some
beds are locally too sandy to be worked.

The iron-content is on the whole distinctly higher than
that of the other British Jurassic ores. Hatch's average figure
from a very large number of analyses is 32·49 per cent. in
the ore as mined and 38·23 per cent. on the dried ore. The
analysis, given in the table on p. 353, shows that the stone is
distinctly siliceous and low in lime. Hence it is usually mixed
with some more calcareous stone, Frodingham or Banbury, for
example.

In this case the difference between fresh and oxidized stone
is considerable, but not so great as in the case of the more
limy varieties elsewhere. Owing to the scarcity of lime there
is of course less loss by leaching and consequently a less well-
marked concentration of iron. Nevertheless the difference of
appearance in the two stages is very striking.

Most of the stone is won by open workings, but there are a
few regular mines, where the stone is good and the cover too
thick for open working. Over very large areas the overburden
is thin and soft, so that steam diggers can be used.

Average Composition of Jurassic Ironstones.
Raw Stone as received at the Works.

	Fe	Mn	SiO$_2$	Al$_2$O$_3$	CaO	MgO	S	P	Moisture
Northampton and Rutland:									
Inferior Oolite (siliceous)	32·5	0·24	14·7	6·1	2·7	0·4	0·10	0·60	15·2
Cleveland:									
Middle Lias (siliceous)	28·1	0·41	11·8	10·2	4·7	3·5	0·26	0·47	6·8
Leicestershire and S. Lincoln:									
Middle Lias (slightly siliceous)	25·2	0·23	10·9	8·0	9·6	0·6	0·11	0·25	16·4
Oxfordshire (almost self-fluxing):									
Middle Lias	24·0	0·27	10·2	7·6	12·2	0·6	0·06	0·23	15·6
N. Lincolnshire (limey):									
Lower Lias	22·7	0·96	8·1	5·1	18·2	1·0	0·16	0·31	10·7

The above figures are obtained by averaging the results of more than 200 analyses made at ironworks; the information was placed at the disposal of the Ministry of Munitions by the owners[1].

The Iron Ores of Lorraine. It is impossible here to discuss the extremely interesting historical, political and economic questions connected with the iron ores of Lorraine. It must suffice to say that of recent years this area has been by far the largest producer in Europe. It is, however, only since the great increase in the manufacture of basic steel that this district has become of first-class importance.

Since the readjustment of territory at the Peace the iron-field lies entirely in France, except for a small area in Luxemburg and a few acres in Belgium. The iron ore forms part of the strata of the Jurassic system, at about the same general horizon as the Inferior Oolite ironstones of the English Midlands, namely, the lower part of the Bajocian series, though some French geologists are inclined to assign them to the top of the Lias. The strike of the rocks is approximately north and south, and the dip is to the west at a fairly gentle angle. The chief northern area, beginning in Luxemburg, has a total length of about 40 miles;

[1] Hatch, *Journ. Iron and Steel Inst.* vol. xcvii, 1918, p. 79. See also the same author, *Average Analyses of British Iron Ores and Ironstones*, published by the Ministry of Munitions, 1918.

south of this comes a barren stretch some 15 miles long, and then the Nancy field, which is about 13 miles long. The northern, or Briey field, is divided into sections by various natural features, the chief subdivisions here recognized being those of Orne, Landres, Tucquegnieux: still further north are the basins of Crusnes and Longwy, the latter adjoining Luxemburg. Different names were given to the prolongations of these fields eastwards into the former German Lorraine, thus:

$$\text{Longwy} \quad = \text{Audun-le-Tiche,}$$
$$\text{Tucquegnieux} = \text{Fentsch,}$$

and so on. These need not be enumerated further.

The iron-bearing formation consists of six or according to some authors seven, or even eight, different beds, which are not all present in any one locality. They are named according to their supposed characteristic colours, but most of these names do not seem to be very apposite descriptions. The general view at present is that they should be divided into three groups, as follows[1]:

Upper: ferruginous limestone and red seam.
Middle: yellow and grey seams.
Lower: brown, black and green seams.

The base of the whole series consists of a green, sandy, pyritic marl, and the thickness of the iron-bearing formation is generally from 80 to 100 feet, but near Nancy rarely more than 33 feet. The unworkable part of the series consists of limestones and marls.

The following may be taken as an average section of part of the series at a point where four beds exist.

			metres
Red Seam	...	...	3·0
Marne	...	...	2·0
Yellow Seam	...	...	2·5
Marne	...	...	1·5
Grey Seam	...	...	4·0
Marne	...	...	6·0
Brown Seam	...	...	1·0

The so-called "Marne" (marl) of this section is in reality a rather hard ferruginous limestone.

[1] Nicou, *Journ. Iron and Steel Inst.* 1921.

The ore, which is locally known as "minette," is always more or less distinctly oolitic, though the grains are sometimes microscopic. The matrix varies in character, being sometimes calcareous and sometimes siliceous. On the whole the ores are more siliceous in the north and more calcareous in the south, though with local variations. The percentage of iron varies from 30 to 40, increasing on the whole southwards; silica varies from 4 to 20 per cent., and lime from 9 to 14 per cent., the most calcareous varieties being self-fluxing. Phosphorus remains wonderfully constant throughout at about 1·8 per cent.

In 1912 and 1913 the total output of the whole area was approximately as follows:

	1912 tons	1913 tons
French Lorraine	17,300,000	19,600,000
Luxemburg ...	6,500,000	7,300,000
German Lorraine	20,000,000	21,100,000
Total	43,800,000	48,000,000

Several careful computations of reserves have been made and the following figures are estimates of the workable ores remaining in the different districts.

	tons
Briey	2,000,000,000
Longwy	275,000,000
Crusnes	500,000,000
Nancy...	200,000,000
Recovered Lorraine and Luxemburg	2,000,000,000
Total	4,975,000,000

The Clinton Ores of the United States. The output of these ores, although far behind that of the Lake Superior region, ranks second in importance in the United States, amounting in 1920 to 5,347,386 tons. The ore exists as beds interstratified in the Clinton series, which is approximately equivalent to the Llandovery series of the British Silurian. The main outcrop extends from New York State to Alabama, a total distance of about 1200 miles, and there are also smaller areas in Tennessee, Kentucky and Wisconsin. The most important areas, industrially, are Birmingham, Alabama and Chattanooga, Tennessee. In New York the beds are nearly horizontal and can be worked for long distances by stripping, but farther south they become involved in the Appalachian folding, and in places the dips are as steep as 30°. The number of ore-beds varies in different localities, and all appear to be lenticular in character, though of fairly wide extent. In some places there are as many as four different beds, usually not more than 10 feet thick, though occasionally one may run up to 40 feet. The ore varies somewhat in character, both primarily and secondarily. The two chief primary types are oolite ore and fossil ore—both of these may occur in the same bed. The fossil ore consists mainly of recognizable fossils, many of which are trilobites, corals and brachiopods. Each of these types may again show hard ore in depth and soft ore near the outcrop. Owing to leaching the soft ores are less calcareous and therefore more siliceous and richer in iron: all varieties are high-phosphorus non-Bessemer ores, best described as red haematite. The percentage of iron ranges from 31 to 57 in different grades. The sedimentary character of these deposits is obvious, but the source of the iron is not clear. This question is discussed later.

The four following analyses show the gradation from a hard limey ore below to a soft siliceous variety near the surface.

	1	2	3	4
Fe	37·00	45·70	50·44	54·70
SiO$_2$	7·14	12·76	12·10	13·70
Al$_2$O$_3$	3·81	4·74	6·06	5·66
CaO	19·20	8·70	4·65	0·50
Mn	0·23	0·19	0·21	0·23
S	0·08	0·08	0·07	0·08
P	0·30	0·49	0·46	0·10

The Oolitic Iron Ores of Wabana, Newfoundland. This very important and interesting occurrence of iron ore affords another excellent example of oolitic ironstone of Lower Palaeozoic age. The deposit consists of several beds of ironstone of varying thickness, interstratified normally with sediments of Ordovician age, containing a good many fossils. The ores crop out on one large and two small islands in Conception Bay, Newfoundland, some 10 miles west of St Johns. The dip is fairly steep towards the north and a large part of the workings are under the sea. There are two principal workable beds, the Dominion bed near the base of the series, about 35 feet thick;

about 215 feet above this comes the Scotia bed of similar character but only about 15 feet thick. The whole succession is distinctly ferruginous and there are three or four other bands of ironstone of minor importance scattered through about 1000 feet of strata.

The ore is a reddish-brown haematite, rather more grey on a fresh fracture, with a fine-grained oolitic structure, consisting of spherules, usually flattened, up to one-tenth of a millimetre in diameter. These are as a rule composed of a mixture of haematite and green iron silicate, often in concentric layers: in some specimens the matrix is chalybite. There are also present grains of quartz (sand grains) and fragments of fossils, often composed of calcium phosphate. Hence the ore is phosphoric. The microscopic structure has been very carefully described by A. O. Hayes[1]. This author concludes that the ore beds are primary sediments, the iron-content having been present in them at the time when they were laid down, in shallow water: he considers that the oolites are minute concretions formed in the fine muddy ferruginous sediment of the sea floor, before the final consolidation of the rock.

Microscopic Structure of Oolitic Ironstones. Some oolitic ironstones consist of haematite or hydrated oxides, which are opaque even in the thinnest sections, and therefore unsuited to microscopic study, but most varieties are more or less transparent and the study of these is both interesting in itself and helps to throw light on the difficult problem of the origin of such ores. Among oolitic ironstones many variations may be recognized, ranging from limestones containing only a few ooliths of iron minerals, to rocks almost wholly composed of iron compounds, with perhaps only a trace of calcium carbonate. The amount of silica also shows much variation. Inclusions of quartz as sand grains are not at all uncommon. Another notable feature in fresh unweathered specimens is a green, usually amorphous or cryptocrystalline material, which is shown to be some form of iron silicate.

The most striking feature of all true oolites is the presence of round or oval bodies, which under the microscope generally

[1] Hayes, "Wabana Iron Ore of Newfoundland," *Canada Geological Survey Memoir* 78, Ottawa, 1915.

show a distinct concentric structure, sometimes also more or less radial. There is usually a sharp contrast between ooliths and matrix, which may consist of quite different minerals. In most varieties also fossils or fragments of fossils are clearly recognizable. These are often of some importance in that sometimes the phosphorus-content of the ore is mainly present in them, *e.g.* Wabana, where there are abundant brachiopod remains, consisting of calcium phosphate. The minute structure of the

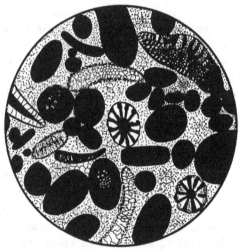

Fig. 67. Oolitic Ironstone, Lower Lias, Frodingham, Lincs. Opaque ooliths of iron oxide and organic remains partly replaced by iron oxide in a ground-mass of crystalline calcite (× 10 diameters).

ooliths is described in detail by many writers: if composed of carbonate they consist of crystalline prisms regularly arranged either radially or tangentially, giving a black cross between crossed nicols. When largely made up of silicate or its oxidation products the structure is less regular. Here another point of interest must be mentioned: by treatment with acid it can be shown that each oolith has a skeleton of silica, which can be isolated[1]. It is often to be observed that the ooliths have been built up around a sand-grain or shell-fragment as a nucleus:

[1] For excellent figures of these see Stead, *Proc. Cleveland Inst. Eng.* 1909–10, No. 4.

sometimes they are compound, *i.e.* two or more small ones, with separate centres, may be enclosed in a continuous outer layer. In some cases the ooliths have been re-crystallized and the minute structure is lost. Oxidation of the green silicate also often renders them partly or wholly opaque and obscures the details. The matrix is as a rule a rather finely crystalline aggregate and in many of the more calcareous ores it consists of calcite. In other instances it may be silicate or chalybite.

The Origin of Oolitic Ironstones. The origin of oolitic ironstones of various types presents some very obscure problems, which cannot yet be regarded as solved. The chief difficulty arises in connection with the presence in some varieties of large quantities of green iron silicates. The controversy mostly centres round the question of the primary or secondary nature of the iron.

The simplest explanation, and one that long held the field, is as follows: the ironstones were originally oolitic limestones, consisting of calcium carbonate, in the form of calcite or aragonite, or both. (The ooliths in limestones are believed to be always formed as aragonite, although they may be converted into calcite later.) The limestones are acted on by percolating water carrying iron salts in solution, and the calcium carbonate is replaced by ferrous carbonate, the aragonite portions being attacked before the calcite. Later on the ferrous carbonate undergoes various processes of oxidation, with or without hydration, forming haematite, turgite or limonite, or very rarely, magnetite. This simple theory, however, does not account in any way for the presence of the green silicates, which in many unweathered ores form the chief constituent.

In consequence of the difficulties in the theory of replacement, it is now believed by many authorities that the iron, either as carbonate or silicate, is an original constituent of the sediment, having been precipitated from iron-bearing solutions, either by some physico-chemical process or by the action of bacteria or other organisms. This idea is quite consistent with the physical structure and chemical composition of the oolite, but certain other facts are hard to reconcile with it. In the first place, these oolitic ironstones possess very close resemblance to oolitic limestones, and above all they are often highly fossiliferous, containing a perfectly normal and well-developed fauna. It is hard to see how these animals managed to live in a solution of iron salts sufficiently concentrated to precipitate the great quantities of iron actually found, more especially as these are always marine sediments, and are interstratified with other normal rock-types of obviously open-sea character. The presence of green iron silicate is not an insuperable objection to the replacement theory since the siliceous skeletons of ooliths and other structures consist of amorphous or colloidal silica, which is attacked by solutions with comparative ease, unlike quartz. In spite of much discussion and many able investigations, this question cannot yet be considered as settled[1].

Bog Iron Ores and Lake Ores. In the more primitive phases of the iron-smelting industry great use was made of ores belonging to the most recent geological formations: such are in fact superficial deposits and can sometimes be seen in the making at the present day in suitable localities. Such ores were used mainly because they were accessible, not because they were of good quality. Hydrated oxides of iron are extensively precipitated in the stagnant waters of marshes and lakes in many parts of the world, and they frequently form also in masses a few inches below the surface of the soil in badly-drained localities, giving rise to what is known by agriculturalists as an *iron pan*. Such pans are commonly found at the lowest limit of free air circulation in the soil and are thus apparently due to a process of reduction to an insoluble state of soluble iron salts held in the water. It is still an open question to what extent this precipitation is due to bacteria or other low organisms: there is, however, much evidence in favour of the idea. The action of iron-bacteria in reservoirs and water-pipes is well known and it is natural to suppose that the same thing occurs in natural waters. Moreover, there can be no doubt that the bedded ironstones of the Coal-measures and similar formations are in their origin essentially similar to the modern bog ores and lake ores.

At one time the deposits of lake ore in Sweden and Finland

[1] See especially an admirable discussion of the whole matter by Hayes, "The Wabana Iron Ore of Newfoundland," 1915. This author favours the theory of primary deposition of the iron.

were of considerable importance. These deposits are found chiefly in shallow water in sheltered bays of the lakes, where there is usually a good deal of vegetation. They are worked during the winter when the lakes are frozen: a hole is made in the ice and the ore broken up and lifted by means of rakes. It is stated that when such a deposit has been thus removed, a workable quantity of ore forms again in twenty or thirty years. An output of 5000 tons of lake ore was reported from Finland in 1913.

Lateritic Iron Ores. These also are modern superficial deposits. The origin of laterite is dealt with under the heading of aluminium (see p. 449). It need only be said here that some Indian laterites form iron ores of quite good quality, often with well-marked pisolitic structure. There is, however, among laterites a wide variation in composition, some varieties being ores of aluminium (bauxite) or manganese respectively. More or less ferruginous laterites have a very wide distribution in many parts of the tropics, forming in fact one of the most characteristic features of weathering in hot climates.

CHAPTER XVI

NICKEL, COBALT, MANGANESE, CHROMIUM

THE technical applications of nickel have increased greatly in importance of late years. Previously to 1890 its chief employment was as a constituent of a group of alloys of which "German silver" may be taken as the type. The introduction of nickel-plating gave a great impetus to the production, but of still more importance was the discovery of the valuable properties of nickel steel. At the present time nickel is very largely used for all the purposes named, also for many forms of electrical apparatus, and when alloyed with copper for coins of small denominations in place of bronze. It was estimated in 1913 that the total annual production of simple nickel steel was about 50,000 tons and of nickel-chromium steel about 100,000 tons. The useful range of nickel in alloy steels for various purposes is from 2 to 45 per cent., but the proportion most commonly employed is 3 or 4 per cent.; when used along with chromium the limit is about 3·5 per cent. An alloy called Monel metal, possessing very remarkable properties, especially high tensile strength and resistance to corrosion, consists of about 67 per cent. nickel, 28 per cent. copper and 5 per cent. iron and manganese. Consequently the annual output of nickel ores is now very large; in 1913 it amounted to 31,977 tons, calculated in terms of metal, and in 1918, at the height of the demand for munitions, to 46,487 tons[1]. Almost the whole of this output comes from two countries only, Canada and New Caledonia. The geographical distribution of nickel ores was thus of great advantage to the Allies. A considerable amount of natural nickel-chromium steel is made in America direct from ores found in the Mayari district of Cuba: the pig iron produced by smelting of these ores contains from 1 to 1·5 per cent. of nickel and 2·5 to 3 per cent. of chromium.

Nickel Minerals. Nickel is widely distributed in the earth's crust in unworkable forms: it is a constituent in very small amount of many igneous rocks and of many iron ores. Definite

---

[1] *Imperial Mineral Resources Bureau, Statistical Summary*, 1921, p. 63.

nickel minerals are also fairly numerous, but for the most part not commercially important. Nickel forms a series of sulphides and arsenides rather similar to the corresponding iron compounds, while many of them form mixed crystals with iron and cobalt in isomorphous series. From these a certain amount of nickel has been obtained, as for example in Norway and Germany. At the present time by far the most important source is the nickeliferous sulphide of Sudbury, Ontario, Canada, while the second source is the oxidized nickel ore of New Caledonia, a very different type of occurrence. Both of these are of great interest from the theoretical point of view, as well as commercially, and will need somewhat detailed description. Nickel is also produced in some quantity as a secondary product from the silver-cobalt ores of the Cobalt district in Canada, which are dealt with elsewhere (see p. 485). It should be stated that nickel and cobalt are nearly always found in association, but in very varying ratios.

The chief nickel minerals of the sulphide-arsenide group are: millerite, NiS, and niccolite, NiAs (rhombohedral); chloanthite, $NiAs_2$ (cubic-pyritohedral), rammelsbergite, $NiAs_2$ (orthorhombic), and pentlandite (Fe,Ni)S (cubic). A very large part of the production, however, comes from the so-called nickeliferous pyrrhotite, which is in part at any rate a mechanical mixture of sulphides of iron and nickel, chiefly pyrrhotite and pentlandite. Nickel is also obtained in considerable amount from oxidized ores, of which the most important is garnierite, a somewhat complex nickel-magnesium silicate, of varying composition. Annabergite is a hydrated nickel arsenate, $Ni_3As_2O_8 . 8H_2O$, found in Germany and in Canada. Breithauptite, nickel antimonide, NiSb, has been found in some quantity in the Harz Mountains.

On a mineralogical basis therefore the important ores of nickel can be classified into three principal groups, as follows:

(a) Sulphides, e.g. Sudbury; Norway.

(b) Silicates and oxidized ores, e.g. New Caledonia; Oregon.

(c) Arsenical ores, e.g. Cobalt, Ontario.

Other sources of less importance are:

(d) Electrolytic refining of blister copper.

(e) Earthy manganese ores.

(f) Nickeliferous iron ores, e.g. Cuba.

The sulphide ores also yield as by-products gold, silver, platinum, palladium and sometimes other metals: more or less cobalt is nearly always present, and most of the occurrences are also commercial sources on a large scale of copper as well as nickel. The arsenical ores of group (c) are as a rule of greater importance as sources of silver, cobalt and arsenic, the nickel being rather a by-product.

Nickeliferous Sulphide Ores. The study of the highly important ores of this type carried out of late years in Canada and in Norway has raised some general questions of ore-genesis of the highest possible interest. These occurrences have been most carefully investigated, but unfortunately much difference of opinion still exists as to the actual mode of origin of the nickeliferous sulphide masses. Some authorities believe that they are direct products of the differentiation of igneous magmas separated mainly by gravity, while others consider that they are, though primarily of igneous origin, yet in their present position mainly due to hydrothermal processes. The special points at issue will become more intelligible when a description has been given of one or two typical examples. From this point of view the most important are the occurrences in Norway and at Sudbury, Ontario.

Nickel-bearing Sulphide Deposits in Norway. Deposits of this class, some of them of considerable technical importance, are known in many localities in Norway: in all, about fifty have been enumerated. The first point to be noted is that all of these are in connection with basic igneous rocks, and especially with those characterized by rhombic pyroxenes; that is, with norites of the usual classification. These deposits have been comprehensively studied by many geologists, above all by Vogt, who has dealt in a masterly manner with the theoretical, as well as with the practical side of the subject[1].

---

[1] The literature is now very large. Several important papers by Vogt will be found in the *Zeits. für prakt. Geol.* 1893–5. This author has summarized his most recent results in a convenient manner in "Die Sulfid-Silikatschmelzlösungen," *Norsk Geol. Tidskrift*, vol. IV, 1917, separately published by Norges Tekniske Høiskoles Geologiske Institut, as Meddelelse No. 7, 1917, with further references. This is a summary of a larger work with the same title, the second part of which is as yet (1922) unpublished.

Most of the Norwegian occurrences are found, as before stated, in basic igneous rocks of subalkaline character, in which iron and magnesium preponderate over lime, leading to the formation of the rhombic pyroxenes, hypersthene or bronzite $(Mg,Fe)SiO_3$, rather than diopside; the chief felspar is a rather basic plagioclase, commonly labradorite, while quartz is rare or absent. These rocks generally contain about 50 per cent. of silica, and are thus typical norites. Furthermore, Vogt has shown that the sulphides tend to concentrate in those types which contain more hypersthene than corresponds to the eutectic ratio between hypersthene and plagioclase, that is, in the types specially rich in ferromagnesian minerals. This of course means that the sulphides are more soluble in ferrous and magnesian silicate melts than in lime-silicate melts. Some of the very basic norites, approaching pyroxenites, contain large proportions of disseminated sulphides, in exceptional cases even up to 50 per cent. Such extreme cases, however, must indicate incomplete differentiation and imperfect equilibrium in the molten state, besides perhaps in some cases, remixing of already separated fractions. The ores are commonly accompanied by dykes of acidic composition, which may be regarded as being of a complementary nature, representing the more acid portion of the primary magma, which has split into an acid and a basic sulphide fraction. Such dykes are considered to be valuable indicators of the possible presence of workable ore-bodies.

Most of the norite masses, which appear to be of late pre-Cambrian age, are intruded into crystalline schists and gneisses, and are often of a lenticular form in ground-plan, conformable to the foliation of the older rocks. Most of the richer masses of ore occur near the margin of the norite, though always inside the intrusive rock. However, when examined in detail they are always of very irregular form and usually distinctly streaky: in fact the whole mass of norite often shows well-marked flow-banding, due to movement subsequent to partial differentiation, and the streaks rich in sulphides then conform to this foliation. In nearly all cases the nickel is accompanied by copper, in amount sometimes up to half the quantity of nickel: cobalt is in much less amount, often only about one-tenth. Vogt has estimated that if the nickel-content were spread evenly all through the mass of norite it would in most cases amount to less than 0·1 per cent. of the whole.

At the well-known occurrence at Romsaas, in Smaalenene, the nickel-pyrrhotite is found in patches irregularly distributed in an intrusion of quartz-norite, forming an oval mass some 400 yards long by 150 yards wide; the ore is largely associated with a peculiar type of orbicular structure in the norite: the surrounding rocks are gneiss, with foliation striking N.W.–S.E. The richer streaks of ore contain no plagioclase at all but consist of sulphides and hypersthene, with a little biotite and hornblende. The Meinkjär ore-body in Bamle, though small, is of much interest since the geological relations are very clear[1]. The norite forms a mass about 100 yards long by 60 yards wide, lying in gneisses and schists. The sulphide ores are marginal and appear to dip under the norite, so that from Vogt's figure it may be concluded that the sulphides form a layer at the base of the intrusion. The outcrop of sulphides is not quite continuous all round the intrusion, but may thicken in depth. This ore is said to contain 20 per cent. of nickel. The richest nickel mine in Norway is that at Flaad in Evje, which lies in a large mass of somewhat decomposed norite of basic character: the ore forms a series of large, irregular, but generally parallel shoots, having a dip of about 40°, lying near the margin of the norite and accompanied by many acid dykes, which may be called aplite. It has been worked only to a depth of some 300 feet, but shows no sign of exhaustion at that depth. The yield of nickel usually runs to about 2 per cent. of the ore actually mined. Other important nickel fields in Norway are Erteli and other mines in Ringerike, and some occurrences in the neighbourhood of Tvedestrand and Arendal, on the west side of the Kristiania Fjord; also on the island of Hosanger, near Bergen, and various places in the Skjäkerdalen district near Trondhjem: all of these are of the same general character.

The Nickel Ores of Sudbury, Ontario. As already mentioned, the Sudbury area is now by far the largest producer of nickel, and since the mode of occurrence is also of great scientific

---

[1] Vogt, *Zeits. prakt. Geol.* 1893.

interest, it will be necessary to describe the geology in considerable detail. The Sudbury district, using the term in the broadest
sense, forms part of the great "Canadian Shield" of pre-Cambrian rocks. The general succession is shown in a somewhat
simplified and condensed form in the following table[1].

Keweenawan igneous rocks

intrusive contact ∿∿∿∿∿∿∿

Animikie Series ⎰sandstones, conglomerates,
9500 feet   ⎱slates and tuffs.

unconformity ∧∧∧∕∨∖

Greenstones

intrusive contact ∿∿∿∿∿∿∿∿∿

Timiskaming or ⎧
Sudbury Series ⎱quartzites, greywackes,
29,000 feet   ⎰arkose and conglomerate.
                ⎩

unconformity ∧∧∧∕∧∕∖

Grenville Series ⎰gneisses, schists, quartzites
no base seen   ⎱and limestones.

From the economic point of view the Keweenawan Series is
the most important, since the nickel ores belong to this group.
In this area the whole series is igneous and mainly intrusive,
including granite, gabbro, norite and quartz-norite (often called
norite-micropegmatite), besides numerous large basic dykes.

The general structure of the Sudbury area is a syncline or
basin of oval form, elongated in a N.E.–S.W. direction, and
measuring about 36 miles by 16 miles. This basin consists of the
Animikie Series, which may be subdivided into four groups as
follows:

⎧Chelmsford Sandstone.
Animikie ⎪Onwatin Slate.
Series   ⎨Onaping Tuff.
⎩Trout Lake and Ramsay Lake Conglomerates.

This series rests unconformably on all older rocks, while the
Keweenawan rocks, and especially the nickel eruptive, have been

[1] Miller and Knight, "Nickel Deposits of the World," *Report of Royal
Ontario Nickel Commission*, 1917, p. 105.

intruded more or less along the unconformity. It follows from this structure that the ore-bodies occur along two lines, on the north-west and south-east sides of the basin respectively, forming the southern and northern nickel-copper ranges. The ore does not form a continuous sheet, but occurs as more or less definite shoots, which are sometimes quite sharply bounded against the country rock, and sometimes grade into it. It is to be noted that the ore always occurs at or near the outer and therefore structurally lower margin of the norite, or occasionally as dykes running out from this margin, and therefore so far as surface indications go, the field relations suggest that the ore represents a heavy sulphidic layer that settled to the bottom of the norite sheet by gravity differentiation. (It is not clear

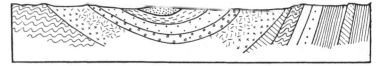

Fig. 68. Theoretical cross section of the Sudbury syncline (after Miller and Knight, *Nickel Deposits of the World*, 1917, fig. 11, p. 121).

whether the intrusion of the norite took place before or after the folding of the syncline, or perhaps concurrently with it, but the point is of minor importance.) However, mining operations in depth have disclosed facts that throw some doubt on this simple theory. In the first place the dip of the ore-bodies proves to be greater than would be expected if they followed the base of the norite exactly, and moreover some of the ore-bodies actually lie within the older rocks, including granites and greenstones. There is also a great deal of disturbance and brecciation along the ore-bearing zone, some of the ore-bodies being really breccias of country rock cemented by sulphides. It is also uncertain to what extent actual chemical replacement of the country rock has occurred. Taking all these facts into account, it has been suggested by some writers that the mineralization was brought about by hydrothermal solutions rather than by direct magmatic differentiation. However, all are agreed that the norite intrusion is in some way the source of the ores.

The Creighton ore-body is one of the most important, and

being geologically comparatively simple, it may be taken as a typical example. It forms at present one of the largest metalliferous mines in the world, and is said to have a reserve of at least 10,000,000 tons of workable ore. Here the relations of the different rock-types are as follows: the norite was intruded into and chilled against an older greenstone; then a mass of granite broke in along the norite-greenstone contact and penetrated the norite in hundreds of dykes along the contact; then a great zone of crush-breccia was formed along the same line; and after all this followed the period of ore-deposition, which is thus seen to be separated from the solidification of the norite by a period of granite intrusion and by considerable crust disturbance. The ore-body occurs at the junction of the granite and norite, but is said to lie largely in the granite foot-wall, the boundary towards the norite being somewhat sharp, although the norite is "spotted" with sulphide for nearly 2000 feet from the margin. The ore consists of a mass of rock-fragments cemented by sulphides; these rock-fragments are of all sizes, even up to 15 feet in diameter, though usually much smaller: when possible they are hand-picked from the ore. The ore-body has a known depth of about 2000 feet measured along its average dip of 45°, while its maximum horizontal dimensions are 1000 and 180 feet respectively: the form is very peculiar and irregular[1]. The hand-picked ore, which still contains up to 20 per cent. of silica, averages 4·44 per cent. of nickel and 1·56 per cent. of copper. Figures given in the publication cited above show the general relations of the ore to granite and norite: the brecciated character of the ore is obvious.

The Levack mine is situated near the westerly end of the northern range, and is notable from the fact that the ore-body lies wholly within the granite gneiss, though generally parallel to the base of the norite, from which it is distant from 40 to 220 feet. The ore-body at a depth of 350 feet has a somewhat lenticular form, measuring about 500 feet by 150 feet in plan: diamond drill cores show that it extends at least 1300 feet down the dip, which is about 40°.

[1] A model to scale is figured in the *Report of the Royal Ontario Nickel Commission*, 1917, facing p. 142.

The Crean Hill ore-body in Denison township is in some ways intermediate in character between those just described. At the surface it lies entirely in greenstone, but below the fifth level it is between greenstone and norite, much of the ore still, however, being in greenstone. The whole is cut by basic dykes of later date. The main body has been worked by nine levels connected with an inclined shaft to a vertical depth of about 750 feet: the average dip is about 70°, becoming steeper in depth. On the surface, where it was at first worked in a large open pit, it has a maximum extension north and south of about 300 feet, but on the ninth level it measures only about 100 feet. Two independent and smaller masses of ore at the greenstone-norite contact have also been worked. The reserves are estimated at about 2,000,000 tons. The hand-picked product still contains from 28 to 35 per cent. of silica, due to included rock-fragments; it carries 2·14 per cent. of nickel and 2·91 per cent. of copper[1].

Altogether something like fifty masses of nickel ore are known to exist in this district and there may be many more concealed beneath drift and swamps. However, they are all of the same general character, though varying much in detail. In general the ore consists of massive sulphides, enclosing blocks of rock of very various sizes, as before described: hence even the hand-picked ore is very siliceous, and in some cases as much as 50 or 60 per cent. of rock has to be picked out. The ore consists almost wholly of three minerals, pyrrhotite, chalcopyrite and pentlandite. The nickel exists as fine threads and strings of pentlandite traversing the pyrrhotite, and possibly segregated from solid solution in the latter after solidification.

In considering the genesis of these ore-masses the following points have to be taken into account:

(a) The occurrence of the ore at or near the base of the norite.

(b) The sheared and brecciated character of the ore-bearing zone.

(c) The intercalation in some places in point of time of granite intrusions between the injection of the norite and of the sulphides.

(d) The scattered distribution and peculiar shapes of the ore-shoots.

---

[1] The foregoing descriptions are taken almost verbatim from the *Report of the Royal Ontario Nickel Commission*, before cited.

The basal occurrence of the ore is easily explained on the theory of magmatic differentiation by the formation of two consolute liquid phases with limited mutual solubility, as worked out theoretically and experimentally by Vogt and explained in the general section of this book (see Chapter II). But the other three points above enumerated have been brought forward in some quarters as evidence showing the inapplicability of this simple theory. With regard to some of the other points, it may be suggested that sufficient attention has not been paid to the fact that the freezing point of a sulphide solution is lower than that of the corresponding silicate solution separated by differentiation: hence the solid norite would at one stage rest on a liquid sulphide layer. Any earth-movements occurring at this stage might undoubtedly lead to any amount of brecciation and injection of sulphide into any planes of weakness thus formed. Remarkable effects might then be produced in a liquid and very mobile layer thus intercalcated between two solid rocks. It is known that molten sulphides are very mobile liquids, in strong contrast to the viscosity of silicate melts at temperatures not far above their fusion points. It is not difficult to imagine that the time-interval between the solidification of the norite and of the sulphide might be long enough to allow of the injection and crystallization of a granite intrusion in places.

The irregular distribution of the ore-bodies has been explained as being due to the settling down of the heavy sulphide differentiate into inequalities in the original floor, which at first sight seems quite a plausible explanation. It is said, however, that some of them lie not in hollows, but above projections of the older rocks. This, however, presents no real difficulty if we adopt the idea of a crust disturbance acting on a liquid layer, as above described. Much stress has also been laid on the frequent occurrence of spots or blebs of sulphide about the size of peas, both in the norite and in the greenstones and granites. The blebs in the norite are just what would be expected on the differentiation hypothesis, by analogy with the blebs of metal and sulphide mechanically entangled in artificial slags, but those in the other rocks are more difficult to explain. It does not seem to be proved however that these spots really have the same composition as the principal ore-masses, and they may have a different origin.

At any rate, whether the nickeliferous sulphides are direct products of dry differentiation, or are more or less of the nature of pneumatolytic or hydrothermal deposits, it is at any rate clear that they are genetically connected with the norite, thus agreeing in all essential particulars with the similar ores in Norway and elsewhere. This is one of the best examples we possess of the undoubted association of a definite type of ore with a particular well-defined variety of igneous rock, namely, nickel-copper-iron sulphides with basic rocks characterized by abundance of rhombic pyroxenes.

The Nickel Ores of Insizwa, Griqualand East. Of much geological interest and with potential commercial value are the sulphide ores of the Insizwa region, on the borders of Griqualand East and Pondoland, South Africa. This occurrence shows both geologically and mineralogically a close analogy to the nickel deposits of Sudbury. The Insizwa range and some neighbouring mountains, running up to just over 6000 feet above sea-level, consist of the dissected portions of an enormous curved basin-shaped sheet of intrusive basic igneous rock, forming part of the dolerites of the Karroo series. This sheet is exceptionally thick, so that on the ground it shows some resemblance to a laccolith, but the upper surface as well as the lower is concave upwards. In the Insizwa range itself part of this great sheet consists of a lens-shaped mass, between 2000 and 3000 feet thick, of an exceptionally basic rock, best described as augite-picrite; this appears to have been segregated from the normal dolerite magma by gravity differentiation; the upper part of the sheet is of more acid composition. The ores lie at and near the bottom of the picrite mass, which dips inwards towards the centre of the basin at angles of from 20° to 25°. There is therefore every prospect that the mineralization will increase in amount when followed down the dip.

The sulphide ores occur partly as disseminations in the picrite, forming a great mass of rather low-grade ore, and partly as rich veins and stringers, which make a considerable show, but are in reality of only subordinate importance compared with the total

bulk of low-grade ore. The chief minerals are pyrrhotite, chalcopyrite and pentlandite, with locally, niccolite and bornite.
Platinum, gold and silver are all present, but it appears still to
be somewhat uncertain with which of the minerals the platinum
is associated.

From the foregoing brief description the analogy with Sudbury is obvious; it may be said, however, that at Insizwa the
evidence in favour of a direct magmatic origin of the ores is
even more conclusive. The bulk of the ore is actually in the
basic rock, and especially at the base of it, and the basic rock
itself (picrite) seems to be quite clearly a gravity differentiate

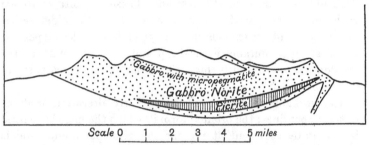

Fig. 69. Generalized section across the Insizwa Range (after du Toit). The
vertical scale is considerably exaggerated and the figure is diagrammatic.

in the lower part of an intrusion of somewhat higher average
silica percentage.

A considerable amount of prospecting has been done with
promising results for the production of copper, nickel and
platinum, and there is always the possibility of similar occurrences at the base of the other large igneous bodies of the
district, especially Tabankulu and Tonti, where pyrrhotite and
chalcopyrite can in fact be seen in many localities[1].

Nickel in New Caledonia. The French colony of New Caledonia is now the second producer of nickel ores, and until the
development of the Sudbury deposits it took the first place. This
island is situated in the Pacific between longitude 164° and 167°
E. and in S. latitude 20° to 22° or thereabouts. It is nearly

[1] du Toit, *The Geology of the Transkei*, Expl. Sheet 27, (Cape) Maclear and
Umtata, 1917, pp. 18–27. Goodchild, *Bull. Inst. Min. Met.* No. 147, 1916.

250 miles long and from 30 to 40 miles broad. Geologically it belongs to the folded arc stretching from New Guinea to New Zealand and forms part of the core of ancient rocks brought up by this line of folds. Hence the longer axis of the island lies N.W.–S.E. About one-third of the island consists of serpentines, more or less altered superficially, and these are the source of important deposits of nickel, cobalt and chromium ores. According to the latest and best geological map by Mr R. H. Compton[1], the largest expanse of serpentine is found in the south-east of the island, with scattered patches mainly near the west coast up to the north-west extremity. The largest mine, that of Tiebaghi, is near the north-west end. These serpentines appear to be intrusive into the crystalline schists of the older series, but may be of Cretaceous age, or even later. The serpentine forms steep mountains with sharp summits and most of the ore-deposits are at fairly high elevations, though not actually on the mountain tops.

The workable ores are in all cases clearly alteration-products of the serpentines: the original form in which the nickel occurred is not quite clear, but at present the principal ore-mineral is garnierite, a hydrated silicate of magnesium and nickel, sometimes of a bright green colour and earthy character, but often of a chocolate colour due to staining by oxides of iron. The average nickel-content is about 7 per cent. in the ore in its natural wet state, or 9 per cent. after drying at 100° C. The chemical processes involved in the decomposition of the nickel-bearing serpentine have apparently not been studied in detail, but may with safety be attributed to the peculiar processes of weathering that occur under tropical conditions, such as prevail in this island, where the temperature and rainfall are high. Much of the ore occurs as vein-like masses and as weathered shells encrusting boulders of more or less unaltered serpentine. The alteration of the rock does not usually extend to a greater

[1] Compton, "New Caledonia and the Isle of Pines," *Geogr. Journ.* vol. XLIX, 1917, pp. 81–103, and coloured map. This map is largely founded on that of Pelatan, *Journal de Génie Civil*, 1892. See also Glasser, *Les Richesses Minérales de la Nouvelle Calédonie*, Paris, 1904, and Colvocoresses, *Eng. Min. Journ.* 1907.

depth than 30 or 35 feet and all the mining is in open workings. In some instances the ore forms a continuous clay-like mass which can be worked in bulk, but usually a good deal of hand-picking is needed, to remove blocks of unaltered serpentine.

During the years 1900–15 the average export of nickel ore from New Caledonia was about 100,000 tons per annum, containing about 7000 tons of metal; from 1910 to 1915 about 26,000 tons of matte were also exported; in 1920 the exports were 90,000 tons of ore and 4437 tons of matte. This was estimated to contain 7400 tons of metallic nickel[1].

Deposits of garnierite ore of character somewhat similar to the above have also been worked on a small scale in Greece[2]. The nickel is here also associated with chromiferous iron ore, and is found in the Lokris and Thebes deposits, irregularly distributed in the upper part of a serpentine where it is overlain by a deposit of chrome-iron ore due to weathering of the serpentine itself. Some of the dry ore contains from 4 to $5\frac{1}{2}$ per cent. of nickel as garnierite and this must be regarded as due to a process of secondary enrichment at the base of the oxidized zone. The nickel, chromium and a small amount of cobalt are undoubtedly original constituents of the peridotite from which the serpentine was formed, just as in New Caledonia. The upper layers of the iron ore deposit also contain anything up to 1 per cent. of nickel as well as chromium, and like the rather similar ores of Mayari, Cuba, can be smelted direct to natural chrome-nickel pig iron suitable for steel making. These are described elsewhere under the heading of chromium.

Arsenical Nickel Ores. Under this heading are comprised a number of deposits, now comparatively unimportant, in which the nickel, usually associated with cobalt, occurs as a constituent of various primary arsenides, such as niccolite, $NiAs$, chloanthite, $NiAs_2$, and gersdorffite, $NiAsS$, as well as annabergite, $Ni_3As_2O_8 + 8H_2O$, and various other oxidation products of the primary minerals. At the present time the chief occurrence of this type is the Cobalt district of Ontario, which is described elsewhere (see p. 485). Ores of this kind have been worked in

[1] *Imperial Mineral Resources Bureau, Statistical Summary*, 1913–20.
[2] Scott, *Journ. Iron and Steel Inst.* vol. LXXXVII, 1913, pp. 447–67.

the past to a considerable extent in France, Saxony, the Harz Mountains, Bohemia and elsewhere. They usually form veins of a perfectly normal type, and do not need any detailed description, as the production of nickel from such sources is now negligible in comparison with the output of ores of the other types.

Cobalt. The technical uses of cobalt are at present somewhat limited, but it is possibly a metal with a future. In many of its properties it has a strong resemblance to nickel, which it may replace for some purposes when supplies of the latter begin to run short. In earlier days the chief use of cobalt was as a colouring matter, since many of its compounds show a beautiful rich blue tint. Like nickel, it can be used for plating metals, although its colour is not quite so good, having a slight pinkish tinge. It has been stated that 3 or 4 per cent. of cobalt confers valuable properties on certain forms of high-speed tool-steel, mainly by increasing the red-hardness of the steel and thus allowing the cutting-tool to run at a higher speed. In this it is unlike nickel, which is deleterious in such special steels[1]. Cobalt-chromium and cobalt-molybdenum steels are actually on the market, but their value cannot yet be regarded as fully established. The magnetic value of cobalt steels is unusually high and they are preferred by some makers to tungsten steel for permanent magnets. At the present time an important use of cobalt is as the principal constituent of a group of non-ferrous alloys, known collectively as "stellite," which possess remarkable properties, resembling some special forms of steel, but possessing the additional advantage of incorrodibility and resistance to acids: a typical alloy of this class consists of cobalt 60 per cent., chromium 25 per cent., and tungsten 15 per cent.; sometimes a little molybdenum is added; in other varieties this metal entirely replaces tungsten.

The principal cobalt minerals are as follows:

Smaltite or tin-white cobalt, $CoAs_2$, a cubic mineral of white or grey colour, usually massive.

Cobaltite, $CoAsS$, cubic; often in forms like pyrites, or massive.

---

[1] Hibbard, "Manufacture and Uses of Alloy Steels," *Bull.* 100, *U.S. Bureau of Mines*, 1916, p. 60.

Erythrite or cobalt bloom, $Co_3As_2O_8 + 8H_2O$, a mineral of a bright pink colour, generally found as an incrustation on primary cobalt ores, owing to oxidation.

Asbolane, a mineral of variable composition, chiefly consisting of oxides of manganese and cobalt, formed by oxidation of primary ores.

The common association of cobalt with silver ores has long been recognized, especially in Germany, where this forms a definite type of most of the older classifications of metalliferous veins. Cobalt is nearly always present in minor proportion along with nickel ores, but at the present time the chief source of the metal is the famous deposits of silver and cobalt in the Cobalt district of Ontario, Canada. The metal is, however, of wide distribution in small quantities, being found as an accessory in many deposits of very varying types.

The Cobalt–Silver Lodes of Saxony. Although of little importance at the present day, these lodes are classical in the history of mining geology. The lodes of Annaberg were discovered in 1492 and soon gave rise to an important mining industry, mainly for their silver content, but all this is now mainly of historical interest, since in Saxony the scientific study of ore-deposits had its origin in the writings of Werner and his followers[1].

The cobalt-silver lodes form one of the main groups of the silver lodes of the ordinary German classification, and can be again subdivided according to the presence or absence of other constituents: three groups are thus obtained[2]:

(a) Silver-cobalt lodes with arsenic and nickel, e.g. Annaberg.

(b) Silver-cobalt-bismuth lodes, e.g. Schneeberg.

(c) Silver-cobalt-uranium lodes, e.g. Joachimsthal.

The Upper Erzgebirge. This region lies on both sides of the crest of the chain, partly in Saxony and partly in Bohemia. The country rock consists of various forms of granitic gneiss and granite, with quartzite, greywacke, conglomerate, hornfels, crystalline limestone and hornblende-schist, all intensively meta-morphosed. The cobalt-silver lodes are arranged in two sets

[1] Mining around Freiberg apparently began seriously about 1160.

[2] Beyschlag, Krusch and Vogt, trans. Truscott, vol. II, p. 655.

striking north-south and east-west respectively: as a rule they are short, not more than half a mile along the outcrop; in the Marienberg district, however, the prevalent strikes are N.E.–S.W. and N.W.–S.E., while the lodes are longer. The most important area is that round Annaberg. The silver minerals include native silver, argentite, cerargyrite and dyscrasite, with cobalt and nickel minerals, while the gangue consists of quartz, barite, fluorite and dolomite. It is evident from the descriptions that we have here a complicated case of the superposition of secondary zones upon primary zones, which are difficult to disentangle. In the sixteenth century silver worth about a million sterling was produced in this district and it is stated that from 1701 to 1850 about 7850 tons of cobalt ore were won, but the metal content is not given.

The most important silver-cobalt-bismuth lodes are in the district of Schneeberg and Neustädtel, due west of Annaberg. Most of these lodes contain but little silver and have been mainly worked for cobalt, in the early days chiefly for blue pigments. Around Neustädtel lodes are extraordinarily numerous, striking more or less N.W.–S.E. or a few degrees on either side. The chief ore-minerals are smaltite, asbolane, erythrite, niccolite, chloanthite, native bismuth and arsenopyrite with a little silver and uranium: the gangue minerals are quartz, calcite and dolomite. The lodes vary from 2 to 10 feet in width. Other lodes around Schneeberg and Schlema are exceedingly rich in silver, but carry only subordinate cobalt, nickel and bismuth in a gangue of barytes. Both types lie in metamorphic schists surrounding granites, in which the cobalt content falls off or ceases altogether. The lodes of the Johanngeorgenstadt area, south-east of Schneeberg, are still of some importance; besides silver and cobalt they contain a good deal of bismuth and uranium, forming a transition to the rich uranium ores of Joachimsthal over the Bohemian border. These are described in some detail under the heading of uranium deposits (see p. 444), but must be mentioned here as they also carry a good deal of cobalt.

Cobalt in Canada. The silver-cobalt-nickel arsenides of Coleman and adjacent townships, commonly known as the Cobalt district, in the province of Ontario, are now the chief sources of the world's supply of cobalt, which is recovered partly as oxides and salts, partly as metal. The total production of cobalt in 1916, calculated as metal, amounted to about 627 tons. The occurrence of these ores is of remarkable interest from the scientific point of view, but as the silver content is of far more importance they are treated under the heading of silver. Deposits of somewhat similar type have also been found in the neighbouring Timiskaming County in Quebec. A considerable amount of cobalt is also contained in the nickel-pyrrhotite ores of Sudbury: in fact nearly all occurrences of nickel-bearing sulphides also carry a greater or less proportion of cobalt.

It is understood that a good deal of cobalt is also recovered as a by-product in the metallurgical treatment of copper ores, and especially from those of Katanga, Belgian Congo.

Manganese. Under modern industrial conditions and especially in view of the recent enormous development of steel-making, manganese has become of first-class importance. Although the pure metal itself has no technical applications, its compounds are essential for the proper working of steel-making processes, and the shortage of manganese was one of the greatest difficulties that German metallurgy had to contend with during the war, no proposed substitutes having apparently proved satisfactory. Manganese also enters into a large number of other technical processes: it is of much importance in glass manufacture, to counteract the green tinge imparted by traces of iron, and as a colouring matter for pottery, tiles and bricks, as well as for paints. It is also employed in various forms of dry batteries and for the manufacture of chlorine, bromine and iodine, and the use of potassium permanganate and similar compounds as disinfectants and deodorizers is well known (Condy's fluid, etc.). Hence the demand for manganese is very large and available supplies on a large scale are strictly local, being mainly confined to three countries, India, Russia and Brazil, though manganese minerals in small quantities are widely distributed; these small occurrences are often very impure and of low grade.

The importance of manganese in the manufacture of munitions of war, chiefly steel, is sufficiently obvious; owing to the cutting off of the Russian supply great development took place in

Brazil, where the output rose from 122,000 tons in 1913 to 532,000 tons in 1917. Russia reached the high-water mark with 1,171,000 tons in 1913 and it is a somewhat significant fact that the German production suddenly jumped from 92,000 tons in 1912 to 330,000 tons in 1913. In that same year the Indian shipments amounted to 832,000 tons, much of which also went to Germany, while from 1915 onwards the American steel trade was seriously handicapped by the shortage of manganese, until home supplies were developed.

Manganese Ores. The number of minerals that can be worked as sources of manganese is unusually large, because the metal itself possesses a considerable variety of oxides, most of which occur in nature: several of the manganese ores show some affinity to the corresponding iron compound.

Hausmannite, Mn_3O_4, with 72 per cent. of metallic manganese, is the richest ore, but is not very common or important. Manganite, $Mn_2O_3 . H_2O$, when pure contains 62·5 per cent. of metal: it is quite analogous to göthite among the iron minerals. The composition of braunite is rather uncertain: it is essentially Mn_2O_3, but always contains a considerable amount of silica, up to 8 or 10 per cent.; hence it is generally regarded as a mixture of Mn_2O_3 and $MnSiO_3$. It is impossible to state definitely, however, in what form the silica exists. Pyrolusite, MnO_2, when pure contains about 63 per cent. of manganese and generally occurs in a massive or powdery form. It is an important ore. Psilomelane is a substance of variable composition, containing various oxides of manganese, often with barium, potassium and water: it grades into an impure earthy substance known as wad.

The manganese silicate, $MnSiO_3$, rhodonite, which is a pyroxene, appears to be in many cases the primary ore, from which the others are formed by oxidation. Manganese carbonate, rhodochrosite or dialogite, $MnCO_3$, is a member of the rhombohedral carbonate group: it is commonly an oxidation product of rhodonite. The silicate and carbonate ores are not of much practical importance for metallurgical purposes. Certain peculiar manganese minerals only found in India will be mentioned later.

In America a large amount of manganese is obtained from the residues of the zinc-manganese ores at Franklin Furnace, N.J.

This occurrence is described in detail elsewhere. Some varieties of Indian laterite are workable manganese ores (see also p. 449).

Most iron ores contain more or less manganese and a complete gradation can be traced from pure iron ores to pure manganese ores. In steel-making the manganese may be added to the charge either as a high-manganese alloy, known as ferro-manganese, or as an alloy much poorer in manganese and richer in iron, called spiegeleisen. Ferro-manganese is usually made from high-grade manganese ores, while spiegeleisen is made from manganiferous iron ores. Hitherto these alloys have been made chiefly in the blast-furnace, but the manufacture of ferro-manganese in the electric furnace is likely to be of importance in the future where water power is available. It usually contains 60 to 80 per cent. of metallic manganese with 6 or 7 per cent. of carbon and not more than 1 per cent. of silicon. In spiegeleisen the percentage of manganese is usually below 25. For some purposes an alloy called silicospiegel, with as much as 10 per cent. of silicon, is employed.

Manganese Production. General. According to the statistics given in *The Mineral Industry*, 1918, in 1913, the last year that can be considered normal, manganese ores were produced on a commercial scale by the following countries: Austria-Hungary, Bosnia, Brazil, France, Germany, Greece, India, Italy, Japan, Russia, Spain, Sweden, United Kingdom and United States. As to many of these little geological information is available and would not be of much value. The figures for the United States are however of interest as showing how in a period of stress low-grade home supplies can be developed to replace high-grade imported ores.

year	tons
1914	2677
1915	9865
1916	27431
1917	114216
1918	304366

The above figures refer only to ores containing more than 40 per cent. manganese (in the case of 1918 more than 35 per cent.). Besides this there was produced in 1918 a total of no less than 1,500,000 tons of manganiferous ores with less than

35 per cent. manganese: this includes 146,000 tons of zinc ore residuum from New Jersey (see p. 305).

The British Isles. The output of manganese ores in the British Isles has always been small: in 1906 it reached high-water mark with 22,762 tons, but is usually between 4000 and 6000 tons: in 1917 it rose again temporarily to 10,000 tons under the auspices of the Ministry of Munitions.

Manganese ores have been worked to a small extent in Cornwall, as at Restormel near Lostwithiel, in the Launceston and Brent Tor district and near Exeter. Most of these mines appear to be worked out.

The most important British deposits are situated in North Wales, in Caernarvonshire, Merionethshire and Denbighshire[1]. The manganese mines of Caernarvonshire are not far from Aberdaron. They consist of beds, mainly of manganese carbonate with some silicate, interstratified with shales and flags containing *Lingula* and graptolites of Ordovician age. The manganiferous bed at Nant, near Llanfaelrhys, varies from 10 to 20 feet thick and contains 30 to 36 per cent. of manganese. At Benallt there are several similar beds folded into a syncline.

In the Llanbedr district of Merioneth the manganese occurs as a bed of mixed rhodochrosite and rhodonite interstratified with the Harlech grits; it varies from 10 to 20 inches in thickness and is altered at the outcrop to black oxides: this is the richest ore, being mainly pyrolusite. The bed is traceable along its outcrop for many miles. A similar bed has also been recognized by Mr T. C. Nicholas in the St Tudwal's Peninsula, Caernarvonshire[2], but this appears to be less rich in manganese.

Manganese ore has also been worked at several localities near Arenig. It fills joints and fissures in a volcanic ash and in lavas and small intrusive masses. Some botryoidal masses of psilomelane contain up to 54 per cent. manganese. Black oxide of manganese was also formerly worked at Nant Uchaf, near Abergele, in the Carboniferous Basement Beds[3].

---

[1] *Special Reports on the Mineral Resources of Great Britain*, vol. I, 1916, pp. 43–54.   [2] *Quart. Journ. Geol. Soc.* vol. LXXI, 1915, pp. 89–91.

[3] "The Geology of Rhyl, Abergele and Colwyn," *Mem. Geol. Survey*, 1885, p. 56.

India. The important manganese deposits of India belong to three types: two of these occur in rocks of Archaean age, while the third type consists of lateritic deposits of uncertain but fairly recent date derived from the older formations. Besides these smaller and less important deposits are found in a great variety of other formations. Only the three important types will here be discussed.

The older of the Archaean manganiferous rocks, commonly known as the Kodurite series, chiefly developed in Vizagapatam and Ganjam, are of igneous origin: the workable deposits being formed by chemical and mineralogical alteration subsequent to intrusion. The second type, known as the Gondite series, chiefly found in the Balaghat, Bandara, Chhindwara, Nagpur and Narukot districts, is in its present form of metamorphic origin and may be described as a series of manganiferous schists and gneisses. Both of these series are of great interest from the scientific point of view, since they constitute well-marked and peculiar petrological types.

The Kodurite Series. The rocks of this series consist of plutonic intrusions which appear to have been injected at a very early date into the still more ancient gneissose and schistose rocks known as the Khondalites. They vary greatly in silica percentage, ranging from ultrabasic to acid, but all are characterized by containing manganiferous minerals, especially various members of the pyroxene group: they are also accompanied by veins of pegmatite and quartz. These rocks therefore constitute an excellent example of a petrographical province, in which manganese is the characteristic and peculiar element. The different types have no doubt been formed by differentiation from one original manganiferous magma. The dominant minerals are quartz, potash-felspar, rhodonite and other manganese pyroxenes, manganese garnet (spandite) and apatite, with accessory biotite, graphite and magnetite. Of these minerals quartz is naturally confined to the more acid members of the series, which are essentially granites, while in the ultrabasic varieties felspar is absent. The typical kodurite, which yields most manganese by its alteration, is a basic rock: it has never been found unaltered, but must have consisted of orthoclase, garnet and apatite, often with pyroxene: thus it is really a garnetiferous gabbro. The ultrabasic forms are extremely rich in pyroxenes, while olivine is not recorded. Analysis shows in the unaltered rocks a considerable proportion of manganese, ranging up to 10 per cent. MnO. Nearly everywhere the rocks are much altered and this alteration has taken the special form of concentration of the manganese; sometimes the result is a mass of crumbly lithomarge containing garnet and apatite, but more commonly the change has gone much further, every mineral having been more or less replaced by manganese compounds. These changes are supposed to be brought about by groundwater containing carbon dioxide, which dissolves the manganese compounds and re-deposits them. A large proportion of the manganese appears to be derived from the garnet.

There are two chief types of manganese ore: (a) a mixture of pyrolusite and psilomelane, (b) a mixture of psilomelane and braunite. Some varieties of the ore contain a good deal of iron and may be conveniently classed as ferruginous manganese ores or manganiferous iron ores. The following table shows the average proportion of the important constituents present in samples of both classes from the Kodur district.

	Manganese ores	Ferruginous manganese ores
Manganese	44·34	36·75
Iron	9·08	15·20
Silica	4·15	5·72
Phosphorus	0·32	0·33

The ores usually occur as long narrow belts interbedded with the gneissose rocks of the neighbourhood: they generally show a steep dip, and as the workings are entirely open-cast and no boring seems to have been done little is known as to their behaviour in depth, and it is uncertain whether the manganese beds cut across the foliation of the gneisses or are parallel to and folded with them: both interpretations have been suggested in certain cases.

The Gondite Series. The most important of the manganese-bearing rock-series of India are of a different character from the foregoing, since it is generally agreed that the present condition of the rocks is due to metamorphism. They are believed to have

originated in Archaean times as manganiferous sediments belonging to the Dharwar system; they consisted originally of a
great thickness of conglomerates, sandstones and clays with
interbedded chemical limestones and ironstones. The whole
series was afterwards strongly disturbed and metamorphosed,
so that these rocks became intimately folded in with the older
gneisses, from which in some areas they are hardly distinguish-
able. For some reason which is not understood parts of this sedimentary series was very rich in compounds of manganese, which
later, owing to metamorphism, gave rise to the present great
development of manganese minerals, as well as the associated
bands of haematite rock and jasper which are so characteristic
of the Dharwars, as of many other pre-Cambrian formations
(compare the Lake Superior region and parts of South Africa).
In different parts of the country, however, the intensity of the
metamorphism has varied a good deal, so that various grades
of alteration may be recognized. The most important of these
grades is that known as the gondite series; this is the most
highly metamorphosed of all.

The gondite rocks are chiefly developed in the following areas:
Narukot State, Bombay Presidency; Jhabua State in Central
India; the districts of Balaghat, Bandara, Chhindwara and
Nagpur in the Central Provinces. Of these the third is the most
important and will alone be dealt with here.

The typical unaltered rock of the gondite series is one composed entirely of a manganese garnet near to spessartite, and
quartz; other varieties contain apatite and manganese silicate
(rhodonite). It is supposed that the manganese ore-bodies have
been formed from these different varieties by various forms of
chemical alteration, especially by concentration of the manganese, similar to that occurring in the case of the kodurite
series. This is not difficult to conceive, since an analysis of a
typical gondite contains 10·36 per cent. of metallic manganese,
while rhodonite, which is abundant in some varieties, contains
over 41 per cent. of manganese. This change was probably
brought about by the influence of ground-water. The process is
essentially one of oxidation, and the resulting ores usually consist of braunite with either psilomelane or hollandite, the latter

being an oxide of manganese, iron and barium of rather uncertain composition.

As before stated, these rocks are really folded sedimentary strata, hence the manganese-bearing beds tend to occur in synclinal forms. Not much is known as to their persistence in depth, but since the alteration is supposed to have occurred at or about the time of the folding, there is no reason why the concentration of ore should not extend to the bottoms of the synclinal troughs, and in all probability they do so, but hitherto there has been little or no underground mining, so that this question has not yet been properly tested. The average analyses of a very great number of samples work out somewhat as follows: manganese, 50–54 per cent.; iron, 6–8 per cent.; silica, 6–8 per cent.; and phosphorus, 0·07 to 0·11 per cent. These figures are for first-grade ores and represent the qualities usually stipulated for by large buyers. Second grade ore runs from 44 to 48 per cent. manganese.

Manganese in Laterite. Compared with the other two, this source of manganese is not now important and may be dismissed briefly. The origin of laterite is discussed elsewhere (see p. 449) and it will suffice to say that some varieties of laterite derived from manganiferous rocks contain a sufficiently high proportion of manganese to form workable ores. The manganiferous laterite of Goa belongs to the low-level group, but is formed by disintegration of rocks in place: it is very similar to the high-level laterites of Belgaum. The other workable occurrences are mostly in a variety of laterite, called lateritoid by Fermor, which is always derived from the rocks of the Dharwar series. The ores usually consist of psilomelane or pyrolusite or a mixture of the two, usually closely associated with iron ores, and grading into limonite and other ferruginous types. Many of these are really manganiferous iron ores. In Mysore (Chitaldrug, etc.) the ores are often oolitic: silica and phosphorus are commonly lower than in the types previously described. The best qualities contain manganese 44–56 per cent., iron 2–10 per cent., silica 1–3 per cent. and phosphorus 0·015 to 0·060 per cent.[1].

[1] Fermor, *Mem. Geol. Survey India*, vol. XXXVII, 1909.

Russia. In the years before the war the Russian Empire was the largest producer of manganese ores in the world. In 1906 and 1907 the yield averaged just over one million tons, and in 1913 the high-water mark was reached with 1,289,370 tons. After this it naturally fell suddenly to almost nothing, and at present little or no information is available, owing to the disturbed state of the country. Manganese ores exist in three different districts: in the Caucasus, in the Urals and in the Government of Ekaterinoslav. Of these the first is by far the most important and is responsible for about 90 per cent. of the total output.

The manganese deposits of the Caucasus lie in the neighbourhood of Chiaturi, 42 kilometres north of Kvrilli station on the Poti-Tiflis (Transcaspian) railway, in the Government of Kutais. The country consists of granitic and syenitic rocks overlain by Cretaceous and Tertiary sediments, forming a plateau dissected by deep valleys. The ore occurs in a brown sandstone of Miocene age at a considerable elevation above the floor of the main valleys: it forms a bed usually 6 or 7 feet thick and the reserves have been vaguely estimated at from 22,000,000 to 110,000,000 tons, but there are few satisfactory data to base a calculation on; the average percentage of manganese is about 45, but this is raised to 51 or 52 per cent. by cleaning and sorting. From Chiaturi the ore is transported by a narrow gauge railway to Sharapan on the main line, and reloaded into broad gauge trucks for transport to Poti on the Black Sea[1].

Over an area estimated at about 8 square miles on either side of Nikopol on the Dniepr in the Government of Ekaterinoslav, manganese ore forms beds averaging from 3 to 5 feet in thickness in Oligocene strata overlying granite and gneiss. It consists of nodules of psilomelane and pyrolusite in a bed of sandy clay and averages 46 per cent. of manganese, with 12 per cent. of silica and 0·25 per cent. of phos-

---

[1] Haenig, *Berg. u. Hüttenm. Jahrb.* vol. LVII, 1909, pp. 140–5. Drake, *Trans. Amer. Inst. Min. Eng.* vol. XXVIII, 1898, pp. 191–208. Harder, *Bull. Amer. Inst. Min. Eng.* Dec. 1916, p. 2223. Scott, *Eng. Min. Journ.* vol. CIV, 1917, p. 647.

phorus. The highest grades run up to 56 or 57 per cent. of manganese[1].

Not much is known as to the geology of the manganese deposits of the Urals, which in the years before the war produced a few thousand tons annually. A new region has recently been discovered in the Gaisinsk district, Government of Podolia, south-west Russia, 220 miles from Odessa, where there are large deposits of high-grade pyrolusite over an area of about 30 square miles[2].

Brazil. Among the manganese-producing countries of the world Brazil now takes the third place and may in time take the first rank, since the supplies of ore, as yet only just touched, are undoubtedly enormous. In many of their geological features they strongly resemble those of India, and it is probable that they are due to very similar processes. In all likelihood many as yet unknown deposits also exist in the less explored parts of this vast country. The stoppage of shipments from India and Russia to the United States during the war gave a great impetus to the manganese mining of Brazil and production rose from 122,000 tons in 1913 to 532,000 tons in 1917, though there has since been some falling off, largely owing to difficulties of transportation and the inefficient management of the State railways.

The most important producing district is in the State of Minas Geraes[3]. There are two chief areas, one near the town of Miguel Burnier on the southern edge of the great iron ore region, the other near Lafayette, or Queluz, about 20 miles further south. The manganese deposits of Miguel Burnier extend from a point a little west of that town to Ouro Preto and form a narrow belt closely associated with the Itabira iron formation of Algonkian age (see p. 329). This consists of an alternation of haematite-schist (itabirite), mica-schist, calc-schist and limestone. The limestone forms lenses in the schists and the lenticles of manganese

---

[1] Harder, *Bull. Amer. Inst. Min. Eng.* May, 1916, p. 770. Haenig, *loc. cit.*

[2] Curtis, *Manganese Ores*, London, 1919, p. 77.

[3] Most of the information as to the manganese deposits of Brazil is taken from Miller and Singewald, *The Mineral Deposits of South America*, New York, 1919, pp. 177–92. See also Harder, *Trans. Amer. Inst. Min. Eng.* vol. LVI, 1917, p. 31.

ore in all probability are alteration products of similar lenticles of limestone, which have been replaced by manganese minerals; some authorities consider that the manganese has been there from the beginning, as in the gondite ores of India. At any rate it is clear that the origins of the iron and manganese ores of this area are parts of a single problem of ore-genesis, analogous to that of the Lake Superior ores and of the Indian gondite type. This question is discussed in detail elsewhere (see p. 326). The ore consists for the most part of amorphous or finely-crystalline manganese oxides, chiefly psilomelane, with pyrolusite and some wad. It averages 50 per cent. manganese, 1 per cent. silica and 0·03 to 0·05 per cent. phosphorus.

The mode of occurrence of the Lafayette deposits is rather different. This locality is outside of the iron-bearing area, and its ores are found associated with the gneisses, schists and granites of the basement complex. The manganese deposits form lenticles in this complex and represent alteration products of an original manganiferous rock, probably sedimentary, and composed of manganese carbonate and silicates. The Piquery ore-body, which is the best known, is described by Derby as presenting the appearance of a gossan resulting from alteration of a vertical rock-mass, some 30 to 40 feet wide: the ore is a hard spongy black oxide, chiefly psilomelane, with included masses of unaltered rock composed chiefly of manganese garnet. There is no doubt that the ore results from alteration of a garnet rock, which in places contains also a good deal of quartz. The ore-bodies at the Morro da Mina mine consist of vertical lenticles of manganese ore, mostly psilomelane, with manganite and pyrolusite: the strike is north-west, and the shoots pitch to the south-east. The average manganese-content is about 50 per cent., with about 3 per cent. of iron, 1·75 per cent. silica and 0·07 per cent. phosphorus. The unaltered rock consists of rhodochrosite, spessartite garnet, rhodonite and tephroite (manganese olivine). This rock is evidently formed by metamorphism of a siliceous and aluminous rock originally largely composed of manganese carbonate: the process being exactly analogous to the metamorphism of siliceous and aluminous limestones and dolomites, with the substitution of manganese for calcium and magnesium in the original rock. The following analysis shows the composition of the original rock at Piquery[1].

			per cent.
CO_2	...	...	22·62
SiO_2	...	...	11·80
Al_2O_3	...	...	7·50
MnO	...	...	47·52
CaO	...	...	3·76
MgO	...	...	6·27
			99·47

Chromium. Although the element chromium enters into the composition of a fair number of minerals, only one of these is of any practical importance as a source of chromium compounds, namely, chrome-iron ore: when pure this is known as chromite, but its composition shows a good deal of variation, since it is a member of the spinel series of minerals which exhibit in a very marked degree the property of isomorphism or the formation of mixed crystals. The composition of the spinel group may be represented by the general formula $R''O . R'''_2O_3$, in which R'' may be replaced by magnesia or ferrous iron, while R''' may be aluminium, ferric iron or chromium: zinc and manganese may also be found, though rarely, in the first group (see p. 305). According to this scheme chromite in the strict sense is $FeO . Cr_2O_3$, while magnetite is $FeO . Fe_2O_3$. It appears, however, that some so-called chromites contain a good deal of the magnetite molecule, while other varieties, often called picotite, contain more or less alumina. A commercial sample of chromite should contain not less than 50 per cent. Cr_2O_3.

Chromite and the similar allied species are essentially minerals of the igneous rocks and are specially associated with the most basic varieties. The majority of the workable deposits of chromite are found in connection with peridotites, or with their alteration products, the serpentines, and it is noteworthy that this mineral is often accompanied by platinum (see p. 393). This is one of the best examples of the definite association of particular metals with special types of igneous rock. Most, if not all, of the larger masses of chromite are clearly segregations due to differentiation of an ultrabasic magma, just as titaniferous iron ores are segregated from basic magmas (gabbro or norite) and

---

[1] Miller and Singewald, *loc. cit.* p. 184.

magnetite from granites. For a further discussion of this subject
see p. 148.

The chief technical applications of chromium are in the manufacture of special varieties of steel, in refractory bricks and in
the preparation of potassium bichromate, a compound used in
a large number of chemical processes and in the construction of
electric batteries. Simple chrome steel is not now largely made,
but chromium is still one of the most important constituents of
the more complex alloy steels. Simple chrome steel possesses
exceptional hardness and is used for burglar-proof safes and
strong rooms, for stamps and rollers of crushing mills and for
the balls of ball mills and for ball-bearings and for small tools,
such as files. Nickel-chromium steels are perhaps the most important of the alloy steels for general purposes and in these the
proportion of chromium is usually from 1 to 1·5 per cent., with
about the same amount of nickel. Chromium is also an important
constituent of the class of non-ferrous alloys known as stellite.

Chromite is a mineral of wide distribution but at the present
time its exploitation on a commercial scale is restricted to a few
countries, namely, New Caledonia, Russia, Rhodesia, Greece,
Turkey and Canada, and during the war a good deal of low-grade
ore was worked in the United States to take the place of imported
supplies, then unobtainable.

From 1828 till about 1850 most of the world's supply of
chrome ore was exported from Baltimore, being obtained from
the peridotites of Maryland and the adjoining districts of Pennsylvania. However, the development of rich deposits in Asia
Minor cut out this industry, which came to an end about 1860,
though a small production of alluvial chrome sand for local use
persisted for many years. The geology of this area was described
in 1899 by Pratt[1], who pointed out that the chromite ore-bodies
always occurred near the margins of masses of peridotite and
recognized clearly that they were due to differentiation in ultrabasic intrusions. The true character of such magmatic segregations of chromite was however first clearly expounded by Vogt
in 1894[2]. In the course of his discussion of the igneous origin of

[1] *Trans. Amer. Inst. Min. Eng.* vol. xxix, 1899, pp. 17–39.
[2] *Zeits. für prakt. Geol.* vol. ii, 1894, pp. 389–94.

ores in general this author gave a good description of the chromite deposits of Hestmandö, northern Norway, which occur in a perfectly fresh peridotite, thus disproving the former idea that chromite was formed by decomposition of chrome-bearing silicates, an idea promoted by its common occurrence in serpentine. Vogt showed that every gradation existed between a peridotite with a little accessory chromite and picotite, through olivine-chromite rocks, to masses of pure chromite in veins, streaks and lumps of all sizes. He furthermore pointed out that in the Röros district of central Norway chromite occurred in an exactly similar manner in a serpentine. In such chrome-bearing peridotites it is often found that the silicate minerals, especially diopside, contain chromium and are of a bright green colour. So far as is known all the workable deposits of the world are of similar type, namely segregations due to differentiation in ultrabasic rocks, either fresh peridotites or more commonly altered to serpentine.

In New Caledonia a magnesia-bearing chromite (magnesio-chromite) occurs in large quantities in serpentinized peridotites, either disseminated in small grains and octahedral crystals, or in segregated lumps and veins. Some of these original segregations are mined, but most of the workable deposits, often erroneously described as alluvial, consist of grains and crystals disseminated in a soft yellow clay formed by weathering of the serpentine, a true residual deposit. At Tiebaghi masses of chromite are found cemented by secondary asbolane, or earthy cobalt ore. The chromite industry of New Caledonia is very important and there are numerous mines and open workings, as at Tiebaghi, Plum, St Vincent, Cape Goulvain, Nakety and others[1]. The Tiebaghi mine is said to be the largest single producer of chromium in the world, and the deposits are closely associated with the nickel of New Caledonia (see p. 373).

Of late years a very important source of supply is the chromite deposits of the Selukwe district in Rhodesia[2]. Here the mineral is found in lenticular masses in a series of talcose schists with

[1] Lacroix, *Minéralogie de la France et de ses Colonies*, vol. IV, 1910, p. 313. Compton, *Geogr. Journ.* vol. XLIX, 1917, pp. 85 and 92.

[2] Zealley, *Trans. Geol. Soc. S. Africa*, vol. XVII, 1915, pp. 60–74.

abundant carbonates. These schists are obviously highly metamorphosed peridotites intrusive into the banded ironstones and
other older rocks of the district and again in their turn invaded
by the Mont d'Or granite. Chromite is found disseminated
throughout the schists, but the important ore-bodies are lenticular
masses, often of great size, varying from 150 to 600 feet in
length and as a rule from 10 to 20 feet wide, each containing
many thousands of tons. Of these large masses about 130 have
been mapped and many smaller ones exist. At the Chrome
Mine a group of eight or ten of them are quarried within a small
area. They consist of nearly pure chrome ore, ranging from 35
to 54 per cent. Cr_2O_3; as the mineral contains a certain amount
of magnesia and alumina it approaches picotite in composition.
The commercial product averages about 50 per cent. Cr_2O_3.
There are also enormous bodies of low-grade chromite ore in the
Umtebekwe valley, north of the Selukwe mineral belt.

The chromite deposits of the Ural Mountains in Russia have
long been known: they have been important producers and are
of considerable interest on account of the close association of
chromium and platinum in serpentines. Both the chromite and
the platinum are clearly original minerals formed during an
early stage of the consolidation of the original peridotite and it
is stated that the chromite crystallized earlier than the platinum,
which is moulded on it; there is also a considerable amount of
gold and all three have been obtained from placers at Nijni-
Tagilsk and Malo-Mostrovska. In Greece also chromite is obtained in considerable quantity from large masses of serpentine
in Thessaly and Magnesia, and in the islands of Euboea and
Skyros. Much chromiferous iron ore of low grade (about 3 per
cent. Cr_2O_3) is also mined in Boeotia and Lokris. In Asia Minor
large deposits have been worked near Mersina on the coast about
100 miles north-east of Cyprus, and also near Brussa. The ore
appears to be of high grade, but little information is available
as to its geological features.

The outstanding fact of geological interest in connection with
the chromite deposits is that it is abundantly clear that they
are in every case original magmatic segregations in rocks of
ultrabasic character. They are always associated with peridotite

or with some form of serpentine derived from peridotite, and chromium is not found along with the pyroxenites or amphibolites without olivine. Even the Rhodesian talc-schists with chromite are no exception to the rule, since these certainly represent peridotites altered by intense dynamic metamorphism; we can thus draw up a comparative scheme showing the degrees of alteration possible, thus:

Hestmandö, Norway. Chromite in fresh peridotite—no change.

New Caledonia and Urals. Chromite in serpentine—hydrothermal change.

Selukwe, Rhodesia. Chromite in talc-schist—dynamic metamorphism.

In each case the chromite is precisely similar, and even in the last-named it is entirely unaltered. It appears therefore that chromite is an original mineral and very stable under any degree of metamorphism, behaving under such conditions like magnetite. In this respect it differs much from nickel and cobalt, which during the serpentinization and weathering of peridotite in New Caledonia, for example, often form silicates and other oxidized ores, while oxidation products of chromite are of trifling importance.

Chromium also occurs to a considerable extent in certain varieties of iron ore, often in association with nickel. These ores have a special value for the manufacture of a particular kind of natural chrome-nickel steel, and considerable quantities have been mined for this purpose in Cuba and in Greece. The most important are those in the eastern part of Cuba, commonly known collectively as Mayari ores, from the name of one of the chief deposits. These ores are obviously decomposition products of a lateritic nature derived from serpentine and generally have a thickness of 18 to 20 feet. The averages of many hundreds of analyses show the following figures:

			per cent.
Iron ...	...	...	40–50
Nickel	...	...	0·5–1·0
Chromium	...	...	1·5
Silica ...	...	...	2–6

The iron is present partly as limonite and partly as haematite. The pig produced by direct smelting of these ores contains chromium 2·80, nickel 1·50 and manganese 0·87 per cent. Part of the chromium is intentionally oxidized and carried off in the slag and the typical Mayari steel contains 0·10 to 0·40 per cent. chromium and 1 to 1·5 per cent. nickel. The total reserve of this type of ore, which forms the surface capping of large plateaux, is estimated at about 3,000,000,000 tons and before the war the importation into the United States averaged about 1,400,000 tons[1]. Deposits of a similar type have also been worked in Greece, where they are derived from serpentines intrusive into Cretaceous sediments. The most important localities are in Lokris and Boeotia, and on the islands of Euboea and Skyros. Some of the deposits also contain nickel (see p. 375). The mineral occurs in some cases in lenticular masses and pockets at the contact of serpentine and limestone or in fissures in the limestone[2]; in other cases it forms superficial cappings to ridges, but most of the ore seems to represent weathered dykes of serpentine. The field relations, however, vary a good deal and it is difficult to give a brief general description. On the island of Seboekoe, off the south-east coast of Borneo, are extensive deposits of iron ore carrying chromium and nickel, very like those of Cuba, also derived from serpentine.

From the brief descriptions just given it is clear that there exists a definite type of chrome-nickel-iron ore, formed as a residual deposit on the surface of serpentines, that have undergone a weathering process of lateritic type, such as is closely associated with tropical climatic conditions. The first stage in the process is the serpentinization of peridotites containing primary chromium and nickel; according to the most recent ideas this may be a hydrothermal change brought about by magmatic waters soon after the intrusion of the serpentine, but the final decomposition to ore, with concentration of iron, chromium and nickel, with concomitant removal of silica, magnesia and lime, is certainly carried on at the surface, and is mixed up with the general question of the origin of lateritic deposits, which is discussed at some length under the heading of aluminium ores (see p. 449).

---

[1] Rumbold, *Chromium Ore*, Imperial Institute Monographs, 1921, p. 47.
[2] Kilburn Scott, *Journ. Iron and Steel Inst.* vol. LXXXVII, 1913, pp. 447–67.

CHAPTER XVII

MERCURY, ANTIMONY, ARSENIC, BISMUTH

Mercury. The geology of the mercury deposits is in many respects peculiar and of special interest from the information afforded as to the genesis of ore-deposits in general. One or two of the occurrences may in a sense be regarded as working models of the formation of ores, since the processes may actually be seen in active operation. Of course it is clear that such modern formations, being necessarily near the surface, cannot be taken as examples of the origin of deep-seated types, to which so many ore-deposits belong, but nevertheless, when considered from this point of view, they do cast much light on certain obscure problems. Another noteworthy feature is the great variety presented by these ore-bodies; although not particularly numerous or widespread, they are recorded from rocks of almost all classes and of a wide range of composition: in fact nearly every occurrence of mercury ores is a special type of its own, and has a different origin, when considered in detail. Fundamentally, however, the great majority of them may be attributed to a volcanic origin, more or less clearly expressed in their geological relations.

The mercury minerals also present some interesting and unusual features. Actually they are very numerous, but comparatively few are of any importance. The native metal is occasionally found, but apparently only as a reduction product of other minerals. By far the most important ore is cinnabar, HgS, the well-known substance commonly called vermilion. This occurs commonly in the massive or granular form, but when crystalline possesses very remarkable physical and especially optical properties. It crystallizes in the plagihedral or trapezohedral class of the rhombohedral system, like quartz, and shows a similar type of polarization. Mercury sulphide also occurs occasionally in a different form, metacinnabarite, a grey or black substance crystallizing in the tetrahedral class of the cubic system, like zinc-blende. Isomorphous with these are also various compounds of mercury with selenium and tellurium, such as tiemannite, HgSe, and coloradoite, HgTe. Kalgoorlite and coolgardite are examples of mixed gold, silver and mercury tellurides found in West Australia: they are not really definite mineral species. This is an interesting example of the isomorphism of sulphides, selenides and tellurides, which form a fairly natural chemical group. Mercury is also found as a constituent of some of the complex sulphides: for example in one variety of tetrahedrite it replaces some of the copper. Mercury has been produced as a by-product from this ore in Hungary.

Among the minerals commonly associated with mercury ores are pyrite and marcasite and very often sulphides of arsenic and antimony. Owing to their colour some of the latter group need careful discrimination. The most usual gangue minerals are silica in some form, sometimes quartz, but more usually jasper, chalcedony or opal, together with calcite and dolomite. As will appear later, one of the most characteristic features of mercury deposits is the prevailing silicification of the surrounding rocks, due to the action of ascending thermal waters of ultimate volcanic origin. However some of the most important of the world's deposits of mercury ores are disseminations in sedimentary rocks, to which the foregoing remarks hardly apply. A remarkable fact is the common occurrence of bitumen and gaseous hydrocarbons in mercury deposits; this undoubtedly has some genetic significance. In several instances the ores are accompanied by deposits of native sulphur now actually in course of formation at or near the surface, and some of the Californian deposits are visibly due to hot springs rich in chlorides, sulphates and boron compounds. In all these instances the volcanic origin is obvious: in fact it may be said that in most cases deposition of cinnabar belongs to the solfataric stage of volcanic activity.

We see then that there are two principal types of mercury ore-deposit; the first, probably primary but of very small practical importance, is the mercurial tetrahedrite, found in veins and lodes of the usual fissure-filling type: the second and by far the more important is the cinnabar deposits, which are nearly always disseminations and replacements of the actual material of the country rock; only occasionally in cavernous limestones does it occur as true cavity-fillings. This kind of deposit belongs to a much higher zone in the earth's crust than the tetrahedrite lodes. Owing to their much greater importance these will be considered first.

The Mercury Mines of Almadén. This district, till lately the largest mercury producer of the world, lies on the northern slope of the Sierra Morena, in the province of Ciudad Real, Spain. It is composed of Silurian and Devonian sediments, mainly slates, limestones and sandstones or quartzites, with occasional beds of a basic volcanic ash. There are also some sheets and masses of a basic intrusive rock, usually described as diabase. The cinnabar occurs as disseminations and replacements in three beds of sandstone or quartzite, which are nearly vertical, and are associated with slates in which graptolites have been found. The three beds together have a total thickness of about 50 feet. The cinnabar either fills the original pore-spaces of the rock, or forms a mass of fine veins and replacements of the sandstone. The actual quartz of the sandstone has been in many places replaced by cinnabar and the ore is unusually rich, running on the average about 8 per cent., as against 2 per cent. in California. The best bed, known as the San Pedro y San Diego lode, runs in places up to 15 per cent. There is no sign of falling off in depth; in fact at 1200 feet the ore is somewhat richer than at the surface. The total value of the mercury production from Almadén is said to have amounted to over £43,000,000; the average annual output before the war was about 1500 tons of metal. This may be taken as a typical example of the sandstone replacement type of deposit. The origin of the mercury is obscure, but may with probability be referred to some form of igneous activity; the presence of beds of basic tuff is suggestive in this respect.

The Mercury Deposits of Idria. As a typical example of cinnabar deposits in limestone we may take those of Idria, in Krain (commonly called Carniola in this country). This locality is situated about 30 miles north-east of Trieste, and according to the latest information it now appears to be in Italian territory. The sedimentary formations here represented include Carboniferous slates (Gailtal Series) with Permian and a well-developed representative of the Alpine Trias consisting largely of limestones and dolomites. The structure is complicated and the mercury deposits are largely associated with an overthrust which brings Carboniferous strata above the Trias. The ore-bodies mainly occur in the Trias limestones (Werfen, Guttenstein, Wengen and Skonza Series) in a zone of disturbance and brecciation below the great overthrust previously mentioned. In some places also the ore occurs in masses of breccia filling inclined fault planes striking E.N.E. and dipping at about 30°. In this instance the ore is mainly to be regarded as cavity-fillings in cavernous limestones, the cavities being partly due to solution and partly to brecciation resulting from the violent earth-movements. The actual source of the ore-bearing solutions, however, is not at all obvious, as there appear to be no volcanic rocks in the immediate neighbourhood. The uprise of the solutions also was probably a direct result of the crust disturbances that caused the overthrust and brecciation. Most of the ore is cinnabar in massive forms, either bright red or stained black or dark grey by bitumen. Native mercury occurs only in some of the upper beds as a reduction product of the primary cinnabar. There are also small quantities of pyrites and fluorite, the latter being indicative of a deep-seated origin, and some secondary sulphates, especially gypsum and magnesium sulphate[1]. The annual production before the war was about 500 tons of metal.

The Mercury Deposits of Tuscany. It has been shown that in Tuscany there were three well-defined phases of ore-deposition, as follows[2]:

(1) In the Eocene, copper ores connected with intrusions of serpentine.

(2) In the Miocene, masses of iron ore with copper veins and various sulphides, connected with granites and quartz-porphyries.

(3) In the early Quaternary, mercury and antimony ores, with eruptions of trachytes and andesites.

The after-effects of the latest phase still persist to a large extent in the form of solfataras, hot springs and emanations of sulphuretted hydrogen gas (putizze).

---

[1] Kossmat, *Jahrb. k. k. geol. Reichsanst.* vol. XLIX, 1899, pp. 259–86.
[2] Lotti, *Zeits. für prakt. Geol.* 1901, pp. 41–6.

The ore-deposits of the third phase are well seen in the region that may be described in a general way as Monte Amiata. Here all the ore-bodies lie along a zone about a mile long parallel to the general trend of the folds of the district, namely, N.N.E.–S.S.W. This zone is an important fault plane, and ten miles away in the same direction is an occurrence of hot sulphurous springs. The ore-bodies are masses of rather soft clayey material, intercalated with limestones of various ages, especially Lias and Eocene; ore is also found in an Eocene sandstone and in trachyte. The ore-bearing clays are considered to represent the insoluble residue of limestones that have been leached by waters holding sulphuric acid in solution, due to oxidation of sulphuretted hydrogen: this action has given rise in places to deposits of gypsum. The clays are impregnated with cinnabar, antimony ores, and pyrites, with sometimes realgar, with here and there native sulphur and a little fluorspar[1]. The form of some of the ore-bodies is peculiar: in the Siele mine the ore-shoots lie at the contact of two masses of limestone with shales, having a steep dip. They have been followed down for about 200 metres; in side view they resemble pipes, but in front view, in the plane of the limestone, they look like a map of a branching river system. These shoots undoubtedly follow the courses of the ascending mineralizing solutions. The clay of which they consist has been formed by the decalcification of argillaceous limestones by the action of waters containing sulphuric acid due to the oxidation of sulphuretted hydrogen. A precisely similar residue has been obtained artificially by dissolving specimens of the limestone in acid[2].

The Mercury Deposits of the Pacific Slope. The mercury deposits of California have been described in detail by many geologists and the literature is fairly extensive. They are found in the Coast Range, over a length of some 400 miles from Santa Barbara to Colusa, with a few scattered deposits in the north of the State. The rocks range in age from Mesozoic to Quaternary, apparently to a great extent underlain by granite and sometimes metamorphosed. The deposits are sometimes veins, with

[1] Lotti, *Zeits. für prakt. Geol.* 1901, p. 141.
[2] Spirek, *ibid.* 1897, p. 369.

definite walls, but more commonly masses of irregular veinlets and stringers, often in brecciated rock merging into disseminations. Some ore-bodies may be described as stockworks. In some instances there appears to be evidence of actual replacement of the quartz of sandstones. Nearly all, if not all the deposits appear to be in sediments, but the association with igneous activity is clear; in the neighbourhood are many occurrences of lavas, chiefly basalts and andesites, but it is uncertain whether the cinnabar is directly due to solutions derived from these or from the underlying granite. At any rate there is no doubt that the deposition is of recent geological date, since some mercury deposits are visibly in process of formation in connection with hot springs[1]. One of the most striking cases is at Steamboat Springs, Nevada, where water issues at temperatures ranging from 75° to 85° C. The vapours coming from the same fissures include sulphur dioxide and sulphuretted hydrogen, while the waters are depositing cinnabar, sulphides of arsenic and antimony, with sulphur, gold and silver. The country rock consists of granite and quartz-porphyry, overlain by sediments and recent lavas. The hot water comes from the granite, which is very rotten, and a good deal silicified with opaline silica. This deposit was at one time worked commercially, but under great difficulties owing to the high temperature. This locality is only 6 miles from the famous Comstock Lode, where ore-formation is also in progress (see p. 489).

A case of great interest from the point of view of vein-formation is to be seen at Sulphur Bank, on the shores of Clear Lake, California. This is a region of extinct but geologically recent vulcanicity, lying in the Coast Range belt, which was strongly folded in Miocene times. There are many very perfect cones in the neighbourhood, as well as solfataras, hot and effervescent springs and some pools containing borax; in fact all the evidences of expiring vulcanicity. Sulphur Bank itself is a low rounded hill, consisting of blocks of what was originally an andesite lava, but now intensely decomposed and converted into siliceous masses of chalky appearance, sometimes with a harder grey core inside white spheroidal decomposition-shells; between these

[1] Becker, Monograph 13, *U.S. Geol. Survey*, 1888.

blocks, near the surface, are abundant crystals of native sulphur. At a depth of a few feet the sulphur disappears and is replaced by cinnabar. The most interesting part of the deposit, however, lies below the pile of lava-blocks, where the underlying sand-stones and shales are much brecciated along certain bands. From these shattered belts water issues at a temperature of 70° C., together with much carbon dioxide, and between the blocks near the outcrop is a mass of blue mud rich in alkaline sulphides and boracic acid. The ore-bodies, which occur also at and below this horizon, are brecciated masses of sediments, angular or subangular, cemented by cinnabar, pyrites, silica and blue mud in varying proportions, the amount of cinnabar seem-ing to increase in depth, where also the cementation is more complete. At some horizons and especially in the lower part of the lava, the cement consists of opaline, gelatinous silica, streaked and stained by cinnabar. It seems clear that the mercury ores are deposited from solution in sulphides and that the overlying sulphur is also formed from these same solutions, by oxidation on coming in contact with air.

The most doubtful point about this occurrence is as to whether the lava flow constituting the mound and the cinnabar deposits are in genetic connection, or whether the lava flow simply happens to overlie the fissures from which the ore-bearing solutions issue from a deep-seated source: the latter seems the more probable alternative, when the high temperature of the water is taken into account[1]. At any rate the connection with recent igneous activity is obvious.

The most important quicksilver mines in California are New Almaden, in Santa Clara County, 13 miles south of San José in the Santa Clara valley, and New Idria in the south-east part of San Benito County, in the southern part of the Diablo range. Both of these have been producers since before the middle of last century; the former was discovered as early as 1824. At New Idria the ore-bodies consist of fissure-fillings in much meta-morphosed Lower Cretaceous rocks, consisting of sandstones, shales and "serpentines." These rocks are much twisted and

[1] Le Conte and Rising, *Amer. Journ. Sci.* Ser. 3, vol. XXIV, 1882, pp. 23–33.

crushed, and in places silicified. The ore-bodies are connected with fault-fissures filled with breccia, which is much silicified and converted into a very hard, chert-like mass: the ore, consisting of cinnabar and pyrite in varying proportions, occurs mostly along cracks and crevices in the breccia. Much inflammable gas is given off from the workings. The ore decreases in metal-content in depth: it now runs under 1 per cent., but it is said that one-half per cent. ore can be worked at a profit[1]. Here the temperature appears to be normal and there is no suggestion of deposition still in progress.

Mercurial Tetrahedrite Ores. It is stated in most books on the subject that mercury in commercial quantities has been obtained in certain cases as a by-product in the metallurgical treatment of certain copper ores, especially tetrahedrite, but on this point little definite information is available. This mineral, schwatzite or spaniolite, sometimes contains as much as 17 per cent. of mercury; it has been worked at Brixlegg and Schwatz in the Tirol, and also at Maškara in Bosnia and at Igló and Dobschau in Hungary. The veins in the latter localities lie in ancient rocks and the ore is superficially altered to cinnabar, amalgam and copper carbonates[2].

Antimony. The technical uses of antimony and its compounds are numerous and rather important. The metal is a constituent of many alloys, which can be classified under four principal headings, namely, bearing metals, type metal, Britannia metal and hard lead. All of these are employed on a large scale, but the demand for the fourth class or hard lead was stimulated to an enormous extent by the war, since antimony is an essential constituent of the ordinary type of shrapnel bullet. Consequently antimony mining was extremely active owing to the demand from munition makers. The object of the antimony is to harden the lead and thus prevent the bullets from losing their shape on the bursting of the shell. Antimony oxides and sulphides are also used to a considerable extent as paints, in the vulcanizing of rubber, in enamelling and in the manufacture of matches. It seems possible that in the future it may be found practicable

[1] McCaskey, *Min. Res. U.S.A.* 1908.
[2] Mieleitner, *Die technisch wichtigen Mineralstoffe*, München, 1919, p. 122.

26–2

to use antimony for some purposes for which the more expensive tin is now employed.

By far the most important ore is stibnite or antimonite, Sb_2S_3, which commonly occurs in long bladed prisms or in massive forms, and possesses the remarkable property of fusing easily in the flame of a candle: it is also extremely soft. Jamesonite is a double sulphide of lead and antimony, or lead sulphantimonide, $2PbS.Sb_2S_3$, which occurs in considerable quantity in Cornwall in lead veins. Antimony is also a constituent of a considerable number of minerals belonging to the class of complex sulphides found in the zone of secondary enrichment of copper and silver lodes; such are for example tetrahedrite (fahlerz), bournonite and pyrargyrite. Antimony and gold are also often found in association.

The greater part of the world's production of antimony comes from China and Japan. In the first-named country the resources are undoubtedly very large, and it appears that most of the antimony exported from Japan is really of Chinese origin, having passed through an intermediate phase of local commercial speculation. As to the geological relations of the Chinese antimony deposits very little information is obtainable.

The Antimony Deposits of China. The largest deposit of antimony ore in China, and perhaps in the world, is situated at Hsi-Kuang-Shan, east of Sing-Hua, in the province of Hunan. Other important antimony mining districts are situated in the north of Kwantung, and in Kwangsi and southern Yunnan.

The Hsi-Kuang-Shan mining district is situated in a rugged region, about 2500 feet above sea-level, and consists of a series of sediments ranging from Silurian to Carboniferous, folded into anticlines and synclines striking N.N.E.–S.S.W., with minor cross folds, giving a dome-like arrangement, which is further complicated by faults. The stibnite occurs in the lowest visible bed, a quartzitic sandstone about 160 feet thick. This rock is a good deal brecciated and contains large caves, sometimes lined with crystals of quartz and stibnite. The main supply of ore is found in open joints and cracks in the quartzite, sometimes forming veins and flats up to a foot thick or even more, of nearly pure stibnite. Up to 1915 about 100,000 tons of metallic antimony have been extracted from this region, and the total available reserves appear to be very great[1].

At Pan-chin, in the Yi-Yang district, stibnite occurs in quantity in quartz veins, and this type of deposit seems to be characteristic of the rest of the Chinese deposits, the Hunan area being quite exceptional in character.

France. In the seven years before the war France was, after China, the largest producer of antimony, the output having averaged about 5000 tons per annum, with a maximum of 6390 tons of metal in 1913. It is said, however, that by that date the home supplies were nearly exhausted, and that the French smelters were importing ore from China. In the early days of the war the necessary supply for munitions was kept up by importation of ore from Algeria; towards the end of the war period the consumption in France was at the rate of 600 tons per month. Most of the French antimony deposits are found in the region of the Central Plateau, especially in the departments of Puy-de-Dôme, Cantal and Haute Loire. They consist mainly of quartz-veins with lenses and masses of stibnite, cutting the granite, gneiss and mica-schist of the pre-Cambrian formations, as well as, in some places, sediments of later date. In general the formations are true veins with a gangue of quartz and sometimes barytes, and the stibnite is associated with more or less pyrites, blende, galena, cinnabar and realgar. Stibnite is also found in small amount as an accessory in some copper and lead veins. The most important area lies mainly in the departments of Haute Vienne, Corrèze, Puy-de-Dôme, Cantal, Haute Loire, Ardèche and Lozère. Some well-known localities are in the immediate neighbourhood of Limoges, in Haute Vienne; at Channac, 10 kilometres south of Tulle, in Corrèze, where veins from 40 to 70 cm. wide lie in black clay-slates; at Malbose, in Ardèche, where lamellar and fibrous masses and fine needles of stibnite are found with barytes and calcite in veins in mica-schist. Splendid crystals have been obtained from the old mines at Lubilhac, near Massiac, on the borders of Cantal and Haute Loire; individual crystals are sometimes 7 or 8 inches long. Veins are common in the mica-schists of this region. At Neronde,

---

[1] Tegengren, Bulletin 3, *Geol. Survey China*, 1921, pp. 1–26.

in Loire, the chief vein lies in Carboniferous slates and lime-stones, which themselves carry disseminated stibnite. The same mineral has also been worked on a commercial scale at Pierre de Crau and La Ramée, in Vendée, which geologically belongs to Brittany. The veins of all these regions are connected with intrusions of acid rocks, commonly granites and microgranites[1].

Bolivia. This country has constantly been a small producer of antimony, but in 1916, during the war boom, the production suddenly jumped up to 27,000 tons of ore: it is not known, however, how much metal this contained. The ores are found in the same general region as the tin ores, but not in the same lodes. The principal producing districts during the boom were the Chuquiutu area, near Uncia, the country about Porco and around Atocha. The stibnite veins with a gangue mainly of quartz fill narrow fissures in the black Palaeozoic shales that are so prominent in the eastern Andean range, and they are worked mostly in a very primitive fashion by Indians. The hand-sorted ore usually contains 50–55 per cent. of antimony metal. Many veins of stibnite are found in the silver mines of Pampa Larga, province of Atacama, Chile, and some ore has also been got near Santa Rosa, San Antonio and Juliaca in the department of Puno, Peru[2]. From the rather scanty descriptions yet avail-able it appears that antimony veins are a not uncommon feature of the Andean mineralization, and some of them appear to be rich, with remarkably little gangue. Most of the veins, however, are rather narrow. The genetic relations of these deposits are not clear: they appear at any rate not to belong to the same phase as the tin-tungsten-bismuth-silver veins of Bolivia, though they may be later differentiation products of the same metal-liferous magma.

. Deposits of antimony exist in many parts of Mexico and there has been a considerable production at various times. The principal mines are in the Sierra Catorce in the States of San Luis Potosí and Queretaro. The ores are partly sulphide and partly oxidized. In western Sonora there are deposits of oxidized

---

[1] Lacroix, *Minéralogie de la France et de ses Colonies*, vol. II, p. 449.

[2] Miller and Singewald, *Mineral Deposits of South America*, New York, 1919, pp. 87, 282, 462.

ores, worked by American companies, while a large deposit of lead-antimony ore (jamesonite) exists at Zimapan, Hidalgo, and shows considerable promise. Other deposits of more or less importance are known in many States.

The statistics of antimony production are not always intelligible to the layman, and sometimes apparently inconsistent with each other, owing to the somewhat unfortunate nomenclature employed in the trade. Thus "antimony regulus" means metal, while "crude antimony," "needle antimony" and "antimony matte" are all used to indicate, not metal, but merely sulphide, fused and cast into moulds and therefore containing only about 70 per cent. of metal.

Arsenic. The technical uses of arsenic and its compounds are fairly important and the output is considerable. The chief uses are in the manufacture of glass and in the preparation of insecticides, especially sheep-dips, and for the purpose of destroying objectionable vegetation, such as the prickly pear in Australia. Metallic arsenic also imparts hardness to lead and is used in the so-called chilling of shot. The metal and its compounds are widely distributed in nature, but in most cases its exploitation is merely as a by-product in the mining, dressing and metallurgical treatment of other ores, so that it is unnecessary to enter at length into the geology of the arsenic deposits.

The chief primary mineral is arsenopyrite or mispickel, $FeSAs$, a mineral of prismatic form and silvery metallic lustre, very common in metalliferous veins: closely allied to this is löllingite, a rather more rare compound. Along with marcasite these form a somewhat closely related series, thus:

Marcasite	FeS_2.
Mispickel	$FeSAs$.
Löllingite	$FeAs_2$.

Mixed crystals containing these three compounds probably exist to some extent. Another important ore of arsenic, probably in some instances of primary origin, is enargite, $3Cu_2S.As_2S_5$. Arsenic is also a common constituent of the complex sulpharsenides and similar minerals of the zone of secondary enrichment in copper and silver lodes: such are, for example, tennantite,

$4Cu_2S.As_2S_3$, isomorphous with tetrahedrite, and proustite, $3Ag_2S.As_2S_3$, one of the ruby silver ores. Arsenides of nickel and cobalt are not uncommon, belonging mainly to the pyrites group, and apparently of primary origin, while the simple arsenic sulphides realgar and orpiment, AsS and As_2S_3 respectively, have also been mined on a small scale in various localities: these are probably of secondary origin. Native arsenic is also found in some localities, as in the lead-silver veins of Freiberg, Saxony and Joachimsthal, Bohemia. This is unimportant.

Arsenopyrite is one of the most abundant and characteristic minerals of the deep-seated tin-tungsten lodes associated with tourmaline granites. In the Cornish lodes it is found in great quantities in the lower levels within the granite, as in the famous Rogers lode in East Pool and other mines of the Camborne-Redruth district. A large output is obtained by roasting the concentrates: the arsenopyrite is decomposed and oxidized, yielding Fe_2O_3 and As_2O_3. The latter is volatile and is condensed in chambers, to be afterwards refined. The total production of white arsenic in Cornwall in 1919 was about 2500 tons.

The average annual production of white arsenic in the United States is about 5000 tons, but in 1918, owing to demands for war purposes, this suddenly jumped to 10,000 tons. It comes mainly from the copper smelters at Butte, Montana, where enargite is one of the most important ores (see p. 241) and from Tintic, Utah. Quite recently Canada has developed a considerable arsenic industry, mainly as a by-product in the smelting of the arsenical silver-cobalt-nickel ores of the Cobalt district, which are described elsewhere (see p. 485). A good deal of arsenic is now produced in Queensland, chiefly from the Stanthorpe mining field and near Gladstone, in response to the large local demand for insecticides and plant-poisons, especially for dealing with the prickly pear pest. The ore is mispickel[1]. Some development has also occurred in the large deposits of mispickel known to exist in Rhodesia, and in the Mutue Fides-Stavoren tinfield in the Transvaal[2]. The gold of the St John del Rey and

---

[1] *The Mineral Industry for 1918*, vol. xxvii, 1919, p. 43.

[2] Wagner, "The Mutue Fides-Stavoren Tinfields," Memoir 16, *Geol. Survey of the Union of S. Africa*, 1921, pp. 64 and 175.

Passagem mines in Brazil is chiefly contained in arsenopyrite (see p. 463).

Bismuth. The chief uses of the metal bismuth are as a constituent of various fusible alloys. Some of the more common forms, *e.g.* Rose's alloy and Newton's alloy, approximate closely to the composition of the ternary eutectic of lead, tin and bismuth, and melt at about 94° C., while by the addition of a small amount of mercury an alloy can be obtained which melts at a temperature as low as 37° C., or about that of the human body. Bismuth salts are also used to some extent in medicine. The total output of the metal is quite small, owing to the limited range of its uses, and it is as a rule a by-product in the mining and metallurgy of more important metals.

Bismuth is found to a considerable extent in nature as the native metal, and most of the production comes from this: it also occurs as bismuthine or bismuthinite, Bi_2S_3, and to some extent as bismuth ochre, Bi_2O_3 (sometimes called bismuthite, whence some confusion has arisen).

At the present time the chief commercial source of supply is Bolivia, where bismuth occurs as a characteristic member of the tin-tungsten mineral assemblage of the veins associated with silver ores in the Cordillera Real (see p. 272). The chief minerals are the native metal and bismuthinite. One of the most important localities is in the Huayna Potosí district, near La Union, the most productive mines being the Carmen, Esperanza and San Alberti, of which the Carmen is the best. The ore, which is very largely native metal, occurs as veins and stringers in quartzite and slate. A mass of native bismuth weighing 100 lb. has been found here. At Tasna and Chorolque on the other hand the ore is mainly the sulphide: the Tasna veins lie in slates cut by acid dykes and contain also a little mispickel and stibnite as well as argentiferous galena, cassiterite and wolfram. Chorolque mountain is a mass of intrusive quartz-porphyrite and sedimentary rocks with rich veins at the contact, the bismuth usually being found at lower levels than the tin and silver[1]. The mines of the Tasna-Chorolque area are chiefly

---

[1] Miller and Singewald, *Mineral Deposits of South America*, New York, 1919, pp. 103 and 130.

worked by the Compagnie Aramayo de Mines en Bolivie, of Geneva, who are probably the largest bismuth producers in the world. In 1915 the bismuth production of Bolivia was 570 tons, worth about a million dollars.

In Australia bismuth is an important constituent of the very remarkable mineralized pipes lying in granites in Queensland and New South Wales, which also yield tin, tungsten and molybdenum. These are described in some detail elsewhere (see p. 423). The annual production, however, is only about 30 tons. There is also an output of a few tons from Tasmania.

The various bismuth minerals have been recorded from a large number of localities, including many of the most important mineral areas of the world and they doubtless exist in many other places, but as a rule they are to be regarded merely as mineralogical specimens and no good purpose would be served by a more detailed enumeration of them.

CHAPTER XVIII

THE MINOR METALS

Classification. Under this heading are included a considerable number of elements, belonging to various chemical categories, which are used for a large and increasing variety of technical purposes. It is of course very difficult to give any exact definition of what is meant by the term "rare metal," since rarity is purely a relative term: as a matter of fact some of the metals here included are really of very wide distribution, but as a rule in small proportions only; such for example is zirconium. Indeed few rocks, either igneous or sedimentary, are entirely free from the mineral zircon. It may perhaps be said that it is the highly specialized character of their technical applications rather than their natural occurrence that is the most distinctive feature of this group, and in fact they are most conveniently subdivided on this basis. According to this scheme we arrive at the following classification: it will be noticed that some elements occur under more than one heading.

I. Metals used for hardening special steels: tungsten, molybdenum, vanadium.

II. Metals used as deoxidizers in steel: titanium, zirconium.

III. Metals used in special illuminants: thorium, cerium, zirconium, tantalum, niobium.

IV. Radio-active elements: uranium, thorium.

Most of these groups are chemically heterogeneous, their members belonging to different groups from the constitutional and genetic point of view, and it is not at all clear why some of them possess similar properties: for example, tungsten, molybdenum and vanadium have little in common and yet they impart very similar properties to ferrous alloys. There is also a large and increasing number of metallic elements, such as yttrium, lanthanum, erbium, ytterbium and others, commonly found in association with thorium and cerium and having very similar chemical and physical properties. These metals are difficult to separate and determine and have not yet found any individual technical applications: it is therefore unnecessary to discuss them in detail.

From the geological point of view the rarer metallic elements show an immense variety in their manner of occurrence, but most of them are primarily associated with igneous rocks. One of the most striking features is the close connection that exists between many of them and the pegmatitic facies of differentiation. Titanium is highly characteristic of basic igneous rocks, while vanadium is altogether abnormal in its occurrence, but the rest of the elements in the foregoing list belong typically to the acid and intermediate rock-groups, especially the granites and syenites, although it would not be true to say that they are exclusively confined to this type.

Tungsten. Of late years the metal tungsten has acquired very considerable technical importance, especially as a constituent of alloy steels, which possess some remarkable properties. During the war the demand for the so-called high-speed tool-steels became very great, for munition-making, and this demand, coupled with a shortage of world-supplies, led to a tungsten boom from 1916 onwards. The ores reached extraordinary prices and in consequence great activity arose in the prospecting and development of tungsten properties in many parts of the world, especially in the United States, Burma, China and Bolivia. As a consequence of this commercial activity much attention was also paid to the scientific side of the subject and a considerable literature appeared on the genesis of tungsten ores[1]. Tungsten is also an important constituent of various non-ferrous alloys, especially those of the stellite group, which are essentially cobalt-chromium-tungsten alloys, and several varieties of acid-resisting metals, and platinum substitutes. Owing to its remarkably high melting point (about 3300° C.) and electric properties, tungsten is much used in electrical apparatus, and in resistance furnaces, while its compounds are employed in various processes in dyeing, and in the glass and ceramic industries. An important application of the metal is in the filaments of ordinary incandescent electric lamps and particularly in the so-called half-watt (nitrogen-filled) lamps, where the consumption of current is

[1] Hess and Schaller, Bull. 583, *U.S. Geol. Survey*, 1914.

extremely low; it is also used for the poles of certain types of arc-lamps.

The two chief sources of tungsten are the minerals wolfram and scheelite. Wolfram is the general name commonly applied to a series of isomorphous minerals, consisting of mixtures of iron tungstate and manganese tungstate in any proportion, forming a typical example of mixed crystals. Both compounds are known in nature in a pure or nearly pure state, the iron compound, $FeWO_4$, being called by mineralogists ferberite, while the manganese compound, $MnWO_4$, is called hübnerite. It has been proposed by Messrs Hess and Schaller[1] to adopt the following convention:

Varieties from 100 per cent. $FeWO_4$ to 80 per cent. $FeWO_4$, 20 per cent. $MnWO_4$ to be called ferberite; from 100 per cent. $MnWO_4$ to 80 per cent. $MnWO_4$, 20 per cent. $FeWO_4$ to be called hübnerite; all intermediate varieties to be called wolframite: however, the ordinary practical man calls the whole lot wolfram (or wolframite). It is not a matter of any technical importance whether the base is iron or manganese, and all varieties sell at the same price[2].

The mineral wolfram, in the broad sense, crystallizes in the monoclinic system, but rarely shows good terminal faces; it usually occurs as prisms with one good cleavage and strong metallic lustre, and a black or very dark brown colour; the varieties rich in manganese are browner. It is hard and heavy, its density being about the same as that of black tin: hence arose much trouble and waste of tin ore before the days of magnetic separation; some rich ores were rendered useless by the impossibility of separating the two minerals except by costly chemical treatment.

The other practicable tungsten ore-mineral is scheelite, $CaWO_4$, which crystallizes in the tetragonal system, but usually occurs as massive aggregates of a grey or yellow colour, with a greasy lustre. When pure it contains about 80 per cent. of WO_3. Its density is about 6 and it therefore cannot be separated from cassiterite by mechanical means, while both minerals are entirely

[1] Hess and Schaller, Bull. 652, *U.S. Geol. Survey*, 1917.
[2] Vogel, *Mining Mag.* vol. xx, 1919, pp. 12–17.

non-magnetic. Any scheelite present in tin-wolfram ores will therefore pass over with the tin. Fortunately this mineral association appears to be uncommon.

Stolzite, $PbWO_4$, is a rare mineral closely related to scheelite, and some varieties of scheelite contain copper in addition to calcium. These are of no importance. In some instances wolfram and scheelite have undergone decomposition near the surface, giving rise to masses of yellow tungstite, which is believed to be amorphous WO_3, but this rarely occurs in commercial quantities.

With regard to the occurrence of these minerals, we may say in general terms that wolfram is found in igneous rocks and in siliceous and argillaceous sediments surrounding them, while scheelite is found where limestones have been mineralized by tungsten-bearing solutions from granitic magmas.

On genetic grounds deposits of tungsten ores may be classified into four groups, as follows[1]:

Primary
- (1) Wolfram ores with tin.
- (2) Wolfram ores without tin.
- (3) Scheelite ores.

Secondary. (4) Residual, eluvial and alluvial deposits.

This classification is mainly one of convenience and it is not possible to draw a hard and fast line between groups (1) and (2). The distinction, however, is of some importance from the ore-dressing point of view: in ores of group (2) magnetic separation is usually not required, unless much iron is present as sulphides, when roasting and magnetic separation may be necessary.

The general geology of the wolfram-tin ores is essentially the same as that of the tin ores, and scarcely needs extended treatment. In a very large number of cases important tin deposits are accompanied by more or less wolfram, which sometimes forms a fairly well-defined sub-zone. This is very clearly the case in Cornwall, as already described (see Chapter XIII). The same association is found well-developed in the Erzgebirge, in Spain and Portugal, in Australia and the Malay States, and the tin region of Bolivia has also produced much tungsten. The

---

[1] Rastall, *Geol. Mag.* 1918, p. 195.

special characters of these areas are dealt with fully in the chapter just referred to.

Tungsten in Burma. One of the most important tungsten-producing areas of the world is the Tavoy district in Lower Burma, and this likewise shows some features of great geological interest. In the later years of the war Tavoy produced nearly a quarter of the whole world's output of tungsten ores. A highly mineralized area extends from Tavoy in Tenasserim, Lower Burma, throughout some intervening Siamese territory into the Malay Peninsula, a distance of about 1000 miles, and is continued into the Dutch islands of Banka and Billiton. This is a region of highly folded rocks forming part of the great Burmese-Malayan arc: the folded rocks are penetrated by vast masses of granite, probably of Permo-Carboniferous or early Mesozoic age, to which the mineralization is due. Generally speaking, in the north tungsten is dominant, gradually giving way to tin towards the south, a feature perhaps depending on relative depth of denudation[1]. Hitherto the greater part of the Tavoy production has actually come from various forms of superficial deposit, shoad and to some degree alluvium, but lode-mining has been developed to some extent and the literature of the subject is now large. One or two errors unfortunately crept into the earlier descriptions and have gained wide currency, but good descriptions will be found in the papers quoted in the footnote[2]. The lodes are all in obvious connection with the granite, which is intrusive into the Mergui schists: they belong to three main types: (1) wolfram-quartz lodes, which are the commonest; (2) cassiterite-quartz lodes; (3) wolfram-greisen—the two last types are rare so far as at present known. The other associated minerals are molybdenite, bismuth, bismuthinite, chalcopyrite, mispickel, galena, blende, iron ores, mica, felspar, chlorite and rarely fluorspar. In places wolfram and tin are also disseminated in the granite in workable quantity, and some parts of the mineralized granites are so much weathered as to be workable, like float deposits, by hydraulic methods. It is quite obvious that these deposits

---

[1] W. R. Jones, *Bull. Inst. Min. Met.* No. 186, March, 1920.

[2] Jones, *op. cit.* Morrow Campbell, *Min. Mag.* vol. xx, 1919, pp. 76–88. Coggin Brown and Heron, *Rec. Geol. Survey India*, vol. L, 1919, pp. 101–21.

belong essentially to the magmatic or pneumatolytic phase of the granite intrusions, and the rich lodes appear to be confined mainly to the upper part of the granite masses. Although fluor-spar is rare, and tourmaline completely absent, at any rate in Tavoy proper, yet there are affinities to the more common type of tin lode, and it has been suggested that a tin-zone may underlie the tungsten zone, on the assumption that tungsten compounds are more volatile than tin compounds; as to this there is still some doubt, and Morrow Campbell maintains that wolfram in these lodes is always formed before cassiterite[1].

Wolfram Lodes without Tin. As before stated a regular gradation can be traced from cassiterite-wolfram lodes to wolfram-quartz veins without tin: this type is perhaps most common in America, although it is known in Cornwall, where a tin-wolfram lode can sometimes be traced into a quartz vein with wolfram and this again into a pure quartz vein without any metallic minerals. Many quartz-wolfram veins are obviously pegmatitic in character.

An interesting Cornish example is the wolfram lode of Castle-an-Dinas. Here a small granite boss, much greisenized, forming probably an outlying cupola of the St Austell intrusion, is cut by a vein of quartz about 3 feet wide, carrying much wolfram in crystals up to several inches in length. Tin is almost completely absent from this lode, though found near at hand. The lode itself likewise contains no pneumatolytic minerals, though these are common in and around the granite[2]. The filling must have been due to a highly differentiated magmatic solution, containing only silica and the elements of wolfram.

From the scientific point of view one of the most interesting cases of quartz-wolfram veins without tin is seen in the Sierra de Córdoba in Argentina. Here the veins, which lie in gneiss, are traceable into a large granite mass of unknown age: they are clearly pegmatitic differentiates of the granite. The wolfram occurs in large crystals in the quartz and in masses up to half a cubic yard in size. The other chief minerals are mica, apatite, fluorspar, molybdenite and chalcopyrite: the abundance of apatite is unusual

---

[1] Morrow Campbell, *Trans. Inst. Min. Met.* vol. xxx, 1921, p. 3.
[2] Davison, *Geol. Mag.* vol. lvii, 1920, pp. 347–51.

in wolfram lodes. Some years ago there was considerable production from these mines which were controlled by German interests[1].

One of the most important tungsten-producing areas of the world is Boulder County, Colorado. This mining field lies on a plateau at a height of about 8000 feet on the eastern slopes of the Rocky Mountains. The country rock consists of biotite-hornblende-granite, gneiss and quartz-mica-schist, all of pre-Cambrian age. The wolfram occurs in a group of veins striking S.W.–N.E. and accompanied by parallel gold-silver veins of two types, sulphidic and telluridic respectively. The wolfram seems to be more closely associated with the telluride veins. The gangue mineral is quartz with rarely a little felspar, calcite and chalcedony. The only other minerals are chalcopyrite, galena and blende in small quantity. The veins are extremely rich in wolfram[2]. In 1917 this area produced 2707 short tons of wolfram concentrates carrying 60 per cent. WO_3. The origin of these ores is still somewhat obscure and some writers consider that they are of comparatively recent origin.

Scheelite Deposits. The mineral scheelite is fairly common under certain more or less well-defined conditions. Some small occurrences are apparently due to alteration of wolfram in the presence of lime compounds, but most scheelite has been formed by the action of solutions derived from tungsten-bearing granites on calcareous rocks, commonly limestones. Hence such scheelite deposits should in strictness be placed in the category of contact deposits. It follows that scheelite can be usefully prospected for wherever granites of the tin-tungsten type are intruded against calcareous sediments, or even highly calcic igneous rocks. Scheelite is therefore sometimes a constituent of skarn rocks (for a definition of this term see p. 422). The mineral is not particularly easy to identify and is not well known, so it is possibly commoner in suitable localities than is generally believed. Since it contains an even higher percentage of tungsten than wolfram, its presence is in no way a drawback in concentrates, except that it cannot be separated magnetically from tin.

[1] Bodenbender, *Zeits. für prakt. Geol.* 1894, p. 409.
[2] Hess and Schaller, Bull. 583, *U.S. Geol. Survey*, 1914. Hess, Bull. 652, *U.S. Geol. Survey*, 1917.

At the present time only one area is of real importance as a producer of scheelite concentrates, namely the Atolia district in California. This is a remarkably arid region with no surface water: consequently mining and still more ore-dressing present special difficulties. These have been in part overcome by deep well-borings, as at Randsburg. In 1917 California produced 2781 short tons of 60 per cent. concentrates, almost entirely scheelite from Atolia. In 1918 the State of Nevada also produced 898 short tons, mainly scheelite.

In the Atolia district scheelite occurs as contact-metamorphic deposits, associated with garnet, epidote and ferruginous calcite, as contact zones up to 3 feet wide at the margins of the granite and limestones, as replacement deposits alongside narrow quartz-veins in the country rock, and occasionally as veins with quartz and calcite in the granite. Some of the veins and contact masses have been followed down the dip to a depth of 700 feet. Scheelite has also been mined extensively near Bishop, in the Owens valley, Inyo County, from low-grade deposits in metamorphic limestones, which can be very cheaply worked in open quarries.

The following general description of the scheelite area of the western States is slightly condensed from a personal communication kindly supplied by Mr F. L. Hess, of the U.S. Geological Survey, the leading authority on this subject.

"In the Great Basin region a broad belt of limestones extending from N.W. Utah across central Nevada, along the Sierra Nevada and round its southern end has been invaded by tungsten-bearing granites, which formed scheelite in the metamorphosed limestones, in deposits of variable size and composition. In places they are largely iron garnet and dark green epidote; in other places these minerals are paler in colour and contain less iron. Hornblende is abundant in some deposits. The ore is patchy, with rich spots, but usually ranges from 1 to 2 per cent. WO_3. Some patches may contain millions of tons of workable ore, though more commonly only a few thousand tons. The tungsten mineral is invariably scheelite, generally white, sometimes grey or yellow, in particles and crystals ranging from microscopic size up to 2 or 3 inches across."

These deposits clearly belong to the skarn type.

Origin of Tungsten Ores. From a general survey of the mode of occurrence of primary tungsten ores, it is quite clear that they are essentially of magmatic origin, being in fact the product of highly mineralized solutions differentiated from special types of acid igneous rocks, mainly rather acid granites, occasionally, however, monzonitic, very often associated with tin. The deposits are often pegmatitic in character and belong to the types of deposits formed at the highest temperatures, and at the greatest depths. The general relations of such veins are sufficiently discussed in the chapter on tin ores[1].

Secondary Tungsten Deposits. Although the primary tungsten ores are the most interesting, nevertheless it must not be forgotten that the major part of the world's production has actually come from various kinds of superficial deposits of secondary origin, some sedentary, some transported. There must also be included here some occurrences of rather exceptional character: under tropical conditions, such as prevail in Burma and the Malay States, it is possible for hard rocks, such as granites, to become so deeply weathered that it is possible to work them by hydraulicking methods, just like true alluvials. One at least of the best-known "alluvial" propositions of this region is actually an undisturbed granite in place, with abundant disseminated primary ore. Still more common are residual and eluvial deposits at the outcrops of lodes, in which the more resistant minerals have become concentrated in the surface deposits. Such accumulations also naturally tend to creep down hill under the influence of geological processes, and grade insensibly into true alluvium. There has been a good deal of difference of opinion as to the stability of wolfram under weathering, and even yet some obscurity prevails[2]. The chemical behaviour of the mineral under oxidizing conditions apparently depends on some factors not yet thoroughly understood. In comparatively rare instances wolfram seems to oxidize directly to tungstite, or to pass rather readily into solution and disappear from alluvial deposits: in other cases, though not very com-

[1] See also, Rastall, *Geol. Mag.* 1918, p. 193, etc. Rastall and Wilcockson, *Tungsten Ores*, Imperial Institute Monograph, 1920, pp. 7–8.

[2] Turner, *Econ. Geol.* vol. xiv, 1919, pp. 625–39.

monly, it is found in considerable quantity along with alluvial tinstone, as in the "head" of Bodmin Moor, Cornwall, a deposit supposed to have been formed under sub-glacial conditions[1]. At any rate it seems to be clearly established that wolfram is less stable than cassiterite, and under similar conditions does not travel so far from the parent source. The disappearance of wolfram in alluvial deposits, in comparison with cassiterite, is probably due in part to the possession by the former mineral of a very good cleavage, leading to flaking and mechanical disintegration. It also seems to be established that tungsten ores do not undergo secondary enrichment in depth.

Molybdenum. Until quite recent times the technical applications of the metal molybdenum and its compounds have been very limited, but in the last few years there has been a noteworthy development and increase in its employment, especially in the manufacture of special steels and as a constituent of some non-ferrous alloys, *e.g.* stellite. During the war the ores were in considerable demand and some new fields were opened up, particularly in Norway, Canada and Australia, while prices temporarily reached a very high level. It seems probable that the use of molybdenum as a steel-hardener will extend in the future, when the processes have become standardized by further research: another important factor is the provision of an adequate and regular supply, which has not existed in the past, and the market for molybdenum ores has always been uncertain. Besides its application in the metallurgy of steel, metallic molybdenum is also in use for various electric purposes, while its compounds are employed to some extent in the dyeing, tanning and ceramic industries; also a considerable amount of ammonium molybdate is used in chemical laboratories for the determination of phosphorus in various technical products, especially iron ores, steel and fertilizers.

The geology of the molybdenum deposits is comparatively simple. The metal is of very wide distribution, but as a rule only in small quantities, and its occurrence is clearly connected with a definite type of igneous rock, namely the acid intrusives, the granites and quartz-porphyries. The primary mineral of this

---

[1] Barrow, *Quart. Journ. Geol. Soc.* vol. LXIV, 1908, p. 384.

type is molybdenite, and the metal therefore belongs to the sulphide group of metallogenesis. The other molybdenum minerals appear to be always of secondary origin.

The Molybdenum Minerals. Only two minerals containing this metal are of any economic importance as ores, namely, molybdenite and wulfenite: others are known, but are mainly mineralogical curiosities. Molybdenite, MoS_2, strongly resembles graphite in appearance and is occasionally mistaken for it, but can be distinguished by its bluish tinge, lighter coloured streak and much higher density. Wulfenite, $PbMoO_4$, is a tetragonal mineral closely related to scheelite and stolzite, and intermediate forms seem to exist, forming an isomorphous series, though probably incomplete. Wulfenite is found in the oxidized portions of lead veins and appears to be due to reactions between decomposition products of molybdenite and galena.

Types of Molybdenum Ore-Deposits. A general consideration of all the important known deposits of molybdenum ores shows that they can be reduced to very few types; these are as follows:

(*a*) Segregations and disseminations in acid igneous rocks;

(*b*) Pegmatites, veins and pipes in connection with granites;

(*c*) Contact deposits at junction of granites and limestones; all these yield molybdenite only.

(*d*) Wulfenite deposits in oxidized zones of lead veins.

Of the molybdenite deposits the second type is the most common, and the greater part of the world's production probably comes from this source. The third type is of importance in Canada, while a large proportion of the molybdenum output of the United States has recently come from wulfenite ores in Arizona, but this particular occurrence appears to be worked out, and in the last year or two the United States production has been mainly from molybdenite, as elsewhere.

From the metallogenetic point of view the behaviour of molybdenite is very similar to that of its frequent associate tungsten. It is most commonly found as an original constituent of pegmatites derived from magmas of highly acid composition, and of quartz veins, which are in their origin essentially similar to pegmatites. It is not necessary here to discuss the mechanism of the formation of such ore-bodies. For a general account of the process see Chapter II.

The chief producers of molybdenite ores are Canada, Norway and Australia. The types of occurrence in these countries show some interesting differences and will be briefly described.

Molybdenite in Canada. The occurrences of this ore in Canada show a considerable amount of variation, belonging to all the three first types as above enumerated. In the No. 1 pit of the Dominion Molybdenite Co. at Quyon, Quebec, is worked a large dome-shaped segregation of biotite-syenite, measuring about 130 by 60 feet, enclosed in a fine-grained hornblende-biotite-granite. The syenitic ore-body contains molybdenite, pyrite, pyrrhotite, tourmaline and fluorspar; a dyke of pegmatite cutting the syenite also contains a little ore, while other segregations of a rather more basic nature (mica-diorite) also carry some molybdenite.

But most of the Canadian molybdenite occurs either in pegmatite dykes of the Laurentian series or in altered limestone in contact with granites. The pegmatitic type scarcely needs any more detailed description; but the limestone type is more interesting. The metamorphic limestones or dolomites of the Grenville series form lens-shaped masses in contact with the intrusive hornblende-gneiss of Laurentian age. There has evidently been a good deal of assimilation in both directions and every gradation can be traced from the hornblende-granite through pyroxene-granite, syenite and gabbro. The ore-bearing rock which comes next in order consists mainly of lime-pyroxenes with scapolite and is described as contact-pyroxenite; epidote and sphene are often present, and the whole facies of the rock suggests a metasomatic replacement by material derived from the granite magma: the ores may therefore be described as contact deposits, and the whole rock is similar to that known to Scandinavian geologists as "skarn." Many occurrences of this type are known in different parts of the provinces of Quebec and Ontario.

In 1918 the total production of molybdenum as metal contained in concentrates from the Canadian mines amounted to 113·5 tons.

Molybdenite in Australia. Although molybdenite occurs in some quantity in most parts of the eastern mineral belt of Australia, from the Cape York Peninsula to Tasmania, the most important area is in the Stanthorpe and Ballandean district in the extreme south-east of Queensland and the adjoining New England district in the north-east of New South Wales: there is also a considerable production from the Whipstick area in the south-east of the last-named State. The general geology of this eastern part of Australia has already been referred to in the sections on tin and tungsten: it must here suffice to say that the occurrences of molybdenite are always associated with granite intrusions; when considered in detail, however, these occurrences show many points of special interest in relation to their manner of origin.

The general mineral association of the Australian molybdenite deposits includes ores of tin, tungsten, bismuth and molybdenum in varying proportions. Those richest in molybdenite are chiefly associated with a rather coarse-textured non-porphyritic granite with from 70 to 73 per cent. of silica, usually described as "sandy" granite by Australian geologists, while the granites richest in tin and wolfram are more acid and porphyritic in structure. The molybdenite occurs in several different types of deposit including pegmatite and aplite veins, quartz veins with some felspar and quartz veins without felspar, and sometimes as contact deposits, but the most important ore-bodies are in the form of pipes in granite, as hereafter described.

In the north of Queensland a good deal of molybdenite is associated with the wolfram veins at Wolfram, but only in some instances in payable quantities: it generally lies in a gangue of clear white quartz in joints and fissures in a grey biotite-granite, with native bismuth and in depth mispickel and pyrite. Circular pipes of quartz with wolfram, bismuth and molybdenite are also known here, but are not so well developed as further south.

The Stanthorpe district of Queensland and the New England district of New South Wales form a single geological unit, lying on both sides of the inter-State boundary, not far from the coast. A brief description of the Kingsgate area, some 20 miles east of Glen Innes in New South Wales, will serve as an illustration of the general character of the deposits. Kingsgate lies on a plateau at a general level of about 3750 feet above sea-level, consisting of "sandy" granite, with peaks of tin-bearing granite rising to 5000 feet. The ore-bodies, lying in the sandy granite, are oval and cylindrical masses of quartz and other silicate minerals, vertical or more or less inclined: when inclined they usually dip towards the granite margin. They usually vary from 10 to 20 feet in diameter, less commonly up to 30 feet and the famous "Old 25" pipe on the Yates property measures 40 by 40 feet in plan. Sometimes the pipes branch downwards, and they often show what is known as "step and tread" structure; that is, a series of right-angled bends in the vertical plane, doubtless related to the major joints of the granite. Some pipes have been followed down for 500 feet vertically.

The structure of these pipes is very peculiar and interesting, in that, when inclined, the arrangement of the different minerals appears to show a relation to the flow of solutions in the pipe, regarded as a hollow cylinder. The molybdenite and bismuth usually lie in what may be called the gutter of the pipe, the natural line of flow in an inclined open cylinder. Thus the ores lie mostly at and near the foot-wall of the pipe. The pipes are, however, in reality filled with varying associations of minerals, chiefly quartz, felspar, often more or less completely altered to sericitic mica, normal mica and sometimes garnet. Quartz is nearly always the dominant mineral. It is clear that these pipes have been formed by pneumatolytic alteration of the original minerals of the granite, with addition of metallic minerals, by solutions escaping along joints from the interior of the granite. They are thus closely related to the tin and wolfram greisens of Cornwall and Saxony. In some pipes bismuth is the chief ore, in others molybdenite. As a typical example we may choose the Wet Shaft pipe worked by the Sachs of Kingsgate Syndicate. This pipe is oval in cross-section and about 8 feet in its longest diameter. The upper part, including rather more than half the section, consists of white quartz with some molybdenite; below this is a flat layer of brown quartz very rich in molybdenite; below this again comes a band of granular cavernous quartz with much bismuth, occupying the gutter of the pipe; the lowest layer or foot-wall consists of granite with highly-decomposed and sericitized felspar. In another pipe in the same area, called Sachs' pipe, the order is somewhat different, since the richest molybdenite ore forms the lowest layer, in hard granular quartz, and most of the bismuth occurs along with large patches of molybdenite above it: the uppermost layer is similar to that before described.

The Whipstick district in the extreme south-east corner of New South Wales, consists of Devonian sediments intruded by granite and overlain by Tertiary basalts. Here molybdenum and bismuth ores also occur in pipes, many of which, however, are largely filled with mica and garnet rock: the general relations, however, are very similar and the origin of the deposits is to be referred to the same cause as in the northern area. Some of the pipes also yield silver and gold. The output of ore from this district has been considerable[1].

In the Yetholme district, not far from Bathurst, N.S.W., molybdenite occurs in contact deposits due to the intrusion of a bathylith of granite into Palaeozoic sediments, including limestone, which is altered to a sugary marble and shows absorption of silica from the granite, giving rise to garnet and wollastonite rocks. The ore occurs chiefly in a garnet zone with amphibole, epidote and wollastonite, from 2 to 30 feet thick. This is a skarn rock, like some of the Canadian occurrences[2].

Molybdenite in Norway. The only other country that has produced molybdenite ores in any large quantity is Norway. The chief area is the Knaben district in southern Norway, east of Stavanger and north-east of the Flekkefjord. Several mines in this area made great profits during the war, and other companies hurriedly put up expensive plants, often on poor and doubtful prospects. Three fairly well-defined types of deposit have been recognized: (1) quartz veins and lodes lying in granite, with dissemination in the granite walls; (2) mineralized fissures in granite, usually parallel to quartz lodes; (3) disseminations

[1] E. C. Andrews, "The Molybdenum Industry of New South Wales," Bull. 24, *New South Wales Geol. Survey*, 1916.

[2] *Ann. Rep. Geol. Survey, N.S.W.* 1915, pp. 176–7.

in granite[1]. None of these seem to show any particular features of interest. The Gursli mine in the Moi district is peculiar in that it works mineralized fissures with ores of copper and molybdenum in a basic rock, probably a norite, and also a dissemination of molybdenite in the same rock. The Dalen mine in Telemarken works a flat-lying quartz vein, apparently an apophysis from a granite, in siliceous slate. From Bröggers' classical description of the pegmatites of southern Norway[2] it is clear that molybdenite is of very wide distribution in these dykes: it is often associated with copper, lead, zinc and arsenic and with rare-earth minerals, but these occurrences are of no commercial importance.

As before stated, wulfenite is found in a good many localities as a secondary mineral where deposits of molybdenite and lead ores have undergone decomposition in proximity. It has been worked to some extent in Spain, where it is found in limestones, but the chief output has been from Arizona, especially from the Mammoth and Collins mines at Schultz, Pinal County, Ariz. The veins lie along faults in granite intruded by quartz-porphyry and have a brecciated filling of country rock with a great variety of metallic ores, including copper, lead, a little gold and silver and a good deal of vanadium. The copper and lead ores are almost entirely oxidized to carbonates and sulphates, but a little unaltered galena has been seen. Most of the ore has recently been obtained by working over the old tailings from the former gold workings. These dumps are now exhausted, but much low-grade ore remains in the lodes.

VANADIUM

By W. H. WILCOCKSON, M.A., F.G.S.

Vanadium is widely distributed both in igneous and sedimentary rocks; in the former it is found chiefly with the intermediate and basic magmas and in the latter associated with iron, aluminium and other elements, the oxides of which are

---

[1] E. R. Woakes, *Trans. Inst. Min. Met.* vol. XXVII, 1918, pp. 184–195. *Mining Mag.* vol. XVIII, 1918, pp. 100–2.

[2] See memoirs cited on p. 28.

precipitated from alkaline solutions; or in close relationship to carbonaceous matter, particularly fossil wood.

The more important occurrences of vanadium have been classified by De Launay[1]; his classification is based on genetic lines and is, with some modifications, quoted below.

I. *Segregations from basic magmas.*

Stony meteorites, titaniferous magnetites—such as those of Taberg (Småland) and the Adirondacks—chromite and rutile of apatite-pegmatites.

II. *Hydrothermal veins* (generally associated with acid intrusions).

(a) Associated with pitchblende, as in the Joachimsthal (Bohemia).

(b) Gold telluride, Boulder County, Colorado; Kalgoorlie.

(c) In gold-bearing silver-lead veins, Province of San Luis, Argentina (both the Boulder County and San Luis deposits are in association with wolfram-bearing pegmatites).

(d) In the form of roscoelite at Vanadium, Colorado.

(e) As mottramite at Mottram St Andrew in Cheshire.

(f) As carnotite at Radium Hill and Mt Painter, S. Australia.

(g) With native copper in the copper deposits of Lake Superior.

III. *Sulphides associated with solid hydrocarbons.*

Patronite of Minasragra, Peru (see p. 430), "coal veins" (asphaltites) of Peru, Oklahoma, Nevada and elsewhere; formed by the evaporation of vanadiferous petroleum deposits.

IV. *Oxidized zones of lead, zinc and copper lodes.*

Minerals, all vanadates, with other oxidized compounds of lead, zinc and copper, especially phosphates and arsenates (for list of typical minerals see p. 432).

Important localities, Wanlockhead and Leadhills in Scotland; the Rhodesian Broken Hill Mines; Spain; the Cordilleran region of the south-western United States and northern Mexico; and Tujam yun in Russian Turkestan.

V. *Deposits in Sedimentary Rocks.*

All composed of oxidized compounds, as a rule accompanied by iron, aluminium, copper, lead, zinc, cobalt and carbonaceous matter. Of these the carbonaceous matter is the most frequent and is variously regarded as the source and as the precipitant of the vanadium.

Examples of these deposits are: the carnotite-bearing sandstone of Colorado, French bauxites and laterites, the "Minette" ores of Lorraine and Luxemburg, the shales, clays and coal ashes of New South Wales, the Kupferschiefer of Mansfeld and the abyssal muds and clays of the deep oceans.

[1] De Launay, *Gîtes Minéraux et Métallifères*, pp. 719–29.

Since many of the varieties of occurrence quoted above are of little economic importance and, moreover, imperfectly understood some typical examples will be selected for description.

Division I. These are all normal instances of magmatic segregation following the laws of differentiation for igneous rocks and do not require further discussion here.

Division II. Though the vein deposits, such as those of the Joachimsthal and of Boulder County, Colorado, are of some theoretical interest as showing the association of vanadium minerals with acid igneous rocks, and though, in past times, the Joachimsthal supplied the vanadium required for ink manufacture and dye making, the only occurrence of any great importance is that of the Vanadium District in Colorado[1]. In this district the vanadium-bearing mica, roscoelite, is found impregnating sandstone belonging to the La Plata formation of Jurassic age. The region, which is part of the Uncompahgre Plateau, is composed of flat lying sediments (sandstones and shales) of Triassic and Jurassic age, through which the rivers have cut deep gorges. The La Plata Sandstone varies from 35 to 60 feet in thickness and is divided by a limestone bed into two parts. The vanadium deposits are confined to the lower part and are found cropping out in the sides of the gorges of the San Miguel river and its tributary creeks over a distance of about 10 miles near the town of Vanadium.

The sandstone is fine-grained, mostly composed of quartz with a few felspar grains, and has a calcite cement. In the impregnated areas the calcite has been converted largely to roscoelite and the felspars have also been replaced by quartz, calcite, and roscoelite in many instances. In addition a number of blue or black oxidized vanadium compounds are found, especially in cracks, among which is the sulphate kentsmithite and occasionally carnotite, which has filtered in from the surrounding rocks.

The roscoelite may be either evenly distributed, concentrated along the bedding planes, or in small green spots: the colour of

---

[1] Hillebrand and Ransome, Bull. 262, *U.S. Geol. Survey*, 1905, pp. 9–31. Fleck and Haldane, *Colorado School of Mines Quarterly*, Jan. 1909, pp. 5–36. Hess, Bull. 530, *U.S. Geol. Survey*, 1912, pp. 142–56.

the rock is green to black according to the amount of vanadiferous minerals present, but may be varied by the presence of other impurities. The deposits are lens-shaped or tabular, between 600 and 700 feet wide and from an inch or two to 30 feet thick. They are placed in definite relationship to an unconformity in the lower La Plata sandstone. The plane of this unconformity acted as the channel for the mineralizing solutions and along it there is always a rich seam of roscoelite from $\frac{1}{4}$ to 1 inch thick, while on both sides of it the sandstone is impregnated with vanadium compounds for a varying distance. The other metals commonly present are copper and chromium, and the roscoelite sandstone often passes out into rock with the chromium mica mariposite; the copper is present as malachite or chalcocite.

The origin of the vanadium-bearing solutions is doubtful, but it seems likely that it is connected with some basic dykes which invade the area and which give place where they terminate to a siliceous fault-rock.

The tenor of these ores is not as a rule much higher than 2 or 3 per cent. of metal, though the rich roscoelite seam carries as much as 8·3 per cent.

The mines were worked for many years by the Primos Chemical Company but have lately been bought and closed by the American Vanadium Corporation.

A rather similar· occurrence of vanadium in the form of mottramite is known at Mottram St Andrew, in Cheshire, where the association is lead, copper and cobalt minerals, with the mottramite impregnating Bunter sandstone on either side of a fault.

Division III. These deposits are widely spread in the United States of America and in Peru[1], most commonly they take the form of vein-like fissures, especially in limestones, filled with some hydrocarbon, such as grahamite or impsonite, through which the vanadium, probably in the form of sulphide, is widely disseminated. These asphaltites, as they are called, are occa-

[1] Gale, Bull. 380, *U.S. Geol. Survey*. Hewett, *Trans. Amer. Inst. Min. Eng.* vol. xi, 1910, pp. 274–99. Baragwanath, *Eng. Min. Journ.* vol. cxi, 1921, pp. 778–81.

sionally mined for fuel and their ash is found to contain vanadium in varying proportions from 0·9 to 15 per cent. They seem to have been formed by the upward seepage, to higher levels, of asphaltic petroleum and its consequent evaporation, the vanadium being an original constituent of the petroleum, derived from the vegetable matter from which the petroleum was formed.

At one locality only has the vanadium sulphide been found isolated and that is the Minasragra mine[1], where it occurs as the mineral patronite ($V_2S_5 + nS$).

This mine, owned by the American Vanadium Corporation, is by far the most important in the world and may be said at present to supply the whole of the world's output of the metal. It is situated at a height of 16,000 feet above sea-level, a short distance to the east of the main coast Cordillera of the Andes, 72 kilometres north-east of Callao and 15 kilometres south-west of Cerro de Pasco station on the Cerro de Pasco railway.

The ore-body takes the form of a lens-shaped mass about 30 feet wide and 300 feet long. It occupies a zone of crushing in red gypsiferous shales of Mesozoic age, strikes 20° west of north, the same direction as the country rock, and dips 75° to the west, rather more steeply than the shales.

The lens contains several materials which are, in order of abundance:

(1) *Quisqueite*, a black lustrous hydrocarbon, with conchoidal fracture, of hardness 2·5 and specific gravity 1·75.

(2) *Coke*, a dull black vesicular hydrocarbon, again with conchoidal fracture, hardness 4·5 and specific gravity 2·4; often containing globules of quisqueite.

(3) *Patronite*, a greenish-black mineral, with uneven fracture, hardness 2·5, specific gravity 2·65–2·71 and composition $V_2S_5 + nS$.

(4) *Bravoite*, a reddish-yellow mineral, occurring in small amounts in the patronite: composition (Fe,Ni)S.

(5) An undetermined silicate resembling halloysite.

The ore-body is mainly composed of the first three substances, the last two being only visible under the microscope. The quis-

---

[1] Singewald and Miller, *Eng. Min. Journ.* vol. CII, 1916, pp. 583–7. Miller and Singewald, *The Mineral Deposits of South America*, New York and London, 1919, pp. 487–91.

queite, coke and patronite are arranged in separate regions of the lens, the purer patronite usually in the middle, with the coke and quisqueite outside, in a manner somewhat resembling the successive fillings in a mineral vein. So striking was this resemblance, that Hewett considered that the different materials had been separated by a kind of differentiation and "intruded" separately, the patronite coming last, partly as masses of pure sulphide and partly disseminated through the coke and quisqueite. Further development of the mine has shown that the lens is surrounded by a zone of material called *veta madre*, about 40 feet thick and extending beyond its ends. This is composed of shale with thin veinlets of gypsum and anhydrite, and contains between 1 and 12 per cent. of vanadium, showing a gradual passage from unaltered rock to the most intensely mineralized portions of the deposit. This passage from shale to ore suggests that, while the coke and quisqueite were intruded into cracks in the shatter zone, the following (or accompanying) vanadiferous solutions replaced the shale itself, though the fact that the gypsum was not also affected constitutes a difficulty in accepting the explanation.

In and around the upper parts of the ore-body considerable quantities of oxidized vanadium compounds are encountered, both filling cracks in the rock and also replacing the shale. They are red or brown at the surface, becoming greyish black at a depth of 10 feet, and contain anything between 57 and 70 per cent. of vanadic oxide. They are found as low as the level of the lowest tunnel at 120 feet below the surface. The preservation of these oxidized compounds is a direct result of the rainless climate, since if there were any considerable amount of precipitation they would be dissolved. Originally the workings were an open cut in the oxidized ore, but now the unaltered patronite is won by tunnels driven into the hillside. The original oxidized ore contained about 20 per cent. of vanadic oxide, but the patronite at present mined yields a vanadic oxide to the amount of 35 and 40 per cent. after calcining. In 1921 arrangements were being made to instal plant to produce a concentrate of between 85 and 90 per cent. The concentrate produced at the mine is all sent to Bridgeville, Pennsylvania, for reduction to ferro-vanadium.

The analyses of the various materials are as follows:

	Patronite	Quisqueite	Coke
S (soluble in CS_2)	4·50	15·44	0·64
S, combined	54·29	31·17	5·36
C	3·47	42·81	86·63
H	—	0·91	0·25
N	—	0·47	0·51
O (by difference)	—	5·39	4·64
H_2O (at 105° C.)	1·90	3·01	—
Ash	—	0·80	1·97
V	19·53		
Fe	2·92		
Ni	1·87		
SiO_2	6·88		
TiO_2	1·53		
Al_2O_3 and P_2O_5 ...	2·00		

	Red Oxide	Green Oxide
V_2O_5	67·60	57·33
V_2O_4	tr.	4·76
MoO	2·82	3·28
SiO_2	1·17	0·57
TiO_2		0·07
Al_2O_3		—
P_2O_5	3·31	—
Fe_2O_3		19·53
CaO	4·30	0·70
MgO	?	tr.
H_2O	20·81	13·89

Division IV. In the deposits of this division the vanadium is always found with the oxidized products of sulphidic ores of lead; the lead is often accompanied by other metals, such as zinc, copper, molybdenum and occasionally tungsten, while phosphorus and arsenic are generally present in noticeable amount. Since the concentration of vanadium is always found in the oxidized parts of the lodes the minerals are all oxidized species, such as vanadinite, descloizite, cuprodescloizite, pyromorphite, mimetite and wulfenite. As a rule the richest deposits occur not far removed from the level of ground water.

The source of the vanadium in these ores is obscure: the vanadium minerals have obviously been deposited from solutions which have leached the upper parts of the ore-bodies in which they occur. Vanadium has rarely been reported as present in the unoxidized ore below, and it is suggested the metal must have been derived from some external source, such as the country rock.

An interesting occurrence of ore of this type has been described from the Rhodesian Broken Hill mines, situated on the Cape to Cairo railway, about 300 miles north of the Victoria Falls. Here lead and zinc ores occur in limestone. They are oxidized to a depth of 300 feet, below which they pass to sulphides, and in the oxidized region contain vanadium. The common metalliferous minerals are hemimorphite and cerussite, with some hydrozincite, limonite and wad, and are accompanied by smaller amounts of descloizite, vanadinite, pyromorphite and a number of zinc phosphates, such as tarbuttite ($Zn_3P_2O_8Zn(OH)_2$), hopeite and parahopeite ($Zn_3P_2O_8.4H_2O$). The descloizite occurs in the main ore-body as pockets up to an inch in width and as massive seams and incrustations mixed with limonite. The vanadinite is mainly found along with pyromorphite and the other phosphates on the floor of caves in the limestone, associated with bone breccias which evidently served as the source of the phosphorus[1].

A small output of vanadium ores has been made from this mine.

Division V. This division includes deposits of a very varied character, which cannot be described in detail for lack of space. The American carnotite ores are by far the most important and also the most interesting and will therefore be described, though they cannot be said to serve as a type for the division[2].

These carnotite ores are found over wide areas on either side of the Green River in south-western Colorado and south-eastern Utah. The rocks of the region are all sandstones and shales from Carboniferous to Jurassic in age and contain vanadium ores at several horizons, especially in the McElmo formation of Jurassic age. The deposits are usually lenticular in form, the lenticles often being disposed obliquely to the bedding; they rarely extend more than 100 feet in length or width and vary in thickness from a few inches to as much as 14 feet, with 4 or 5 feet as a more usual figure. As a rule there are only a few inches of high-grade ore but each lenticle yields a few hundred pounds to as much as 1800 tons in exceptional instances of workable ore. Frequently small funnel-shaped pockets of high-grade ore, the "Bug Holes" of the miners, are found to enter the upper side of the ore-bodies inclined at a low angle. They are a few inches wide and may be as long as 30 or 40 feet; the filling is usually high-grade carnotite in the form of a yellow clay, but is sometimes a black vanadium compound and the sandstone close around is usually cemented with gypsum or quartz.

In addition to carnotite, these ores always carry a number of

---

[1] Hubbard, *Eng. Min. Journ.* vol. xcv, 1913, p. 1297; *South African Mining Journal,* 1920, special number, p. 107.

[2] Boutwell, Bull. 260, *U.S. Geol. Survey,* 1904, pp. 200–10. Gale, Bull. 315, 1907, pp. 110–17, and Bull. 340, 1908, p. 259. Curran, *Eng. Min. Journ.* vol. xcii, 1911, pp. 1287–8; vol. xcvi, 1913, pp. 1165–7 and 1223–5. Burwell, *ibid.* vol. cx, 1920, pp. 755–8. Moore and Kithil, Bull. 70, *U.S. Bureau of Mines,* 1913.

other minerals, usually compounds of vanadium, such as hewet-
tite, metahewettite, volborthite, calciovolborthite, a number of
dark coloured vanadium minerals including the sulphate kentsmithite, and often some roscoelite. They are generally associated with gypsum and frequently contain fossil wood, round
which the carnotite is often concentrated. The origin of the
deposits is still a matter of some dispute. Hillebrand and
Ransome suggested that the vanadium and uranium compounds were originally widely distributed through the rocks in
which they now occur and that they were leached out by percolating waters containing sulphuric acid or alkaline sulphates
and deposited wherever the beds contained calcite or fossil
wood, which acted as precipitants, or wherever the solutions
were held up by a bed of shale. This view is upheld by several
other authorities and is supported by laboratory experiments
on the solubility and precipitation of carnotite performed by
Notestein. Hess, on the other hand, thinks that the ores were
deposited contemporaneously, the carnotite being deposited
from solution in sea water by the reducing action of decaying
organic matter, while Fleck and Haldane suggest that the
uranium and vanadium minerals have been carried to the
positions which they now occupy by the ordinary processes of
concentration and sedimentation.

These ores occur both in the Colorado plateau region, on the
east side of the Green River, and in the folded districts of Utah
on the west. A fair amount of mining has been done in the
Colorado region, but chiefly for radium, and the output of
vanadium has not been large.

The only countries whose output of vanadium is large enough
to be noticed here are the United States and Peru. The production
in recent years of these countries, expressed as metal, is given
below in long tons.

	Peru	United States	Total
1913	—	386	386
1914	4	404	408
1915	790	560	1350
1916	760	411	1171
1917	805	432	1237
1918	229	246	475
1919	494	254	748
1920	1095	455	1550

Titanium. This element is widely distributed in nature and occurs in considerable quantity compared with some of the more valuable metals. According to Clarke it amounts to nearly one-half per cent. of the lithosphere, equivalent to about 0·77 per cent. of titanium dioxide, in which form it is usually stated in analyses. Nevertheless its practical applications are very limited and indeed in most cases its presence in ores and economic minerals is a drawback rather than otherwise. Although titaniferous iron ores occur in great quantity in many localities they present considerable difficulties in smelting, and have not hitherto been largely exploited. By the introduction of improved methods however these ores may be of greater value in the future.

Titanium occurs commonly in several different mineral forms. The most abundant probably are ilmenite and titaniferous magnetite, constituting the so-called titaniferous iron ores. These occur usually as disseminations or segregations separated during the early stages of the cooling and crystallization of basic igneous rocks: since these minerals are very stable they also tend to accumulate in detrital deposits derived from basic igneous rocks and ilmenite is an important constituent of "black sands" (see p. 478). It is now generally agreed that "titaniferous magnetite" is a mechanical mixture of ilmenite and magnetite. The mineral titanite or sphene, which is a calcium titanosilicate, is also very widely distributed, especially in acid and intermediate igneous rocks. Of more interest from the scientific point of view are the three crystalline forms of titanium dioxide, rutile, anatase and brookite, and the first of these is occasionally worked for special metallurgical purposes. Rutile is a very resistant mineral and is abundant as one of the heavy minerals in sediments, but the others are much less common, and need not be discussed further.

The principal occurrences of rutile of commercial importance are in the form of syenitic pegmatites in Virginia, U.S.A., and in Norway. At Roseland, Nelson County, Virginia, rutile occurs in considerable quantity along with ilmenite in pegmatites, where it has been mined to a considerable extent[1]. At Kragerö,

---

[1] Watson, *Econ. Geol.* vol. ii, 1907, pp. 493–504.

Risör and other places in south-east Norway it is found in veins of albite, where it is also worked commercially.

The most important large-scale segregations of ilmenite and so-called titaniferous magnetite are in Scandinavia and the Adirondack region of New York, in both of which regions they are of some commercial importance. The great ore-mass of Taberg, in Sweden, which may be taken as a typical example of direct differentiation, is described elsewhere (see p. 321). The titaniferous iron ores of the Adirondacks are segregated from a gabbro which is itself intrusive in a large mass of anorthosite. Mineralogically the ore is a finely granular admixture of magnetite and ilmenite and as mined it also contains felspar, pyroxene, olivine and other definitely igneous minerals: the ore-masses usually shade off gradually into the country rock (gabbro) but sometimes form well-defined dykes[1]. Somewhat similar orebodies also occur in association with the great Duluth gabbro laccolith, and a very remarkable dyke of nearly pure ilmenite, up to 300 feet wide, cuts an anorthosite intrusion. All of these are clearly due to differentiation in basic magmas.

Titanium ores are exploited for the manufacture of ferrotitanium, which is now used to some extent as a purifier and deoxidizer for steel. Other uses are for the preparation of titanium tetrachloride for smoke clouds in warfare, and for various applications in arc-lights and in the leather, textile, pottery and other trades. A recent development is the production of white titanium paint, which is said to be superior to zinc white: this is chiefly made in Norway, from the ilmenite ores of that country, also to some extent in the United States.

Thorium. Until about 1885 thorium minerals were regarded merely as chemical and mineralogical rarities, but the invention of the incandescent gas-mantle by Auer von Welsbach rendered them of considerable economic value and the so-called Rare Earth Industry is now an important branch of chemical technology[2]. The earliest forms of incandescent mantle were

---

[1] Kemp, 19th Ann. Rep. U.S. Geol. Survey, Pt III, 1899, pp. 377–422. Singewald, Bull. 64, U.S. Bureau of Mines, 1913.

[2] Johnstone, The Rare Earth Industry, London, 1915.

composed chiefly of zirconia, but it was soon found that thorium oxide with about 1 per cent. of cerium oxide gave much better results, and this is now the standard form.

Thorium is a constituent in varying quantity of a large number of minerals, mostly rare, including thorite, thorianite, samarskite and gadolinite. Thorite and thorianite have both been found in commercial quantities in Ceylon, where they were chiefly obtained from gravels: they have, however, been found in place in weathered pegmatites associated with a mica-spinel rock[1]. In 1905 the export of thorianite from Ceylon amounted to about 8 tons valued at £4800. Gadolinite and samarskite are also rare minerals found in special types of pegmatite in Norway, Russia, South Dakota and Texas, where they have been exploited to some extent as a source of rare earths in general.

But at the present time the only important source of thorium compounds is the mineral monazite, which contains this element in varying proportions, ranging from nothing to 28 per cent. in rare cases, but usually averaging about 8 per cent. Owing to this great variation it is obvious that careful analysis of samples is indispensable.

Monazite sands have been exploited to a certain extent in Ceylon, but the most important deposits in that part of the world are found on the coast of the State of Travancore, from the southernmost point of India for about 100 miles up the west coast. Along this coast are extensive sand-beaches and sand-bars, while at certain points elevated dunes of ancient date also occur. In these sands the monazite is found naturally concentrated in the same manner as in Brazil. The crude sand is usually black from the presence of dominant magnetite and ilmenite, but sometimes red when garnets are in excess. Zircon is also common, while the light constituents are quartz and calcite. The monazite has been traced to its source mainly in the pegmatites cutting the charnockite and gneissose rocks that make up so large a part of this State: some of it may also occur as an accessory in the gneisses themselves[2].

[1] Sterrett, *Min. Res. United States*, 1907, vol. ii, p. 791.
[2] Tipper, *Rec. Geol. Survey India*, vol. xliv, 1914, pp. 186–96.

Analyses of monazite from Travancore generally show from 8 to 10 per cent. of ThO_2: this is nearly double the amount usually found in the Brazilian mineral, and these deposits must be considered high grade.

In the early days of the incandescent mantle industry a fair amount of monazite was obtained from North and South Carolina, mainly from the outcrops of pegmatites in which it is a common constituent, but this supply was soon undersold by the Brazilian sands. Most of the monazite was found in various residual and alluvial deposits in grains about the size of a pea, but the supply was limited and this source is now hardly worked[1].

Monazite has been recorded as an accessory mineral in granites, gneisses and pegmatites in many parts of the world, and there is always some possibility of its occurrence in quantity in residual and alluvial deposits formed by denudation of such areas: like all other heavy minerals it tends to concentrate at the bottom of such deposits where due to water action, and thus is easily missed on a superficial examination.

Monazite appears to be common as an accessory mineral of igneous rocks, especially granites, in certain parts of the world, but only in small quantity, so that it is rarely identified in microscopic examination of thin sections. It is also found in large crystals as a constituent of certain pegmatites along with other rare-earth minerals. Monazite is essentially a phosphate of the metals of the cerium and lanthanum groups, the thorium being present apparently as a more or less accidental constituent, possibly as minute inclusions of thorite, or in solid solution.

Monazite is heavy, fairly hard and chemically stable, so that it tends to survive and accumulate in residual and alluvial deposits. The chief source of the mineral is in beach-sands which have undergone a natural process of gravity concentration by wave-action. The other important constituents are usually magnetite, ilmenite, zircon and garnet, together with quartz. Since the hardness of monazite is not very great (about 5–5·5) the grains are commonly well rounded and the yellowish colour and lustre are characteristic.

[1] Sterrett, Bull. 340, *U.S. Geol. Survey*, 1908, pp. 272–85.

The most important deposits of monazite sand are found on the coast of Brazil in the provinces of Bahia, Espirito Santo and Rio de Janeiro. The deposits chiefly worked are those forming part of the actual beach, between tide-marks, and these are the property of the Government, being let on lease to contractors on a heavy royalty. Extensive deposits also occur inland, but are little worked as yet[1]. The grains of monazite are of remarkably uniform size and must be derived from the granitic rocks of the Brazilian highlands. The proportion of monazite in these sands is very variable, ranging from 2 to 60 per cent. in the natural condition, before concentration, which is usually effected by magnetic separation, or sometimes by shaking tables of the Wilfley type.

Tantalum. At one time the metal tantalum was in considerable demand as the material for the earliest practicable form of the metallic filament electric lamp[2]: however, for this purpose it was soon displaced by tungsten, and none of the other numerous technical applications of the metal appear to have proved of any value, and there is now no demand for the ores. It is, however, always possible that some practical application may be found in the future for this metal, which possesses some useful properties, rather similar to those of platinum, while its melting point is much higher, about 2900° C. In natural minerals tantalum is nearly always associated with the rather similar metal niobium. The principal ores are the allied minerals tantalite and columbite, which may be represented by the general formula $(Fe,Mn)O.(Ta,Nb)_2O_5$, and are generally found together in varying proportions. Samarskite is a highly complex mineral, containing tantalum, niobium, cerium, yttrium, thorium and uranium, with ferric iron. These minerals clearly belong to the pegmatitic type of segregation from igneous magmas of acid composition. Columbite occurs in some quantity in the remarkable pegmatites of Ivigtut, Greenland, and of Etta Knob, South Dakota, which are described elsewhere, and generally speaking it is a not uncommon member of the tin-tungsten paragenesis. Samarskite belongs mainly to the rare-earth syenitic pegmatites of Scandinavia and the Urals.

---

[1] Freise, *Zeits. für prakt. Geol.* 1909, p. 514.
[2] W. von Bolton, *Zeits. Elektrochemie*, vol. XI, 1905, p. 45.

The small commercial production of tantalum ores about 12 or 15 years ago mostly came from the Greenbushes and Pilbara districts of Western Australia. The mineral has also been worked in the Black Hills of South Dakota (Etta Knob), at the mica mines in Mitchell and Yancey Counties, North Carolina, and at Branchville, Conn. Samarskite has been found in considerable quantity in the Ilmen Mountains, near Miask in the Urals.

Zirconium. The mineral zircon is of very wide distribution, commonly occurring as an accessory constituent of igneous rocks, especially in those of acid and intermediate composition. It is often enclosed in biotite, in crystals of microscopic size, while larger crystals are found as constituents of some syenites, and in pegmatites. Owing to its hardness and chemical stability zircon finds its way in quantity into sedimentary deposits and is one of the most abundant of the "heavy minerals" of sands and sandstones. Large transparent crystals of zircon, chiefly from pegmatites, have been cut as gem-stones, under the name of hyacinth. The hardness and high refractive index render the mineral well adapted for this purpose. The best stones have a yellowish or clear brown colour, occasionally pink or colourless.

Zirconium is also a constituent of a large number of other minerals, most of which are rare, being found chiefly in pegmatites of the rare-earth type, often in connection with syenites, as in Norway: among these are rosenbuschite, lovénite, wöhlerite, eudialyte and eucolite and catapleiite. In most of these the percentage of ZrO_2 ranges from 15 to 40. Cyrtolite is a mineral of rather uncertain composition, supposed to be formed by alteration and hydration of zircon. Baddeleyite or brazilite consists essentially of ZrO_2 with a little iron, and is found in Ceylon and Brazil.

In recent times several technical applications of zirconium and its compounds have been discovered. It was first utilized commercially as the chief constituent of the early forms of the incandescent (Welsbach) gas mantle, but for this purpose zirconia was soon displaced by thoria. Zirconia is also used in the glower of various forms of the Nernst lamp. Owing to its highly refractory characters it is sometimes employed for the manufacture of crucibles for very high temperature work, and the alloy of zirconium and iron, ferro-zirconium, is sometimes employed in the manufacture of special steels, about 1 per cent. of a 20 per cent. alloy being added to the steel to remove nitrogen and oxygen. It has also been found that zirconia gives good results as a lining for steel furnaces, as it is not corroded by the molten metal to any appreciable extent.

Although actual figures are not obtainable, it is clear that the commercial production of zirconium minerals is quite small. Only two minerals are of any practical importance, zircon and baddeleyite.

As before stated, zircon in large crystals is commonly found in pegmatites: one of the most important occurrences is in Henderson County, North Carolina, where a dyke about 100 feet wide cuts through pre-Cambrian gneisses and can be traced for about one and a half miles. The upper part of the dyke is much decomposed and kaolinized and the zircons can easily be separated from the felspar by hydraulic processes. At somewhat greater depths they can be obtained by crushing the rock and washing. Many other minerals containing rare earths are also present. This occurrence has been exploited to a considerable extent in connection with the manufacture of Nernst lamps in America[1]. The zircon from this locality is to a considerable extent altered to cyrtolite, but the two minerals are of nearlv equal value, and do not need to be separated.

The zircons used as gem-stones mostly come from gravels and other alluvial deposits, where owing to their hardness and density they tend to become concentrated at the expense of lighter and softer minerals. Gem zircons chiefly come from gravels in Ceylon.

As to the geological relations of the Brazilian baddeleyite very little appears to be known, although there has been considerable production. The mineral is found as pebbles and broken fragments, often of a glassy appearance, in the gravels of the Serra de Caldas in Minas Geraes. The commercial product averages about 69 per cent. ZrO_2 and contains up to 26 per cent. silica, the rest being mainly ferric oxide[2].

---

[1] Pratt, Bull. 25, *North Carolina Geological Survey*, 1916, p. 15.

[2] Wedekind, *Berichte d. deut. chem. Gesell.* vol. xliii, 1910, p. 290.

Uranium and Radium. The ores of uranium are chiefly valuable for their radium content, the uses of uranium itself being somewhat limited. It has been employed as a steel alloy metal, but without much success, while uranium salts impart a yellow colour to glass and pottery. The uses of radium are too well known to need recapitulation here. It always exists in uranium minerals in extremely small proportion, less than 1 part to 3,000,000 parts of uranium. The processes of extraction are very complicated and costly, so the price of radium is always likely to remain enormously high.

Uranium is a constituent of a large group of minerals, mostly of very complex composition. At the present time, however, by far the greater part of the world's production comes from certain deposits that are also sources of vanadium: the geology of these deposits is dealt with in the section on vanadium; hence no details will here be given (see p. 433, where the occurrence of carnotite, the most important of these ores, is dealt with). Apart from this source the chief uranium minerals are as follows:

Uraninite, cleveite, or bröggerite (crystalline), pitchblende when massive. This is a highly complex mineral, which is generally described as uranyl uranate, with lead, thorium or zirconium, often the metals of the lanthanum and yttrium groups, and also containing nitrogen, argon and helium. Radium was first discovered in this mineral and it has been shown that both it and the helium are products of the radio-active disintegration of uranium. This breaking down passes through many stages, a similar process also taking place in thorium, and the final product is believed to be lead[1]. Massive pitchblende is said to contain no thorium. Gummite is an alteration product of uraninite, of doubtful composition.

Torbernite and autunite are hydrated phosphates of uranium and copper or lime respectively, usually written

$$\text{Torbernite} = CuO.2UO_3.P_2O_5.8H_2O,$$
$$\text{Autunite} = CaO.2UO_3.P_2O_5.8H_2O.$$

Torbernite forms beautiful bright green tetragonal crystals, usually flat plates, with a very perfect cleavage parallel to the

[1] Rutherford, *Radio-active Substances and their Radiations*, 1913.

flat faces, giving a micaceous appearance. Autunite is yellow in colour and forms crystals very like those of torbernite in form and angle, but it is said to be orthorhombic. Similar minerals are also known containing arsenic and vanadium instead of phosphorus, but they are very rare.

Samarskite is a rare mineral containing uranium with cerium, yttrium, niobium and tantalum. The only important source is in North Carolina, where lumps up to 20 lb. in weight have been found in mica workings in pegmatites.

Of these minerals pitchblende is the commonest: it is often a good deal decomposed, being covered with a thick coat of yellow uranium oxide; sometimes it is entirely converted into a mass of this substance. Since uranium has the highest atomic weight of any known element, most of the minerals containing it have a high density: the specific gravity of uraninite crystals varies from 9·0 to 9·7. Some varieties of pitchblende have a density as low as 6·5.

The chief European localities where pitchblende is worked as a commercial proposition are Cornwall and Joachimsthal in Czechoslovakia: cleveite, bröggerite and other crystalline varieties occur in considerable quantities in pegmatites in Norway (Arendal, Anneröd, Elvestad), while samarskite is found near Miask, Ural Mountains.

Uranium in Cornwall. Ores of uranium in considerable variety have been found in many of the Cornish mines, the chief minerals being pitchblende and torbernite. According to the published descriptions they chiefly occur in association with copper, lead, silver, cobalt and nickel, and therefore appear to belong to the upper zones. Small quantitites of uranium phosphates have been noticed in the oxidation zones of copper lodes in very many mines of the district, so the metal seems to be widely distributed.

Pitchblende was at one time found very abundantly in the workings of the Trenwith mine in the St Ives district, where much copper ore was thrown on the dump owing to the presence of this "impurity," which was actually of much greater value than the copper, though this fact was not then realized. Rich specimens were also found from time to time in the adjoining

St Ives Consols mine. The chief uranium deposit in Cornwall is
that of the South Terras mine, situated in the Fal valley, a little
south of Terras Bridge, near Grampound Road station. This
mine has produced copper, tin and iron ores, and one lode,
running N.-S., is composed of quartz, chlorite, magnetite, ochre,
zincblende and a little copper, tin and garnet. It has at its
centre a leader varying in width from a mere film up to a foot
consisting mainly of rich uranium ore. Near the surface this is
composed of green torbernite and yellow autunite with zippaeite,
but below the 30 fathom level is found only unaltered pitchblende[1]. From this lode some hundreds of tons of rich uranium
ore have been produced: hitherto much mystery has surrounded
the operations of this mine. Uranium ores have also been found
in some quantity in the St Austell Consols mine in a lode lying
very near the continuation of the line of the South Terras lode[2].

Uranium Lodes in the Erzgebirge. The best-known locality
for uranium in the Erzgebirge is at Joachimsthal, on the southern
side of the frontier, now in Czechoslovakia. Here the country
rock consists of schists, gneisses and amphibolites penetrated by
granites. The lodes form two systems striking N.-S. and E.-W.
respectively, the last named being the older, but the N.-S. lodes
the richer in uranium, which is usually associated with silver,
cobalt and nickel, and a gangue of calcite or dolomite. The
pitchblende is irregularly distributed in the lodes and seems to
be specially associated with carbonates, or with included fragments of slate[3]. These lodes contain no tin, but in other parts
of the district uranium minerals are found in smaller quantities
in tin-bearing lodes.

However, it appears from a careful study of the subject that
the association of uranium with tin is by no means so close as
is often believed. In Cornwall in particular the pitchblende and
derived minerals are chiefly found in the copper-bearing portions

[1] Personal communication from Mr E. H. Davison, B.Sc.

[2] MacAlister, "Geology of Falmouth and Truro," *Mem. Geol. Survey*,
1906, p. 179 (with further references). Collins, "Observations on the West
of England Mining Region," *Trans. Roy. Geol. Soc. Cornwall*, vol. XIV,
1912, pp. 241 and 342. Penrose, "The Pitchblende of Cornwall, England,"
Econ. Geol. vol. x, 1915, pp. 161–71.

[3] Krusch, *Zeits. f. prakt. Geol.* 1911. p. 83.

of the lodes and in silver-lead veins belonging to a still higher zone. The same argument applies to the Erzgebirge, and the uranium deposits of Gilpin County, Colorado, are associated with gold-silver lodes of a still shallower type, and containing some tellurides. Uranium, therefore, is not a characteristic metal of the deepest zone, but belongs to a region of lower temperature and pressure, namely the upper part of the copper zone and the silver-cobalt zone of the usual German classification. It has not hitherto been worked so far as is known at very great depths anywhere, and does not appear to occur in any quantity in the real deep-seated tin lodes, but only in those containing tin as a minor and somewhat fortuitous constituent beyond the radius of its usual occurrence in quantity.

CHAPTER XIX

ALUMINIUM

Occurrence and Properties. Although aluminium is actually the most abundant metal in the accessible part of the earth's crust, amounting according to Clarke to 7·85 per cent.[1], nevertheless by far the greater part of this is unavailable by our present metallurgical processes; no means has yet been discovered of preparing the metal on a commercial scale from the common igneous or sedimentary rocks in which it occurs in such vast quantities, as silicates: the oxide Al_2O_3, corundum, is a rather uncommon and very refractory mineral, and the only workable source is a class of substances resulting from the weathering and disintegration of rocks containing alumina, combined with silica when unaltered, but converted by natural chemical processes into various hydroxides and hydrates of aluminium; of these bauxite and various forms of laterite are the most common and important. Aluminium also occurs combined with sodium and fluorine as the mineral cryolite, which is only found in quantity in Greenland. The preparation of aluminium on a large scale is a quite modern development, but one of rapidly increasing importance, owing to the remarkable physical properties of the metal, which combines lightness with strength in an extraordinary degree. From the foregoing it will be readily apparent that the geology of the aluminium ores is markedly different from that of most other metalliferous deposits, since they are for the most part superficial accumulations of fairly recent date, and must be described geologically as rocks rather than minerals, showing, however, considerable resemblances to certain types of ironstone.

A large quantity of aluminium is also used commercially in the form of various soluble salts, especially sulphates and double sulphates (alums). These compounds are manufactured from various natural deposits, such as alunite and shales.

[1] Clarke, "The Data of Geochemistry," Bull. 616, *U.S. Geol. Survey*, 1916, p. 34.

Bauxite. The composition of this substance is variable and uncertain, and it is doubtful whether it should be regarded as a definite mineral. According to the opinion of the best authorities it appears to be mainly a mixture of diaspore, $Al_2O_3.H_2O$, and gibbsite, $Al_2O_3.3H_2O$: it always contains some iron as well as silica, and sometimes a notable amount of titanium. In structure bauxite is always massive, concretionary or oolitic; it may often be described as earthy, and is probably to be regarded as a precipitate of aluminium hydroxides from colloidal solution. The more ferruginous varieties grade imperceptibly into laterite.

It appears probable that bauxite has originated in a variety of ways: the typical bauxite of Les Baux, in the south of France, is associated with Cretaceous sediments and has been explained as a deposit from hot springs. In most cases, however, bauxite has clearly been formed by the decomposition of igneous rocks, either acid or basic. In Germany (Vogelsberg) it has been clearly derived from basalt[1] and in the north-east of Ireland Cole has described bauxite derived from rhyolite or rhyolite-ash: he supposes that the lavas were attacked by acid vapours of volcanic origin, in fact a solfataric action[2]. Certain deposits of bauxite in Georgia and Alabama have been regarded as being formed in a manner analogous to bog-iron ore. They appear to be essentially similar to laterite in their mode of origin. Of late years it has been generally recognized that the more aluminous portions of the laterites of India and other tropical countries are similar to the bauxites of Europe and America and have a similar origin. The following table shows the composition of four typical Indian bauxitic laterites[3].

Indian Bauxites.

	1	2	3	4
Al_2O_3	67·88	64·64	58·23	57·50
H_2O	26·47	24·00	28·10	26·94
SiO_2	0·93	1·79	2·01	2·35
TiO_2	1·04	3·30	6·49	6·61
CaO	0·36	0·04	0·45	0·15
Fe_2O_3	4·09	6·21	5·48	6·53
	100·77	99·98	100·76	100·08

[1] Kilroe, *Geol. Mag.* 1908, pp. 534–42. Münster, *Zeits. f. prakt. Geol.* 1905, p. 242.　　[2] *Trans. Roy. Dubl. Soc.* vol. VI, 1896, p. 105.

[3] Warth, H. and F. J., *Geol. Mag.* 1903, p. 155.

All of these are bauxites of very good quality with only about 6 per cent. of iron oxide at most and therefore comparing very favourably with the Irish, French and German products, most of which contain a considerably higher proportion. Many of the Indian bauxites consist approximately of three-fourths gibbsite and one-fourth diaspore.

The bauxites of southern France are described as follows by Prof. Lacroix: in Provence the deposits are found over a length of 40 km., following an unconformity in the Cretaceous series due to the absence of the upper Aptian and Albian (Greensand and Gault); at Villeveyrac, Hérault, bauxite rests on Upper Jurassic limestones and is covered by Albian strata; in Ariège it rests similarly on a Jurassic dolomite, and pebbles of it are found in the overlying strata, hence it was formed soon after the deposition of the underlying rocks. In France bauxite deposits always accompany an unconformity, and Prof. Lacroix regards it as in all cases a colloidal lateritic deposit, corresponding closely to the secondary or detrital laterites of India. It cannot therefore be regarded as a definite mineral, but corresponds most nearly to a mixture of gibbsite and diaspore.

The chief localities in France are: Corbières, Pyrenees; Les Baux, Bouches du Rhône; and Var in Provence; Saintonge (dep. Charente) and Berry (deps. Cher and Indre). In all of these the geological relations are very similar.

In the United States bauxite is now worked on a very considerable scale and the output of aluminium metal in that country now amounts to about 100,000 tons per annum, besides a certain quantity imported from Canada. This makes up about one half of the world's total. The production of bauxite on a commercial scale is confined to the States of Alabama, Georgia, Tennessee and Arkansas; of these Arkansas is now the largest producer. The Arkansas bauxite area lies south-east and south of Little Rock; there are two types of deposit: (a) beds of primary bauxite, derived from syenite by lateritization, and (b) detrital deposits in Tertiary sediments. Some of the deposits of the first type grade down into kaolin, which is an intermediate stage in the alteration of the syenite. The bauxite deposits of Georgia and Alabama form a belt about 60 miles long, from

Jacksonville, Ala., to Carterville, Ga. The ore forms lenses and
pockets in the residual clays derived from the Knox dolomite
of Silurian age. The chemistry of the process that gave rise to the
bauxite is rather obscure, but appears to be essentially similar to
lateritization. The Tennessee deposits resemble the foregoing and
are mainly found in the neighbourhood of Chattanooga.

Laterite. The more aluminous varieties of laterite are now of
much value as a source of aluminium, many occurrences being
essentially similar to bauxite in composition[1]. Much discussion
has arisen as to the origin of laterite in general and the literature
of the subject is very voluminous. Typical laterite is chiefly
developed in tropical regions and undoubtedly results from a
peculiar type of weathering and decomposition of rocks rich in
alumina and iron. Every gradation exists from a nearly pure
aluminium ore to a high-grade iron ore and in some localities
laterite is also worked as an ore of manganese. Although we
are only concerned here strictly with the aluminous varieties, it
will be well to discuss the whole subject of the origin of laterite.

Indian geologists distinguish two types called high-level and
low-level laterite respectively[2], and there is a real difference
between these, in that the low-level variety has been formed
by denudation, transport, and deposition of rock fragments with
cementation by lateritic material: it is consequently much less
pure, being mixed with material derived from many sources.
Discussions as to origin are chiefly concerned with the primary
high-level variety, or laterite in place.

True laterite covers immense areas in many tropical countries,
being largely developed in India, the Malay Peninsula, the
Dutch East Indies, Western Australia, East, West and South
Africa, and South America. This distribution at once suggests
the influence of a climatic factor in its formation, since it is in
fact the most characteristic superficial deposit of tropical regions,
while it is almost unknown at present in the temperate zones,
though ancient laterites probably exist in what are now temperate regions. From the chemical point of view the most
striking feature of lateritic deposits is the scarcity of silica, lime,

[1] Holland, *Geol. Mag.* 1903, pp. 59–69.
[2] Lake, *Mem. Geol. Survey India*, vol. xxiv, 1891, p. 239.

magnesia and the alkalis, with a countervailing abundance of alumina, iron and sometimes manganese and titanium. This peculiarity is suggestive of some kind of selective solution during the weathering of the original rocks, and Sir T. H. Holland supposes that bacteria may have played some part in this special type of decomposition. There is, however, no absolute proof of this. The general opinion is that the formation of laterite is closely connected with a well-marked seasonal variation of rainfall; that is to say, with monsoon conditions. During the rainy season chemical weathering and solution are active, favoured by the high temperature: much material goes into solution and is carried downwards by the meteoric waters. When the dry season sets in the dominant movement of the ground-water is upwards, largely owing to capillarity. The dissolved material travels upwards again and is oxidized and precipitated in a colloidal form at the surface, thus giving rise to the prevailing concretionary and irregular structure of laterite. The relative proportions of the different metals present, aluminium, iron and manganese, will obviously depend on the character of the rocks from which they are derived: thus we should expect a laterite derived from a basic rock such as basalt to be rich in iron, while from a granite or syenite an aluminous variety poor in iron would be formed. On the whole the distribution of the different types appears to be in agreement with this, but it must be remembered that the basic plagioclase felspars, labradorite and anorthite, contain a large proportion of alumina.

It is clear, however, that many of the deposits in different parts of the world described as laterite are really more correctly designated residual clays and residual ironstones: they have been formed in place by simple weathering processes, and represent insoluble residues: many of them contain a considerable proportion of lithomarge, which is essentially the same as kaolinite, being in all probability a colloidal form of that mineral. In these forms the proportion of combined silica is much higher than in true laterite.

In a comprehensive discussion of the various forms of laterite found in India, Dr Fermor has shown that ordinary laterites are a mixture of hydrated oxides of alumina, iron and manganese, the true "lateritic" constituents, with varying proportions of clay (lithomarge) and sand. He suggests that the term laterite should be restricted to varieties containing over 90 per cent. of the first group; other varieties with a lower proportion he would call lithomargic and quartzose laterites, according to the proportions of clay and sand present[1].

The greater part of the Indian high-level laterite is found on the Deccan basalts, but it is also derived from gneiss and other rocks, such as the granites and diorites of the charnockite series. In West Africa it is found on granites and norites[2], and in East and South Africa on sediments, on granite and gneiss and on volcanic rocks. Its origin seems in fact to depend much more on climate than on the character of the underlying rock.

One of the most remarkable and useful features of laterite is its property of hardening on exposure to air: when first uncovered it is soft enough to be cut with a spade, but rapidly becomes quite hard and solid: thus laterite forms a very good and useful building stone, much used in India even for the construction of bridges and other work requiring considerable crushing strength.

Bauxite in Ireland. Important deposits of aluminous clays, commonly called bauxite, exist in the north-east of Ireland in Co. Antrim, especially in the neighbourhood of Larne, Ballymena and Glenarm. The bauxite locally forms part of a curious and interesting series of beds lying between the lower and upper divisions of the Tertiary basalts, being associated with bole, lithomarge, iron ore and lignite[3]. In some localities bauxitic material has also been described as derived from igneous rocks of acid composition, rhyolite or quartz-porphyry[4].

From a consideration of the geological relations of the deposits interstratified with the basalts it is clear that they represent a lateritic formation, analogous to the Indian laterites and developed during an interval of quiescence and weathering between the eruption of the earlier and later basalts. In places this series is cut by dykes that formed the feeders of the later

[1] Fermor, *Geol. Mag.* 1911, p. 514.
[2] Dixey, *Geol. Mag.* 1920, p. 211.
[3] Expl. Mem. to Sheet 20, *Geol. Survey Ireland*, 1886, pp. 11–16.
[4] Cole, *Trans. Roy. Dubl. Soc.* vol. VI, 1896, p. 105.

basalt sheets. It is difficult, however, to say definitely how much further alteration these deposits have undergone since their formation. The typical succession consists of lithomarge, iron ore and clay with basalt above and below: the bauxite when present replaces the iron ore in the succession, especially the upper, pisolitic portion of it. The lithomarge is a remarkable substance, usually of a pale violet colour with white spots, and as before stated, is essentially similar to china clay in its composition. The following analyses of the bauxite are given by Kinahan[1] (under the name of alumyte).

	Glenarm	Ballintoy
Alumina ...	42·50	52·37
Ferric oxide ...	1·54	1·29
Lime	0·46	0·48
Potash and soda	0·04	0·06
Silica	27·50	12·15
Titanic acid ...	9·40	5·20
Sulphuric acid...	0·08	0·35
Water	18·53	27·13
	100·00	100·00

As will be seen, the first of these is decidedly high in silica and appears to be lithomarge rather than true bauxite.

Here and there the upper part of the ironstone is overlain or replaced by inconstant beds of lignite, and the general character of this upper portion is suggestive of a lake or swamp deposit. A good many plant remains, including stumps of good-sized trees, have been found at various levels, and there is no doubt that the whole represents an old land-surface or succession of land-surfaces of Tertiary age.

Cryolite. In the electrolytic preparation of metallic aluminium the mineral cryolite plays an important part as a solvent for alumina in the electrolyte. It is also used in the manufacture of white enamelled iron-ware.

Cryolite is a double fluoride of sodium and aluminium, having the composition $3NaF.AlF_3$, crystallizing in the monoclinic system as pseudo-cubic crystals with three good cleavages: when pure it is colourless, but often stained brown, grey or nearly black.

Cryolite occurs in workable quantity only in one locality in the world, at Ivigtut, in Greenland, where it is exploited by the

[1] *Trans. Manchester Geol. Soc.* vol. XXII, 1892, p. 461.

Kryolith Mine og Handelsselskabet, A./S., under license from the Danish Government. This deposit is of remarkable character and appears to be essentially a pegmatite of unusual composition, carrying besides minerals rich in fluorine also tin, molybdenum, tungsten, niobium and sulphides of lead, zinc and copper with a little silver and gold. There are also large quantities of chalybite intergrown with the cryolite. The country rock of the district consists of gneiss invaded by non-foliated granite and the cryolite deposit appears to be a pneumatolytic modification of the latter. It forms a nearly oval mass about 500 feet long and 100 to 160 feet wide, sharply bounded against the granite, with steep junctions, on one side nearly vertical, on the other at about 45° or 50°: thus it is of a somewhat dome-like form, widening downwards. It encloses blocks and fragments of gneiss and granite, hence it is evidently the latest intrusion. Along the contacts the granite is more or less converted into greisen, containing yellow gilbertite mica with cassiterite. The cryolite forms large crystalline masses, sometimes measurable by many feet, but usually without definite crystal faces. Sometimes it is intimately intergrown with granitic pegmatite or with chalybite, and average material appears to contain about 25 per cent. of impurity of one kind or another.

This deposit has been extensively worked by open-cut for many years and the average output now amounts to about 10,000 tons per annum; some of this goes straight to the United States, the rest to Öresund, Denmark, for the manufacture of pure alumina.

So far as the genesis of the Greenland cryolite is concerned, its close association with cassiterite, wolfram, molybdenite, lithium minerals and fluorspar is decisive: it is clearly a pegmatitic intrusion belonging to the tin-fluorine paragenesis: it is in reality an exaggeration of the Cornwall-Erzgebirge type: the apparently complete absence of boron minerals, such as tourmaline, however, serves to distinguish it clearly as a separate sub-class. It is the extreme case of fluorine mineralization, and possesses affinities to the pegmatites of the Black Hills of Dakota.

The occurrence of cryolite has also been indicated in the Ural Mountains, in Colorado and in the Yellowstone Park, but none of these appear to be in workable quantity.

CHAPTER XX

THE PRECIOUS METALS

Gold. The annual value of the world's production of gold in the last normal years has been approximately £100,000,000, and of this the British Empire has been responsible for about 60 per cent., the United States coming next with about 20 per cent.: the largest single producer is the Transvaal with nearly 40 per cent. of the total.

Gold is a metal of very wide distribution, being found in more or less quantity in nearly all parts of the world. After iron ores, gold is economically the most important metallic product of the earth, and from the very earliest times it has been the object of active search and exploitation: its discovery in hitherto undeveloped regions and in remote parts of the earth has led to many very remarkable industrial phenomena, and has had notable effects on the history of the world in both ancient and modern times: it is impossible, however, to pursue here this fascinating subject, which would require a whole book to itself.

The greater part of the output of gold comes from mines and workings exploited entirely or mainly for that metal, but a not insignificant fraction of the total is obtained as a by-product in the mining and metallurgy of other metals. Such occurrences will not be dealt with in detail here, attention being confined mainly to the geology of gold mines in the strict sense. A notable feature of the exploitation of gold is the importance of various superficial and secondary deposits: in this gold resembles tin and platinum. The oldest gold workings undoubtedly belonged to this class, and the extensive exploitation of ores worked underground is a comparatively modern innovation, largely made possible by the discovery of new and improved methods of concentration and extraction from low-grade ores on a large scale, especially the cyanide process, which has greatly extended the sphere of profitable gold-mining.

Ores of Gold. The number of naturally occurring compounds of gold is very small, and by far the most important source is the native metal. Compounds of gold and tellurium, known collectively as the telluride ores, occur in quantity in certain districts, and in one or two places gold is associated with selenium, although it is still uncertain whether a true compound exists. Much gold is also obtained from sulphides, occurring in veins or otherwise, and worked primarily or solely for their gold-content. The metal is also recovered in considerable amounts from the treatment of ores worked primarily for other metals, especially sulphides, both simple and complex: in some cases it accumulates in considerable quantity in the oxidation and cementation zones of sulphide deposits, being concentrated by weathering at the outcrop and in the gossan, or by secondary precipitation in the enrichment zone. The gold of the superficial and alluvial deposits is always native. The auriferous conglomerates of the Witwatersrand are a special type and need separate treatment. Natural gold is always more or less alloyed with silver: when much of the latter metal is present the alloy is called electrum. Natural amalgams of gold and mercury are also known, though rare.

The tellurides include a number of minerals often of very variable composition, containing gold, silver, mercury and tellurium. Some of the most important members of the group are calaverite $(Au,Ag)Te_2$, sylvanite $(Au,Ag)Te_2$, petzite $(Au,Ag)_2Te$, and nagyagite, which contains sulphur and antimony as well as gold, silver and tellurium[1]. The condition of the gold in the minerals of the sulphide and arsenide group is still somewhat uncertain: it may exist as a mechanical mixture or in some form of solid solution, in any case probably not as definite chemical compounds.

Types of Occurrence. Reference has already been made to the working of gold as a by-product in the metallurgical treatment of other metals: for example a considerable amount is

---

[1] Kemp, *Mineral Industry*, 1898, p. 295. Some doubt has been cast on the real existence of tellurides as definite chemical compounds, the suggestion being that they should be regarded as alloys of gold or silver and tellurium. There does not seem to be much foundation for this idea (see Lenher, *Econ. Geol.* vol. IV, 1909, p. 544, and Simpson, Bull. 42, *W. Australia Geol. Survey*, 1912, pp. 161–7). The supposed minerals, kalgoorlite and coolgardite, are mixtures (Spencer, *Min. Mag.* vol. XIII, 1903, p. 268).

obtained from the sludge remaining in the tanks in the electro-lytic refining of copper, and many similar occurrences exist. These are not here dealt with further. Including these under one heading we can recognize the following principal types:

(1) Gold-quartz veins.
(2) Telluride ores.
(3) Auriferous sulphides.
(4) Auriferous conglomerates.
(5) Alluvial and superficial deposits.

Of these groups the first, second and third are undoubtedly primary, while the superficial and alluvial deposits are obviously secondary: the origin of the gold in type 4 is still to some extent a matter of discussion: it must be regarded in any case as an intermediate type.

Quartz Veins with Free Gold. The occurrences in this category are the simplest of all, so far as structure and contents are concerned, although the explanation of their origin is not in all cases quite easy. They are probably also the most widely spread: in fact occurrences are so numerous that it is difficult and somewhat invidious to select examples for special treatment. Although of immense importance they are all really very much alike and little would be gained by a repetition of detailed descriptions. From the mass of examples available we select only four; the Kolar Goldfield of Mysore; the lode-system of the Sierra Nevada, California; Porcupine, Ontario; and the saddle-reefs of Bendigo, Australia. It must not, however, be sup-posed that the significance of this group is unduly underestimated.

With regard to the position of the simple gold-quartz veins in the general scheme of ore-deposits, more information is needed: we do, however, possess a series of facts that enable us to form some conclusion on the matter. In the first place the common association of such veins with igneous rocks is well known and significant, and indeed gold is known to occur as an apparently primary mineral in granites and other similar rocks of plutonic origin and it is also very commonly associated with volcanic phenomena. Furthermore there can be no doubt that some gold-quartz veins show relationship to pegmatites, since they often carry felspar and other minerals of similar associations. It seems clear therefore that some of these veins, at any rate, belong to a fairly deep-seated type of mineralization. On the other hand, many gold-quartz veins are structurally related to volcanic rocks, which have been formed at and near the surface; in some gold deposits, *e.g.* the Comstock Lode (see p. 489), much of the silica is chalcedony rather than quartz, and gold is also deposited in the siliceous sinter of hot springs in New Zealand. It appears therefore that gold-silica deposits have a wide range of depth-formation, and cannot definitely be assigned to any particular set of physical conditions.

The Kolar Goldfield, Mysore. As a typical example of an auriferous quartz vein of high productivity we select the well-known Champion Reef in the Kolar goldfield of Mysore, southern India. This lies in an area of highly-metamorphosed rocks, belonging to the Dharwar system, probably of early pre-Cambrian date. These rocks form a long narrow strip of synclinal form, apparently folded into Archaean gneisses and in their turn invaded by later granites. These rocks occupy an area about 50 miles long and from one to four miles wide, striking N.–S., and consisting mainly of hornblende-schists, which were originally basic igneous rocks, with a band of jaspery quartzite on the west and then a belt of mica-schist in contact with the gneisses. On the eastern margin is a belt of crush-conglomerate composed of fragments of granite in a matrix of schist. This series is cut by numerous unaltered basic dykes of later date, generally striking E.–W.

There are several quartz veins carrying more or less gold, but the only one of importance is the Champion Reef, which supports several large mines of world-wide reputation, such as Mysore, Ooregum, Champion Reef, Nundydroog and Balaghat. This great lode strikes N.–S., parallel to the foliation of the schists, and dips to the west at 50° to 55°. It varies in thickness from a mere stringer to 40 feet, swelling out into lenticular masses and being occasionally folded, with considerable thickenings in the crests and troughs of the folds. It consists of a dark bluish-grey quartz of vitreous appearance and often also contains actinolite, pyroxene, chlorite, epidote, calcite and tourmaline in small quantities. Native gold occurs throughout, though not usually in visible form: however, rich specimens with conspicuous gold are often to be obtained, especially on slickensided surfaces and shear planes. Other metallic minerals present are pyrite, pyrrhotite (sometimes nickeliferous), arsenopyrite, galena, blende and cuprite; none of these are of economic importance.

The payable ore mostly occurs in shoots, pitching north in the plane of the lode. Some of these are very large: for example,

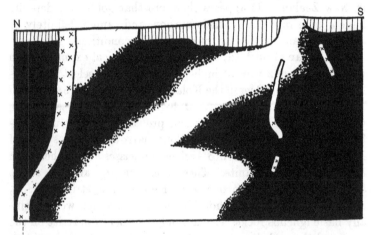

Fig. 70. Longitudinal section of Champion Reef gold mine, Kolar Gold-field, Mysore, India (after Hatch, *Mem. Geol. Survey India*, vol. XXXIII, 1901, pl. 8). Vertical lines indicate portions worked out; dykes are marked with crosses, ore-shoots black.

the Champion shoot and another adjoining it together form an ore-body 700 feet wide and 35 feet thick. Below the 1620 feet level this becomes thinner, but has been followed down for 6000 feet. The gold content is, however, somewhat less in depth.

The average yield, taking the whole lode over a number of years, is over 1 oz. per ton; the Mysore mine for 17 years averaged 27·5 dwt., the Champion Reef for 9 years 26·6 dwt. and Ooregum for 13 years 22·4 dwt. Attempts have been made to work other veins in the district, but with much less success[1].

[1] Hatch, "The Kolar Gold Field," *Mem. Geol. Survey India*, vol. XXXIII, 1902, pp. 1–81.

This may be taken as an excellent example of a single vein of auriferous quartz of high grade, yielding native gold and lying in ancient rocks. With regard to the conditions of origin of such a vein we have not much information, but the highly metamorphic character of the country rock and the presence of actinolite and tourmaline suggest deep-seated conditions with high temperature and high pressure.

The mines of this region have paid many million pounds in dividends and certainly rank among the great goldfields of the world.

The Gold Lodes of California. The total production of gold from the State of California has been enormous, but it must be remembered that a very large proportion of this, especially in the early days, came from placers, deep leads and other forms of secondary deposit. Nevertheless the primary source of the gold in these deposits is to be sought in veins and lodes, and in modern times lode-mining has gradually acquired greater importance as the superficial deposits are worked out.

The auriferous slate belt of California begins in the south of Mariposa County and extends along the western flank of the Sierra Nevada to the northern limit of the State, disappearing under great flows of lava in the adjoining States. Altogether it is about 250 miles long and from 20 to 70 miles wide. Nearly all of this area is more or less mineralized, but there are three specially rich regions: the Grass Valley and Nevada City district, the eastern belt in Tuolumne and neighbouring counties, and the Mother Lode belt. The last is in some ways the most typical and is selected for description here.

The Mother Lode proper is a belt of auriferous quartz veins, extending for a distance of about 100 miles in a north-west direction through Mariposa, Tuolumne, Calaveras, Amador and Eldorado counties. The crest and main mass of the Sierra Nevada is a great bathylith of granodiorite and similar rocks, probably of Cretaceous age, intruded into much folded and metamorphosed Palaeozoic and Mesozoic sediments. Few veins actually lie in the bathylith, most of them being in the metamorphic series on the west. The richest veins are not near the bathylith contact, but at some distance from it, often in close association with a series of small basic intrusions. The Mother Lode forms a belt of more or less parallel and connected veins lying within a very straight belt of black slate about a mile wide, bounded on either side by slates, amphibolites and greenstones. It is itself actually a big reversed fault, striking N.N.W. and dipping about 60° to the east. Individual veins are not usually traceable continuously for more than a mile or two. Some are simple veins up to 20 feet wide or even more, while others are composite veins or belts of brecciated slate with quartz stringers. Some veins are closely associated with albite dykes and in some the gangue is a mixture of quartz and albite. Ferruginous carbonates are also present. The richest ore occurs in steep shoots which usually pitch to the left of an observer looking down the dip. Besides free gold there is much pyrite, and sometimes chalcopyrite, blende and galena. Patches of tetrahedrite and tellurides are sometimes met with. Some shoots have been followed down the dip to a vertical depth of 4000 feet, and the average yield of gold is about 4 dollars per ton[1].

According to Emmons these lodes belong to an intermediate (deep to moderate) zone of ore-formation. Since the ore is very persistent in depth it is clearly due to ascending solutions and genetically related to the igneous rocks: it is to be regarded as a differentiation product of the great granodiorite bathylith[2]. The presence of albite in the gangue clearly shows relationship to the accompanying dykes.

The Porcupine District, Ontario. One of the most promising goldfields of recent development is the Porcupine district, northern Ontario, which lies about 150 miles due north of Sudbury, and north-west of Cobalt. After some unsuccessful attempts the first rich strike was made at the Golden Stairway on the present Dome property in 1909 and a rush immediately began. Development was rapid and the output is now large. For 1919 it was reported as 481,000 oz., and promises to be much greater in the near future.

---

[1] California Mines and Minerals (California Miners Association), San Francisco, 1899. Lindgren, *Bull. Geol. Soc. America*, vol. VI, 1896, pp. 221–40.

[2] Emmons, *The Principles of Economic Geology*, New York, 1918, p. 430.

The general rock succession of the district is as follows[1]:

KEWEENAWAN. Quartz-diabase, olivine-diabase.
ALGOMAN. Granite-porphyry, felspar-porphyry.
PRE-ALGOMAN. Lamprophyre, serpentine, quartz-porphyry.
TIMISKAMING. A series of conglomerates, greywackes, slates and carbonate-rocks, more or less schistose.
KEEWATIN. A complex of basic to acid volcanics, diabase, serpentine, iron-formation and carbonate-rocks, all largely altered to schists.

The principal ore-bodies lie in the pre-Algoman quartz-porphyry, in the Timiskaming series and in the schistose volcanic rocks of the Keewatin, and it is very noticeable that the richest and most extensive ore-bodies lie in the most schistose areas. In the Pearl Lake and Dome group the pre-Algoman quartz-porphyry is very schistose, while elsewhere it is more massive and less altered.

The most common type of workable ore-body takes the form of irregular lenses of quartz, usually with ill-defined boundaries. These vary in size from a few feet long and broad, up to masses hundreds of feet long and 30 to 40 feet wide: these are generally arranged in strings and series, apparently along crush-belts and many of them are known to extend to great depths. Some of the mines are from 1300 to 1500 feet deep and there is no sign of a falling off in yield. The ore-shoots are highly inclined or commonly vertical. Many other deposits are even less well-defined, being just masses of highly mineralized schist. Although quartz-veins and stringers do occur, clean-cut walls are rare, and the ore generally merges more or less gradually into country rock. Besides gold, the following minerals are also recorded: chalcopyrite, blende, galena, pyrrhotite, arsenopyrite, and scheelite. Felspar and tourmaline are also found and some of the masses of quartz and other minerals are of a distinctly pegmatitic character. The gold is irregularly distributed: it is often visible and sometimes very coarse: at the Golden Stairway it formed in places streaks $\frac{1}{4}$ inch wide. Most of it is free-milling and easily amenable to treatment. In the Pearl Lake group it

---

[1] Mackintosh Bell, *Min. Mag.* vol. XXIII, 1920, pp. 139–49. *The Mineral Industry of the British Empire* (Imperial Mineral Resources Bureau), Gold, 1913–19, pp. 87–92. Hore, *Eng. Min. Journal-Press*, vol. CXV, 1922, p. 359.

averages about 10 dollars per ton and in the Crown Porcupine property a large block ran to 30 dollars per ton.

The alteration of the gold-bearing rocks is of considerable interest. Where much shattered, they have also undergone extensive silicification, carbonatization and pyritization; in fact the alteration processes in the schists are essentially of the same nature as those seen in so many of the auriferous volcanic rocks of Hungary and the western United States.

As to the origin of the ores there is a general consensus of opinion that they are due to magmatic segregations derived from the Algoman intrusions. These solutions, of hydrothermal nature, permeated the older igneous rocks and sediments, especially where they were most schistose and shattered, depositing the ores and producing a concomitant alteration of the country rock. This process probably took place at a considerable depth below the surface, since the accompanying Algoman intrusions are of coarse texture and of a general granitic character.

The Saddle Reefs of Bendigo. A peculiar type of auriferous quartz reef, apparently confined to Australia, is shown by the so-called Saddle Reefs of the Bendigo district in Victoria. This is an area of Ordovician rocks, alternations of sandstones and shales, folded into numerous parallel anticlines and synclines of rather small wave-length, striking approximately north and south. When such rocks are strongly folded in this way there is of necessity a tendency to gape in the crests and troughs of the folds; the hollows thus produced have been filled up by deposition of large masses of quartz, in form analogous to the phacolithic type of intrusion. In some instances ten or more of these saddles are known to occur in vertical succession, and three parallel series lie within a horizontal distance of about a mile. The area is cut by several dykes and there is an acid bathylith in the neighbourhood. The ore consists of quartz and some albite, with free gold, pyrite and arsenopyrite. The reefs have been worked to a great depth, in the New Chum mine to 4500 feet, but the gold-content falls off somewhat below 2500 feet[1]. The character of the ore, especially the presence of felspar,

[1] Rickard, *Trans. Amer. Inst. Min. Eng.* vol. xx, 1891, p. 463.

and the association with the dykes clearly indicate an igneous origin for these interesting deposits. It may perhaps be mentioned here that most if not all of the gold-bearing veins of Victoria contain more or less felspar and thus clearly show pegmatitic affinities[1].

The Morro Velho Mine, Brazil. This is selected as a typical example of an auriferous sulphide vein worked primarily for its gold-content. It also possesses several special features of interest. In the first place it is certainly the deepest mine in the world: at present the lowest workings are just over 6700 feet vertical below the surface: secondly there is practically no change in the gold-content from top to bottom, a very unusual feature. There appears to be here no trace of any zonary arrangement of minerals. The mine, which is the property of the St John del Rey Mining Co., Ltd, an English concern, is situated in a hilly district at Villa Nova de Lima in the State of Minas Geraes, in a region of tropical climate and thick vegetation. The country rock consists of much sheared rather calcareous schists with mica and chlorite. The payable ore-body takes the form of an immense shoot, pitching down a vertical plane at about 45°. The stope length is about 700 feet and the width varies from 1 to 30 feet, averaging 10 or 12 feet. In some levels the stope length is nearly 1000 feet. The gangue consists of quartz, chalybite, dolomite and calcite, with arsenopyrite, pyrite, pyrrhotite and a little chalcopyrite; most of the gold is in the arsenopyrite, there being little free gold. Some silver is also produced. In the year 1915 the average return was about $11\frac{1}{2}$ dollars per ton in gold and silver together. It is worthy of note that the lower levels are absolutely dry and the roads must be sprinkled to keep down the dust. The greatest difficulty to be contended with is the high temperature. At 6000 feet the average rock temperature is about 115°, and ventilation is difficult. When the already warm surface air is pumped down to the bottom it becomes so much heated by compression that the cooling effect is very small.

The Passagem Mine, Brazil. This interesting occurrence, worked by the Ouro Preto Gold Mines of Brazil, Ltd, is situated

---

[1] Personal communication from Mr D. J. Mahony.

about 8 miles east of the town of Ouro Preto, in Minas Geraes. The country rock consists of quartz-schists and quartzites, overlain by the iron-ore bearing rocks known as itabirite and jacutinga (see p. 329). The ore-body lies at the contact of the quartzite and itabirite and dips south-east at 15° to 20°. There are two types of ore: white lodes, up to 36 feet thick consisting of white quartz with pyrrhotite and arsenopyrite, the last-named carrying most of the gold, and black lodes, with much tourmaline, quartz and the same sulphides. The black lodes are much smaller than the white lodes and lie below them, either in contact or separated by several feet of quartzite. The black lodes are the richer. Both types appear to be replaced quartzite,

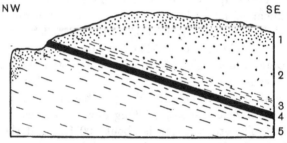

Fig. 71. Section of the lode, Passagem Gold Mine, Brazil (after Hussak, *Zeits. für prakt. Geol.* 1898, fig. 99, p. 345). 1, canga; 2, itabirite; 3, schist; 4, quartzite and lode; 5, quartzite.

but the genetic relations are not obvious: the abundance of tourmaline in the black lodes suggests pegmatitic affinities and formation at a considerable depth. The average yield of gold is about 7 dollars per ton.

In connection with both the cases just described it is to be remarked that this part of Brazil is on the whole an area of highly metamorphosed rocks affording evidence of deep-seated conditions of temperature and pressure with abundant plutonic intrusions: presumably the rich ore-deposits of the region form part of the same series of phenomena, which may date from a time when the earth's crust as a whole was much hotter than it is now, and conditions were specially favourable to chemical and mineralogical processes of replacement and mineralization, conditions then being more or less uniform down to great depths, in contrast with the more rapid temperature gradient of later times.

Gold–telluride Ores. Although tellurides of different metals are not very uncommon in many ore-deposits, there are only two or three areas where gold tellurides are of commercial importance. The chief of these areas are Western Australia, the Cripple Creek region of Colorado and parts of Hungary and Transylvania. The mineral character of the gold tellurides and their chemical composition have already been referred to.

The Coolgardie Goldfield, Western Australia. Although comparatively a recent discovery, this is one of the important goldfields of the world. Western Australia consists for the most part of a massif of ancient rocks, igneous and metamorphic, of undoubted pre-Cambrian age, although there is no absolute proof of this fact: it includes several goldfields, of which the most important is that of Coolgardie, and especially the Kalgoorlie area: one part of this is so auriferous as to be commonly called the Golden Mile. When considered in broad lines the geology of this region is simple. The oldest rocks are a series of sediments, including shales, sandstones, grits and conglomerates with perhaps some lava-flows, later invaded by a series of igneous rocks, basic and ultrabasic, with some still later acid types. These old rocks have been folded into a series of synclinal strips, striking north and south and surrounded on all sides by granites and gneisses. The mineralization is closely associated with movements of shearing and faulting later than the folding just mentioned. As a result of all these disturbances the rocks have been largely converted into schists and amphibolites.

The auriferous deposits belong to two types, which however exhibit intermediate forms. One type is found mainly in the altered igneous rocks, especially in a variety often described as quartz-diabase. These ore-bodies consist partly of replacements by dark cherty quartz, occasionally traversed by veins of white quartz with sulphides, and partly of schistose shear-zones rich in carbonates. The quartzose ore-bodies are essentially lenticular masses of sheared rock, often several hundred feet long along the strike and as much as 100 feet wide; these masses have been highly mineralized, often obviously by percolation from a visible crack or leader. As a rule the rocks are gold-bearing wherever they are sufficiently sheared, and the actual ore as supplied to the mills is in reality for the most part a silicified pyritous schist.

The second type of deposit, found in the calcareous schists, is not quartzose, but consists of shear-zones of the schists rich in carbonates, especially of magnesia and iron. These are of lower grade than the quartzose bodies, and often rather patchy.

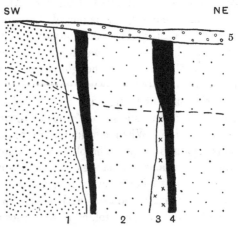

Fig. 72. General section across the north end of the Kalgoorlie goldfield (after Feldtmann, Bull. 51, *Geol. Survey W. Australia*, 1913, pl. XIII, simplified). 1, amphibolite; 2, talc-chlorite rock (altered amphibolite); 3, porphyrite (intrusive); 4, lodes; 5, laterite. The broken line is the base of the zone of oxidation.

Some gold-bearing lodes also occur in the more acid amphibolites: though small these include some very rich veins and leaders, but they are not generally important.

The normal type of sheared ore-body generally strikes N.N.W.–S.S.E. in conformity with the general structure of the district. Another subordinate set, striking nearly north and south, cuts the first set obliquely and good lenses of ore are sometimes found at the crossings. Faults are fairly common, striking north and south, and usually dipping west at about 45°.

Near the surface all the rocks have undergone a good deal of oxidation, the schists being converted into an aggregate of sericitic mica, kaolin and limonite. In this oxidized zone the gold is always free and generally in a spongy or granular form

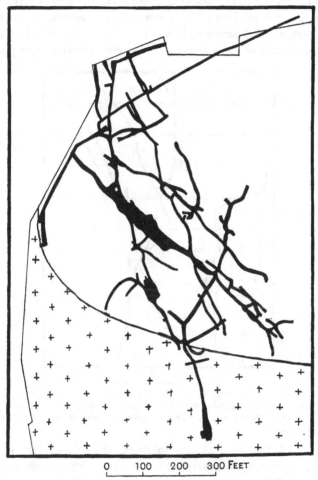

O 100 200 300 FEET

Fig. 73. Plan of level 4, Stratton's Independence Mine, Cripple Creek, Colorado (after Lindgren and Ransome, Prof. Paper 54, *U.S. Geol. Survey*, 1906, fig. 59, p. 454). Crosses, granite; black, veins; white, breccia.

(mustard gold). In the primary zone the gold exists almost wholly as tellurides, and a considerable number of other metallic minerals are also found, including galena, blende, chalcopyrite,

pyrite, tennantite, haematite, ilmenite, magnetite and asbolane. The non-metallic minerals are very numerous[1].

With regard to the origin of these ores, it is pretty clear that they have been formed at a considerable depth, under high temperature and pressure: it is not however clear, which, if any, of the visible igneous rocks is to be considered as the parent

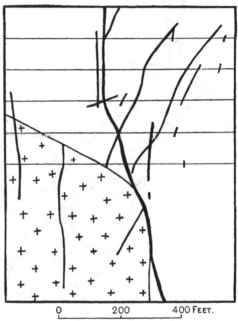

Fig. 74. Section through Stratton's Independence Mine, Cripple Creek, Colorado (after Lindgren and Ransome, Prof. Paper 54, *U.S. Geol. Survey*, 1906, fig. 62, p. 459). Crosses, granite; black, veins; white, breccia.

magma of the mineralizing solutions. The deposition of the ores is obviously connected with the shearing, which is later than the folding of the prevailing basic rocks. The quartzose nature of many of the replacement ore-deposits, especially of those in the amphibolites, suggests a relationship to an acid magma.

[1] Simpson and Gibson, "The Geology and Ore-deposits of Kalgoorlie," Part I, Bull. 42, *W. Australia Geol. Survey*, 1912. Feldtmann and Farquharson, *ibid.* Part II, Bull. 51, *W. Australia Geol. Survey*, 1913.

Cripple Creek, Colorado. Very different in age and general character is the occurrence of telluride ores at Cripple Creek. This district lies a few miles south-west of Pike's Peak, the highest point in the Rockies, and is itself from 9000 to 11,000 feet above sea-level. It is an area of schists and granites, surrounding a great more or less circular plug of igneous rocks, forming the central core of a Tertiary volcano. This plug, composed of tuffs and breccias and intrusions of alkaline rocks, shows a steep wall against the older rocks and is some 2 or 3 miles in diameter. It is roughly circular in plan and the mineralized veins are arranged in a radial fashion around its central point; they are essentially cracks and shear-planes due to the settling of the volcanic mass. Along these cracks there are deposits of quartz and fluorite with some dolomite, carrying abundant gold telluride (chiefly calaverite) with pyrite, blende, stibnite, and molybdenite, with here and there a little wolfram. Besides these fissure veins there are also replacements of the granite. In the oxidized zone near the surface is some free gold formed by decomposition of tellurides. The total gold production of this area to date has been over $300,000,000. Mining has been rendered somewhat difficult by the presence of great quantities of water in the porous volcanic rocks of the plug, while the surrounding schists and granites are dry. A deep-level tunnel has now been completed to drain the whole area down to about 8000 feet above sea-level, or about 1000 feet below the deepest workings.

The Transylvanian Erzgebirge. The Carpathian mountains form part of the main northern folded chain of the Alpine system, and almost encircle the sunken plain of Hungary (Fig. 27). Inside the chain and bordering the Hungarian depression is a discontinuous zone of ancient volcanoes, now deeply dissected and highly mineralized. These include the well-known mining regions of Schemnitz-Kremnitz, Nagybánya-Felsöbánya-Kapnik and others. The most southerly area of this kind, known as the Transylvanian Erzgebirge, is characterized by the occurrence of gold tellurides, especially at Offenbánya and Nagyag, in addition to native gold. The element tellurium was first discovered here. The lodes are fissures and fault-lines in Tertiary volcanic rocks, chiefly dacites and andesites with hornblende. The gangue minerals are chiefly quartz and calcite with manganese carbonate. The gold occurs either as tellurides (sylvanite, nagyagite, petzite and krennerite), as a mixture of tellurides and native gold, or as native gold only, and is usually accompanied by silver sulphides, galena and blende, with antimony and arsenic minerals. At Nagyag the lodes are arranged vertically in the decomposed andesitic lava filling the neck of a volcano which has penetrated schists and Tertiary sediments.

The Auriferous Conglomerates of the Transvaal. At the present time the Witwatersrand goldfield of the Transvaal is the world's largest gold-producer, its output in 1913 having been nearly £40,000,000, or approximately 40 per cent. of the total. This type of deposit, though by no means unique, is of a special character, the gold occurring in a finely disseminated condition in an ancient conglomerate of pre-Cambrian date. There has been much controversy as to the precise manner of origin of the gold, and it is necessary to enter in some detail into the geology of the area. The general succession of the rocks is as follows:

| Karroo System | { Ecca Coal-measures.
 { Dwyka Conglomerate. |
| Transvaal System | { Pretoria Series.
 { Dolomite Series.
 { Black Reef Series. |

Ventersdorp System, mainly volcanic.

| Witwatersrand System | { Upper Witwatersrand Series.
 { Lower Witwatersrand Series. |

Old granite and schists of the Swaziland System.

The Swaziland system constitutes the basement formation of this part of Africa, and includes a great thickness of crystalline schists and intrusive rocks of pre-Witwatersrand age. The next three systems, which are separated by unconformities from the basement rocks and from each other, are totally unfossiliferous and probably also of pre-Cambrian age: the Karroo system contains fossils and is of Permo-Carboniferous age. In parts of the Transvaal, the Waterberg beds, possibly of Devonian age, but perhaps older, are intercalated between the Transvaal and Karroo systems.

The gold-bearing conglomerates (Banket), are found mainly in the Upper Witwatersrand, especially at its base, but the Black Reef conglomerate, which is very similar lithologically, is also auriferous and has been worked on a small scale.

The general structure of the district, as will be seen from the map, Fig. 75, is a broad syncline pitching towards the E.N.E., the northern outcrop running nearly E.-W., from Randfontein

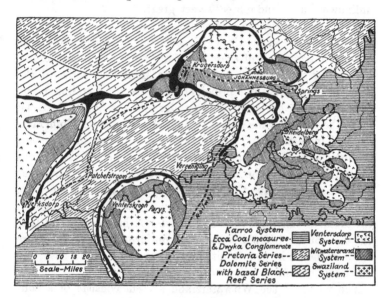

Fig. 75. Geological map of the southern Transvaal, by Dr F. H. Hatch, *Types of Ore Deposits*, 1911.

to Springs, while the southern limb runs N.E. The central part is occupied by Ventersdorp and Transvaal beds, and on the east the whole disappears under a spread of Karroo Coal-measures. The most remarkable physical feature is the northern outcrop of the Lower Witwatersrand series, which forms the Rand proper[1], a long escarpment rising to about 6300 feet O.D., and overlooking the lower ground of the old granite to the north. The northern suburbs of Johannesburg lie on this rising ground. The gold-bearing conglomerates outcrop near the

[1] The Dutch word *rand* means an edge (cf. the *Edges* of Derbyshire and West Yorkshire).

southern base of the dip-slope in a long line which is interrupted here and there by faults. At the outcrop the dip is steep, about 50°, but flattens out in depth. The southern limb of the syncline, near the Vaal river, is in places inverted. The underground extension of the syncline on the East Rand has been traced by borings, some of which are very deep, as much as 6000 feet. The depth to the gold-bearing strata in the centre of the syncline is unknown, but is undoubtedly very great.

The Lower Witwatersrand series consists largely of quartzites and slates, while in the Upper division conglomerates come in at various horizons; all of these latter are more or less auriferous, but by far the most important is the Main Reef series at the base. This includes three principal members, in ascending order; the Main Reef, the Main Reef Leader and the South Reef. Of these the second is the richest in gold. The higher conglomerates of the Kimberley, Bird and Livingstone series are unimportant economically. The Main Reef series is not constant throughout the whole length of the Rand, but shows considerable variations along the strike, thus the Leader disappears towards the west, the other two towards the east; only in the central portion of the Rand are all three fully developed[1]. It has been found that the distribution of gold is not quite uniform, but the richer ore tends to occur in shoots pitching towards the south-east; the meaning of this fact will be discussed later.

The whole of the Witwatersrand system consists of a continuous and conformable series of sediments, shales, quartzites and conglomerates, and the "reefs" are normal beds of a rather coarse conglomerate, composed mainly of quartz pebbles, with some of quartzite and slate. In the cement of this conglomerate so-called pyrites, in reality mainly marcasite, are abundant, and the gold is closely associated with it, partly as fine scales and crystalline flakes and perhaps also to some extent in solid solution. The rest of the cement is a matrix of very finely crystalline quartz and chalcedony with crystals of chloritoid and occasionally rutile and zircon. Visible gold is extremely rare. The gold-content averages about 7 dwt. per ton, and it

[1] Hatch and Corstorphine, *The Geology of S. Africa*, 2nd edition, London, 1909, pp. 116–157.

seems to be well established that the larger the average size of the pebbles the higher is the yield of gold.

With regard to the origin of the gold three principal theories have been put forward in the course of a long discussion. The first, which may be described as the *placer theory*, supposes that the gold was introduced in the solid form at the time of the formation of the original gravels: in fact it regards the reefs, to put the matter shortly, as fossil placers. In its simplest form this theory relies largely on the presence of rounded pellets of pyrites, which are considered to be water worn. However, it has been shown that these supposed pebbles are in reality either radial or concentric concretions and that their roundness is not due to rolling. As a modification of this idea it was suggested by Becker that the original alluvial gold has been dissolved and reprecipitated. This is supported by Gregory, who considers that the gold and pyrites may have at first existed in the form of "black sand" as in so many modern alluvial deposits. The *precipitation theory* supposes that the gold was precipitated from solutions derived from some external source during the actual formation of the original gravel-beds. This is largely based on the crystalline character of the gold, but the physico-chemical theory of the precipitation presents difficulties. The *infiltration theory*, supported among others by Hatch and Corstorphine[1], is based on the fact that much of the secondary cementing material shows a high degree of mineralization, due to subsequent percolation of solutions which may have likewise deposited the gold. The gold is limited to the conglomerates, which would be much more pervious than the finer quartzites. The solutions concerned may be connected with the intrusion of the numerous basic dykes of the area.

The whole question has been elaborately discussed by Mellor[2], who comes to the conclusion that the Banket conglomerates are deltaic deposits, laid down along the coast line of an ancient land. He lays much stress on the lenticular form of the workable bodies on the East Rand (Nigel mine, etc.). These were stream

---

[1] Hatch and Corstorphine, *Trans. Geol. Soc. S. Africa*, vol. VII, 1904, pp. 140–5.

[2] Mellor, *Trans. Inst. Min. Met.* vol. XXV, 1915–16.

deposits formed by rivers running from a rising land mass gradually approaching from the north-west during a period of uplift. He points out that more or less gold occurs in all the conglomerates, and that the conglomerates increase in importance towards the top of the system. The Main Reef Leader, with its unusually high gold-content, is regarded as being due to a single flood. All the deposits show analogy with the auriferous beach gravels of Nome, Alaska. Mellor recognizes that all the gold has been dissolved and redeposited in place.

The latest and most detailed authority on the subject, R. B. Young[1], after a comprehensive discussion and description of the phenomena from every point of view, abandons the idea of contemporaneous precipitation and infiltration and accepts the placer theory.

As before stated, the ore is of low grade, averaging about 7 dwt. per ton, and it is only by large-scale working with carefully standardized methods that profitable exploitation is possible. The mines may in general be assigned to two categories, the outcrop mines and the deep mines. These terms really explain themselves. The outcrop mines mostly work by inclined shafts down the dip, and the deep mines by vertical shafts sunk some distance from the outcrop, in the direction of the dip, that is to the south of the outcrop mines. Thus along the central Rand there is in general terms a double row of mines, extending at intervals along the fifty-mile stretch of country that constitutes this great goldfield. Where the same beds come to the surface along the south-eastern limb of the syncline, although the conglomerates are present they are generally unpayable, owing to excessive distance from the primary source of the gold in the old land to the north-west. Only in a few cases, such as the Nigel mine in the Heidelberg district, has profitable working been found possible[2].

Auriferous conglomerates with a strong superficial resemblance to the Banket of the Rand are found in various other parts of the world, notably in Rhodesia, West Africa (Ashanti) and

---

[1] R. B. Young, *The Banket*, 1917.

[2] Rogers, *Trans. Geol. Soc. S. Africa*, vol. XXIV, 1921, p. 17; Hatch, *Geol. Mag.* vol. LIX, 1922, p. 249.

Western Australia. In fact this may be regarded as one of the
definite types of gold deposit: most of the examples already
mentioned are doubtless of pre-Cambrian age. In the Black
Hills of Dakota there is a very interesting example of an auriferous conglomerate forming the base of the Cambrian. Many
of the pre-Cambrian schists of the area carry a good deal of gold:
depressions in the erosion surface of the schists are occupied by
fragmental deposits with much gold, now more or less cemented to
a hard conglomerate, with a good deal of infiltrated pyrites in the
cement. There can be little doubt that these deposits represent a
series of marine placers of lowest Cambrian age, although the
possibility of some infiltration of gold is not excluded. Many of
the grains of gold are, however, clearly water worn.

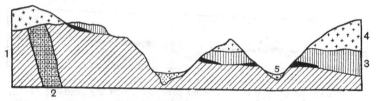

Fig. 76. Generalized section through the Homestake district, South Dakota
(after Irving and Emmons, Prof. Paper 26, *U.S. Geol. Survey*, p. 102,
slightly modified). 1, schist; 2, Homestake vein; 3, Potsdam series
(Cambrian); 4, porphyry; 5, recent placers. Black shading shows payable portions of basal Cambrian conglomerate.

Secondary Gold Deposits. The class of deposits coming
within this geological category are of exceptional importance in
the case of gold. In ancient times a very large proportion of the
gold supply of the world was obtained from alluvial and other
superficial deposits, and the great development of gold mining
in the middle of the last century took its rise in the discovery
of gold-bearing sands and gravels, the more recent and more
stable development of lode-mining occurring at a later date in
the newly-prospected regions. This same process has been constantly in operation up till the present time as more and more
remote auriferous regions are opened up, sometimes in the most
inhospitable parts of the globe, such as the Yukon, Klondike
and Alaska. The only other industry at all comparable in importance with alluvial gold-mining is alluvial tin-mining and
even this is far behind in value.

The class of occurrences generally lumped together as alluvial gold deposits include as a matter of fact a somewhat varying group from the geological standpoint. Some of them, perhaps the majority, are truly alluvial; nevertheless they also include a variety of superficial and residual deposits which are not primarily due to water action. It is unnecessary here to discuss at length the general principles underlying the formation of such deposits: a full account of the processes involved will be found in any text-book of physical geology. The preservation and accumulation of gold in quantity in such deposits are due primarily to two factors, its chemical stability and its density. From the very high specific gravity of gold an important consequence follows, namely, a tendency to work down to the bottom

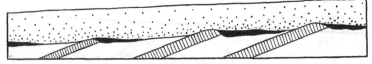

Fig. 77. Longitudinal section of a stream deposit, lying on denuded edges of hard rocks. The upper dotted part represents low-grade or barren overburden. The dark part is rich "pay-dirt" accumulating specially behind bars and riffles of hard rock.

of incoherent deposits of any kind which are kept more or less in movement by water action or any other extraneous disturbance: it is in fact a sort of natural panning effect. With regard to the great stability of gold under normal conditions, it is well known that it is not affected by any of the ordinary chemical agents: nevertheless it is clear that under certain circumstances gold can be dissolved and reprecipitated, because sometimes in gravels nuggets are found that are much heavier than could possibly have been transported by the streams that formed the gravel: in many other instances alluvial gold shows sharp crystal outlines with no evidence of rolling—a state of affairs quite inconsistent with its well-known softness. It must therefore be regarded as established that gold can somehow be dissolved and redeposited in natural waters. The chemistry of the process is not understood, but it is believed by many writers that compounds of manganese play some part.

It is obvious that secondary gold-bearing deposits may be

formed at any period of the earth's history, since at all times during the formation of the stratified rocks similar conditions have recurred again and again, in accordance with the principle of uniformity. For example it is now generally believed that the auriferous conglomerates of the Witwatersrand are really fossil alluvial deposits of pre-Cambrian age. Again in California and Australia rich gold-bearing gravels are known among the Tertiary deposits laid down by river-systems that have long passed away, some of them having been deeply buried under basalt flows, and now again brought to light by modern denudation, with an entirely different system of hills and valleys. Some examples of these so-called "deep leads" are described in a later section.

Owing to the great diversity of conditions that may prevail during the formation of such deposits, of whatever age, it is manifestly impossible to generalize profitably over their characters, form and extent; every individual case must be treated on its own merits, but one or two may be selected for somewhat detailed description, as examples of the rest. Besides the superficial deposits of the land, important quantities of gold are found in some localities in the beach gravels of the sea-shore, as for example in Alaska, and at many other points around the Pacific. All such superficial deposits of whatever origin are conveniently included in the general designation of *placers*.

Thus we may classify placers under the following heads:

(1) Residual placers (sometimes called eluvial placers), resulting from weathering of rock *in situ*, with removal of soluble material, but little or no mechanical transport.

(2) Dry placers: residual deposits formed by weathering, in arid regions, the lighter material having been removed by wind.

(3) Stream placers: the normal type of alluvial deposits in the narrow sense of the term; this is by far the commonest and most important type.

(4) Marine placers: gold-bearing sands and gravels laid down on the sea-shore by waves and currents.

Since the concentration of the gold in such deposits is largely due to its stability and density, it is generally accompanied by other stable minerals of high specific gravity, such as magnetite, ilmenite, garnet, zircon, cassiterite, monazite and sometimes platinum. The first-named minerals are dark in colour and when present in quantity they give rise to "black sands" which are highly characteristic of auriferous deposits and sometimes form a useful indicator for their probable presence.

There can be but few parts of the world where gold has not been found to a greater or less extent in superficial deposits; even in such a notably non-auriferous area as the British Isles gold has been found in small quantities in sands and gravels at different times. Such finds on a small scale are recorded, for example, in the Leadhills district in southern Scotland, in Caithness and Sutherland, and in Ireland (Wicklow), and small amounts have been found in Cornwall. Large quantities of gold have been obtained from washings in various parts of Europe

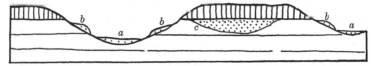

Fig. 78. Ideal section combining different types of auriferous gravel deposits: *a*, modern placers; *b*, high-level terraces; and *c*, a deep lead, the last filling an old valley buried under basalt.

and Asia, as well as from different districts of Africa, but the most famous alluvial goldfields of modern times are those of western North America and Australia.

Placer Gold in California. The great modern development of gold-mining may be said to have begun with the discovery of the rich placer deposits of California in 1849, when a great rush took place to that territory, then only recently acquired by the United States. As is only natural the secondary gold deposits are for the most part found in the same general regions as the primary ores, and to a great extent at lower levels. Many of them are, however, in reality products of denudation and deposition in Tertiary times and therefore they may lie far above present stream levels, while some are buried beneath Tertiary lavas, which may be as much as 200 feet thick, as in Table Mountain, Tuolumne County. Another important consideration in respect of the Californian placers is the very frequent occurrence of gold in raised gravel terraces, of essentially the same nature as the terraces of the Thames, Cam, Ouse and other English rivers. These are often found as isolated patches of gravel of characteristic form lying on the sides of valleys, in some cases far above the existing stream. Similar high-level terraces may have been formed also by streams that have long since vanished, or been diverted into other channels; in fact the possible diversities are endless: in a region of this kind any gravel, lying in almost any position, may carry gold.

In the early days Californian mining, or gold washing as it may well be termed, was very simple and primitive, all the work being done by hand in pans or rockers, or with some elementary form of sluice. Recently, however, various kinds of dredging and hydraulicking have been introduced on a large scale with good results.

Gold Placers in Australia. In the early days of gold-digging in Australia, which began in the early fifties of the last century, most of the output came from different kinds of secondary deposits, ranging in age from Cretaceous to Recent. Pittman has divided the secondary gold-deposits of New South Wales as follows, and those of Victoria are very similar:

(a) Recent and Pleistocene alluvials.
(b) Beach sands.
(c) Tertiary alluvial leads.
(d) Cretaceous alluvial leads.
(e) Permo-Carboniferous conglomerates.

The latest of these, of Pleistocene and Recent age, form deposits of gravel, sand, clay and loam along the courses of existing rivers. Parts of them were extraordinarily rich and it was no uncommon thing to obtain many ounces of gold from a single bucket of gravel. Some very large nuggets were also found, up to 2000 oz. in weight. Most of these rich deposits have been long since worked out and no good purpose would be served by a detailed account of them; they show endless variety of topographical form and composition according to the conditions of formation and the source of the material. In nearly all cases the distribution of the gold could be definitely traced to primary veins, which mainly lie in clearly marked belts, usually running north and south in Victoria.

As in California many of the older alluvial deposits, chiefly of Tertiary age, are now buried under thick flows of basalt, and their disposition is often entirely unrelated to the present topography of the country. These are called "deep leads" and many of them were also very rich, but difficult to work in many cases on account of the great inflow of water. The payable portions are usually confined to the lowest layers of the gravel and much of the gold actually rests on the bed-rock, or on impervious floors at higher levels.

The Yukon and Klondike Gold Fields. The most important alluvial goldfields actively developed within recent times are those of the Yukon River and its tributary the Klondike, on both sides of the Canadian-Alaskan international boundary. The existence of gold in this district has been known since 1869, along the Yukon, but the great rush began with the discovery of the Klondike in 1897. This field lies entirely in Canadian territory, about 50 miles east of the Alaskan boundary, in a region of the most inhospitable climate, and mining there has been attended by terrible hardships. The gravels are of several different ages, and have been classified as High Level Gravels, from 150 to 300 feet above the present streams; Terrace Gravels of irregular extent and at different heights, marking interruptions in the erosion of the present valleys; Low Level Gravels occupying the bottoms of existing valleys and covered by a layer of usually frozen peaty silt up to 25 feet thick. The gold in all cases seems to have been derived from quartz veins traversing the schistose rocks of the regions, which are apparently of pegmatitic origin, since they often contain felspar; some have been followed laterally into normal felspathic pegmatites. The total production of gold from the Yukon province to date has been about 7,000,000 oz. One of the greatest difficulties to be met in the placer mining of the Yukon-Klondike region in general is the permanently frozen condition of the ground. Hitherto much of the gravel has been thawed by hot water or steam, a costly process, but lately it has been found that a rapid circulation of cold water (about 52° F.) is cheaper and even more effective. Most of the richer deposits are now apparently exhausted and production has fallen off to about 70,000 oz. in 1920, from a maximum of over a million ounces in 1900.

The Beach Placers of Nome, Alaska. As an example of auriferous marine gravels we may take the famous beach placers of the Seward Peninsula, Alaska, discovered in 1899. For a distance of 2 or 3 miles east and west of Nome the beaches have been worked over by hand two and even three times and afterwards by dredging. These gravels include two raised beaches, at 79 and 37 feet above sea-level, the highest one being some 5 miles inland. The raised beaches are overlain by frozen tundra (muck) and the gravels and sands themselves are generally frozen. The present beach is made up of sand, fine and coarse shingle, angular and subangular gravel with a few large boulders, the last probably dropped by floating ice. In 1920 seventeen dredges were at work around the Seward peninsula, but this kind of mining has been much hindered by increasing working costs and the output of gold from Alaska has fallen off greatly, having amounted in 1920 to only 380,000 fine oz., of which about one half probably came from placers.

Silver. From very ancient times mining for silver has been an important industry and much romance attaches to the stories of the search for this metal in Europe and still more to the fabulous wealth brought home by galleons across the Spanish Main. But this glamour has now departed and we are faced with the prosaic fact that the greater part of the world's production of silver is obtained as a by-product in the exploitation of other metals. In a good many instances it is true that it is the presence of the additional values in silver that makes the working of the other metals profitable, nevertheless the silver is essentially of secondary importance. From this it follows that it is unnecessary to describe in detail examples of all the different types of silver deposit: many of the important sources are fully dealt with in other sections. In this chapter attention will be confined to certain interesting cases where silver is really the principal or the only product. These, however, do not bulk very largely in the total output. About half the world's production is now obtained as a by-product from lead and copper ores: both of these types are described fully elsewhere. Very many gold ores also yield silver, sometimes one metal, sometimes the other being in excess.

At the present time about 75 per cent. of the world's silver comes from North America, the United States and Mexico each yielding about 30 per cent. and Canada 15 per cent. Other important producers are South America (Peru, Bolivia and Chile) and Australia. Silver is also mined to some extent in Europe, in Great Britain, Spain and Germany, and in some of the component States of the former Austro-Hungarian Empire. Of these sources the Bolivian deposits are partly dealt with under tin, the Australian (Broken Hill) under lead-zinc ores, and the British occurrences in the same chapter. We shall deal here mainly with a few typical examples chosen from America, and one or two European occurrences that are now mainly of historic interest.

Silver Minerals. Minerals containing more or less silver are very numerous and often of complex composition, and as before stated a great deal of silver is often present in ores of gold, copper and lead, in various forms of combination. With gold it often occurs as a natural alloy (electrum) and is also present in tellurides. In the copper minerals silver replaces part of the copper in isomorphous mixture, as for example in tetrahedrite. In galena it may occur apparently either as a mechanical admixture of silver sulphide, or in some kind of solid solution.

Among the true silver minerals in the restricted sense of the term the following are the most important.

	per cent. of metal
argentite, Ag_2S	87·1
stephanite, $5Ag_2S.Sb_2S_3$ ...	68·5
tetrahedrite, $4(Cu_2,Ag_2)S.Sb_2S_3$	variable
pyrargyrite, $3Ag_2S.Sb_2S_3$ ...	59·9
proustite, $3Ag_2S.As_2S_3$ ...	65·4
cerargyrite, $AgCl$	75·3
native silver	—

Besides the above there are other complex sulphantimonides and sulpharsenides, as well as bromine and iodine compounds corresponding to cerargyrite. Argentite is certainly in many cases a primary ore and possibly some of the antimony and arsenic compounds are also primary. Among silver deposits the prevalence of zones of secondary enrichment is noteworthy, and in many lodes native silver is extraordinarily abundant in the upper portions, being formed by chemical processes acting on sulphides. Unlike the copper lodes, however, carbonates and sulphates do not occur in the oxidation zone.

Silver–Lead Ores. The extraction of silver from lead is a very old industry: it was certainly practised by the Romans, for example in Britain and at Lavrion (Ergasteria) in Greece, and has always been a very important source of the metal. The proportion of silver found in galena is curiously variable, even under what appear to be very similar conditions. Thus, for example, there is very little silver in the lead ores of Yorkshire and Derbyshire, whereas it is quite abundant in North Wales, Cumberland, the Isle of Man and Scotland. On the whole silver appears to occur in larger proportion in the higher parts of the lodes, though there are exceptions to this rule, and many miners believe that finely granular galena is more argentiferous than large well-developed crystals. The rich ore of Lavrion, which runs to 60 oz. silver per ton of lead, belongs to the granular type. The manner of occurrence of silver in galena has been investigated microscopically by several observers[1]: Nissen and Hoyt have found experimentally that the limit of solid solution of Ag_2S in PbS is certainly below 0·60 per cent. and probably below 0·20 per cent., the former figure being equal to about 175 oz. of silver per ton of lead. Any amount of silver beyond this exists as a mechanical and visible admixture of argentite in the galena. In some secondarily enriched ores native silver has been observed in cleavage cracks in galena[2]. We may conclude then that in low grade primary ores with only a few ounces of silver to the ton of lead the silver exists in solid solution. In the argentiferous lead-zinc deposits it is clear that the silver is more closely associated with the galena than with the blende, even when galena and blende are actually in contact. Several typical occurrences of argentiferous lead-zinc ores are described in the chapter on those metals and the subject need not be further pursued here. Specially interesting examples are afforded by Broken Hill, N.S.W., and Bawdwin, Burma (pp. 307–312).

[1] Nissen and Hoyt, *Econ. Geol.* vol. x, 1915, pp. 172–9. Guild, *ibid.* vol. xii, 1917.

[2] Finlayson, *Econ. Geol.* vol. v, 1910, p. 727.

The Kongsberg Silver Deposits. One of the most famous silver-mining districts in Europe is Kongsberg in Norway, some 50 miles west of Kristiania. Here the geological and mineralogical relations are altogether exceptional and peculiar. The deposits are found over an area about 20 miles long and from 3 to 6

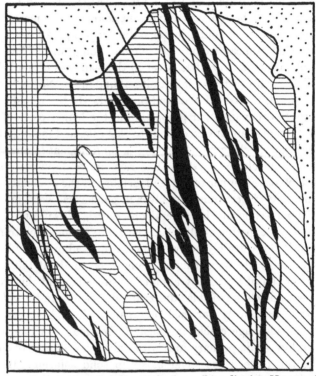

Fig. 79. Geological map of the Kongsberg silver district, Norway (after Krusch, *Zeits. für prakt. Geol.* 1896, fig. 29, p. 95). The district consists of granite, gneiss and schists, with fahlbands (black). The veins are nearly at right angles to the fahlbands, but are too narrow to be shown (scale approximately ¾ inch to 1 mile).

miles wide. The silver was discovered in 1622, and a large number of mines have been worked at different times, certainly over a hundred: the deepest shaft, the Kongens mine, had in 1911 reached a depth of 3000 feet and several others are over 2000 feet. The total production has been about 1000 tons of metallic silver.

Stated in the most general terms the area is composed of crystalline schists invaded by a very great bulk of more or less foliated granites and basic intrusions so that the schists are subordinate in amount to the intrusives. But the ores are largely, though not exclusively, associated with the schists. These include several varieties characterized by hornblende, mica or chlorite. The special characteristic of this region is the occurrence, mainly in the schists, but also to some extent in the foliated granites, of bands parallel to the foliation containing a considerable proportion of sulphides, including pyrite, pyrrhotite and chalcopyrite: these are called "fahlbands." Crossing these fahlbands, approximately at right angles, are veins with a calcite gangue, always very narrow, not often exceeding 4 inches in width, very rarely as much as a foot, but remarkably rich in native silver, with some argentite. The silver is concentrated mainly at the intersections of the veins with the fahlbands, where great slabs of solid metal are sometimes found: most of it however takes a peculiar and characteristic form, generally described as "wire silver." These masses of silver are sometimes irregular fibrous bunches and networks, sometimes mossy or dendritic, and often slightly curved horn-like structures of varying size. The latter were certainly formed from argentite, since a little knob of this mineral is now and then found attached to the end of the horn. Argentite also occurs as a primary mineral, occasionally in very large masses. Besides calcite, a little quartz and fluorite are found in the gangue, as well as silicates, including zeolites, and as a curious subordinate mineral there is some anthracite, which may have played a part in reducing the silver sulphide to metal. At present these lodes are mainly of academic interest, as the production is very small.

The Silver Veins of the Cobalt District, Ontario. Although a recent discovery this area is now the largest silver producer in Canada, and an important factor in the world's silver market. The Cobalt district is near the eastern boundary of northern Ontario, a few miles west of Lake Timiskaming, and about 300 miles due north of Toronto. The silver veins, which were discovered during the making of the Timiskaming and Northern Ontario railway, lie in the ancient rocks of the Canadian shield. The formations here represented are the Keewatin below, overlain with a great unconformity by the Cobalt series of Middle Huronian age: both series are invaded by great sills of "diabase," one of which has been the principal agent in the formation of the rich silver-cobalt deposits. These consist of great numbers of narrow veins, rarely exceeding 8 inches in width and usually 4 inches or less. They occur in all three formations, but seem to congregate especially about the footwall of the diabase. The gangue mineral is calcite, with a little quartz, but the richer parts of the veins are nearly solid masses of ore-minerals. Of these the most abundant are arsenides of cobalt and nickel, including smaltite, chloanthite, cobaltite and nicco-lite, with much native silver. In some places native bismuth is

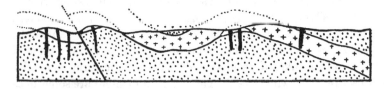

Fig. 80. Ideal geological section of the Cobalt district, Ontario (after Whitehead, *Econ. Geol.* vol. xv, 1920, p. 107, fig. 15). Dotted, Kee-watin; white, Cobalt series; crosses, Nipissing diabase; black, veins.

found, as well as complex silver-bearing sulphides. Near the surface the cobalt and nickel arsenides are oxidized to erythrite and annabergite, the pink colour of the former being very characteristic and a good prospecting indicator. Some rich veins are near the surface mainly an intergrowth of native silver and pink erythrite, while others are mainly calcite and silver.

The distribution of the veins is closely related to the diabase sill. They are found in all the rock-types, but four-fifths of them lie in the Cobalt series, either above or below the sill. Some good veins are found in the hanging-wall of the sill, but un-fortunately most of this part has been denuded away. The best type now lie in the footwall of the sill when this is composed of Cobalt rocks, while some rich veins have been found where the footwall is of Keewatin rocks. Up to 1911 about 112 veins had been active producers.

The origin of these veins is obscure, but the geological relations certainly suggest that they are differentiation products of the "diabase" magma, constituting a special type of mineralization connected with basic intrusions, analogous to the nickel-pyrrhotite type, but specially characterized by silver, cobalt and arsenic. However, on the magmatic hypothesis the calcite is difficult to account for, although this difficulty is lessened by some recent observations on calcite and dolomite differentiated from syenites in Norway[1]. In the latter instance the intrusive calcite is due to absorption of limestone by the magma, with subsequent segregation, and something of the same sort may have happened here, the absorbed limestone acting as a flux for the ores, and helping to differentiate them from the silicate melt of the basic magma.

This region has been extraordinarily productive: the high-water mark was reached in 1911, when the output was about 31,500,000 oz. of silver, as well as cobalt, nickel and arsenic. Since then there has been a steady falling off, and in 1920 the output of the whole province of Ontario was only 9,500,000 oz. from all sources. Some of the largest mines of the Cobalt district are the Nipissing, Coniagas, Kerr Lake, O'Brien and Buffalo. The second of these names is of some interest, being made up of the chemical symbols of its four products, cobalt, nickel, silver and arsenic, Co, Ni, Ag, As.

The Gowganda mining district lies about 50 miles north-west of the town of Cobalt. The geological relations are very similar: the veins, which mainly lie in sills of quartz-diabase, are wider than at Cobalt, but less rich. The mineral content is much the same. The chief producer is the Miller Lake-O'Brien property, which has some high-grade ore.

Tonopah, Nevada. This is one of the most important areas of precious metal mining in the United States. From 1910 to 1915 it produced annually over 10,000,000 oz. of fine silver and over 2,000,000 dollars worth of gold, the total value of the output up to 1916 having been about 85,000,000 dollars.

---

[1] Brögger, "Die Eruptivgesteine des Kristianiagebietes, IV: Das Fengebiet in Telemarken," *Vid. Selsk. Skrifter*, No. 9, 1920. See summary by C. E. Tilley, in *Geol. Mag.* vol. LVIII, 1921, p. 549.

The rocks of the area are entirely Tertiary volcanics, the most important members being a rhyolite and two flows of intermediate composition, the earlier now called the Mizpah trachyte and the later the Midway andesite: there are also some still later rhyolites and basalts. All these rocks are flat-lying, but much faulted. Some authors believe that some of them are intrusive. Most of the ore lies in the Mizpah trachyte, though some veins go down to the rhyolite below. The ore-bodies are for the most part quite typical veins consisting of replacement bodies along fault fractures. The number of silver-bearing and other minerals found here is very large. Bastin and Laney[1] have

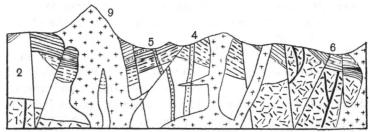

Fig. 81. Generalized section of the mining district of Tonopah, Nevada. 1, Mizpah trachyte; 2, Midway andesite; 4, Dacite breccia; 5, Tonopah rhyolite-dacite; 6, lake beds; 9, rhyolite with veins. (After Spurr, Prof. Paper 42, U.S. Geol. Survey, fig. 10, p. 71, simplified and somewhat modified.)

recognized here two groups of primary ore-minerals: the first or α group consists of minerals still in precisely the same condition as when first formed. These include galena, blende, wolfram, chalcopyrite, arsenopyrite, pyrargyrite, polybasite, argentite, electrum, and probably stephanite and argyrodite. The second or β group includes minerals that have undergone changes during the actual process of primary mineralization; among these are argentite, polybasite, chalcopyrite and electrum. The chief minerals of an undoubtedly secondary nature are gold, silver, argentite, polybasite, pyrargyrite, silver-haloid compounds, malachite and zinc silicate. The gangue minerals are carbonates, quartz, chalcedony, barytes and gypsum. The secondary minerals show by their distribution a very interesting feature,

[1] Bastin and Laney, Prof. Paper 104, U.S. Geol. Survey, 1918. Spurr, Prof. Paper 42, ibid. 1905. Econ. Geol. vol. x, 1915, pp. 713–69.

namely the former occurrence of periods of denudation and oxidation related to old land-surfaces now buried deeply beneath later volcanic rocks.

It is to be noted that the number of minerals here regarded as truly primary is unusually large and includes some species, such as pyrargyrite, polybasite and stephanite of complex composition and commonly considered to be characteristic of the zone of secondary deposition.

The Comstock Lode. One of the most remarkable ore-deposits in the world and one of the richest is the Comstock Lode in Nevada, which has yielded about 380,000,000 dollars worth of precious metals, about three-fifths of this sum representing silver values and two-fifths gold. This great lode is essentially a fault-plane, evidently descending to great depths, since in some parts there are springs of salt water that issue at about 170° F. In such places miners work in fifteen minute shifts only, in a blast of cooled air with a spray of cold water. Here the connection of mineralization and vulcanicity is obvious and ore-deposit is perhaps even now in progress. The fault-plane, which can be followed for a distance of 4 miles or so, dips at about 40° and runs approximately along the intrusive junction of a diorite with a mass of andesite and diabase. The throw of the fault in the middle is about 3000 feet. The country rock is intensely altered, resulting in the formation of sericitic mica, chlorite and pyrites. The ore-body is in the main a mass of shattered rock and quartz several hundred feet wide with the ore in rich shoots and patches. Smaller veins extend upwards into the hanging-wall along secondary fault-cracks. The ore minerals are gold, native silver, argentite, stephanite and galena, with a subordinate amount of secondary minerals, such as pyrargyrite and horn silver. The silver ores are made difficult to smelt by the presence of much manganese oxide, of rather uncertain origin. There are also a little copper, zinc and iron, so that the ores altogether are exceedingly complex. Much gold and silver have been obtained from placers in the streams draining the lode area.

This is an excellent example of extensive mineralization at a shallow depth by solutions of obvious volcanic origin, and may be taken as a working model of this type of ore-formation. In all probability there is a mass of highly-heated igneous rock, perhaps a recent intrusion, at a comparatively small distance from the surface.

Silver in Bolivia and Peru. On grounds of historic interest it is impossible to omit a reference to the once world-famous silver deposits of these countries, from which so much of the fabulous wealth of the Spanish conquerors was derived. However, at the present time the chief metals yielded by the old mining districts are not silver and gold, but tin, tungsten and bismuth, metals mainly disregarded by the earlier miners. From 1553 to the present time Bolivia is estimated to have yielded nearly 50,000 tons of silver, of which 30,000 tons came from Potosí. Other well-known silver-mining areas are Oruro, Colquechaca, Colquiri and Pulacayo, with a host of smaller ones.

The principal mineralization of Bolivia lies in the eastern Cordillera and in isolated hills scattered over the elevated plateau of the Altaplanicie, lying between the two chief ranges. In this area the metalliferous character of the deposits undergoes on the whole, with certain local exceptions, a gradual change from north to south, in that tin is more abundant in the north and silver in the south. As pointed out in the chapter on tin this change is connected with the character of the prevailing igneous rocks. In the most general terms it may be said that silver is the prevailing metal south of Araca, and tin to the north of that latitude. South of latitude 21° S. there seems to be no silver at all, while over the Argentine border the age and character of the mineralization are entirely different.

The ore-deposits of Bolivia are clearly connected with the igneous activity that accompanied the great orogenic disturbances of late Cretaceous and Tertiary times. The richer silver deposits belong to a somewhat late stage of this sequence and are connected mainly with hypabyssal intrusions or even with volcanic rocks. Many of them were probably derived from intrusions of magma itself essentially similar to the great superficial flows, but unable to reach the surface owing to the great thickness of material already ejected. Hence it is unimportant whether the actual rock by which the deposits are formed is called a quartz-diorite, a quartz-dolerite or a dacite; genetically they are all the same. Many of the veins actually lie in sediments which are for the most part of Palaeozoic age, especially Silurian and Devonian, as shown by the fossils, but in the same area rocks of almost the same appearance have been found to contain Tertiary fossils, so the rock-sequence must be of very varying age. As is usual the veins are generally better developed and richer in sandstones and quartzites than in soft shales. The most common width for important veins is about 3 or 4 feet in favourable country rock. Most of the important mines lie at very high elevations, usually over 13,000 feet, and some as much as 17,000 feet.

A very noticeable feature of all these deposits is the great depth of the oxidation zone: indeed in most of the Andean mining regions it is very doubtful whether a true primary zone has ever been reached. With this extensive oxidation is coupled a very high degree of secondary enrichment, with many characteristic minerals. Among these perhaps the most significant is cerargyrite or horn silver, AgCl. The iodine and bromine compounds also occur in some quantity. The reason for the presence of these minerals in such amount is to be sought in the climate and the resulting character of the ground-water. As is well known the west coast of South America is a region of very scanty rainfall and in the past this condition may have been even more accentuated, when the great deposits of guano and nitrate of soda were formed. In arid regions the ground-water, what there is of it, is always rich in salts, especially chlorides, and the silver salts formed by oxidation would be at once precipitated as the insoluble chloride, cerargyrite. A similar phenomenon is seen in the case of copper, forming the insoluble basic chloride atacamite, instead of the ordinary basic carbonates, malachite and azurite, usual in more normal regions. Among other secondary ores ruby silver also occurs in very large quantities, as well as secondary argentite. Much silver also exists in isomorphous mixture or in solid solution in tetrahedrite and jamesonite and other mixed sulphides, sulphantimonides and sulpharsenides, as well as great masses of native silver. The accompanying tin, tungsten and bismuth minerals are described elsewhere, as well as the rare germanium minerals. There are also considerable quantities of lead and zinc minerals, not usually saved owing to transport difficulties, the cost of carriage being too high for such low-priced products. The commonest gangue mineral is quartz, but calcite, barytes, and siderite are also known. Some of the veins rich in tin carry fluor-spar and tourmaline.

As a typical example of these deposits we take the famous Potosí district, from which the Spaniards obtained so much silver in the early days, and which has been worked continuously since 1544. The present output, however, is chiefly tin. The Cerro Rico de Potosí, as it is called, is a peak about 16,000 feet high, composed of sedimentary rocks and volcanic ashes surrounding what appears to be a neck or plug of acid lava, which may be called either rhyolite or quartz-porphyry, and in many places is much silicified. The veins lie chiefly in the quartz-porphyry, but some are in the sediments and tuffs. The total number of veins is very great, with many fine stringers and disseminated patches of ore, so that most of the mountain is mineralized. The veins in depth consist mainly of tetrahedrite and cassiterite with a little quartz and barytes. The oxidation zone is very deep and yields most of the silver; at the bottom of it the chief ore is ruby silver; in the upper levels cerargyrite and native silver. There is also a little copper and still less gold. The ores are very patchy, rich bonanzas alternating with unpayable stretches.

The silver deposits of Peru belong to a different type, being mostly in the category of argentiferous copper, or argentiferous lead. Some few yield mainly silver and the majority of them are worked primarily for the silver that they contain, the lead and copper being of subordinate importance. They are chiefly found in the departments of Cajamarca, Libertad, Ancachs, Junín and Huancavelica. The veins lie in Mesozoic sediments and Tertiary intrusives and usually have a quartz gangue, with sometimes calcite, barytes or chalybite. A very large part of the silver is obtained from argentiferous tetrahedrite, or other complex copper-silver sulphides, pure argentite being less common. In the veins rich in lead and zinc the silver seems to occur in the usual way, in the galena. The oxidation zones do not generally present any very special features, horn silver and native silver being common, as usual in the Andes. Copper may also occur as chalcopyrite and sometimes as enargite[1].

Silver Mines in Germany. In the Middle Ages and until quite recent times an important quantity of silver ore was worked in various parts of Germany; perhaps the most interesting of these areas, especially from the historical point of view, are the Freiberg and Clausthal districts. These were really the first mineralized areas to be studied scientifically. The Freiberg Mining School has always been deservedly famous from the time of Werner and much good scientific work has also been done at Clausthal.

In the neighbourhood of Freiberg in the Saxon Erzgebirge, there are said to be no less than 1100 lodes which have yielded silver in appreciable quantities. The chief rocks of the neighbourhood are an older grey biotite-granite-gneiss, which has been invaded by laccoliths and other masses of a red muscovite-granite. Both of these appear to have been intrusive into an older series of sediments, now in large part metamorphosed to schists, but also including quartzite, greywacke, limestone and conglomerates. These are probably Lower Palaeozoic. The lodes have been divided by German geologists into two series, one supposedly younger than the other, but it is doubtful whether this classification possesses much real significance. The lodes vary much in character and contents; those containing most of the silver belong to three types:

(a) Quartz-silver lodes, with argentite, ruby silver, native silver and mispickel in a gangue of quartz and chalcedony.

(b) Silver-lead lodes, with argentiferous galena, blende, ruby silver, native silver, fahlerz and argentite, in a gangue of quartz, dolomite and rhodochrosite.

(c) Barytes-lead-silver lodes, with argentiferous galena, blende, pyrite and some copper minerals in a gangue of barytes, quartz, chalcedony and fluor-spar. Some of these contain a good deal of cobalt.

---

[1] *Note.* Nearly all the information in this section is taken from Miller and Singewald, *Mineral Deposits of South America*, New York, 1919, the latest and most reliable general account of its subject. Here full details will be found of the most important occurrences.

Each of these types occurs most commonly in one particular district, while the whole mineralized region covers a large area and adjoins the tin-field of the Upper Erzgebirge.

In parts of the area last mentioned, in the tin district of Annaberg, Schneeberg, Johanngeorgenstadt and Joachimsthal, there are many lodes carrying silver and cobalt. These have been worked near Annaberg since 1492, and gave rise to a considerable industry. The Schneeberg lodes were discovered some 20 years earlier and are still of importance, though mainly for cobalt and bismuth rather than silver. A short account of the minerals of this district has already been given in the chapter on metallogenesis (see p. 170).

The Harz Mountains form part of the core of the ancient Hercynian or Variscan fold-range, contemporaneous or nearly so with the Armorican folding of southern England and western France. They consist of Devonian and Carboniferous strata, invaded by granites, the latter forming for example the well-known Brocken. A great number of lodes, mainly striking W.N.W.–E.S.E., are scattered over the plateau of the Upper Harz around Clausthal and south of Goslar. Some of these lodes are traceable for 5 or 6 miles and the larger ones may be as much as 100 feet wide. The most important ore is argentiferous galena, with blende and a little copper, mostly in a gangue of quartz, sometimes calcite, with barytes only in the south-western area. The lodes of the St Andreasberg district, only 10 miles away, are quite different. This is a region of overthrust Devonian rocks in the form of a wedge in which are situated numerous veins of calcite, with occasional quartz, carrying galena, blende and many silver minerals, such as native silver, argentite, pyrargyrite and dyscrasite, with some nickel, cobalt and arsenic. A peculiarity of these lodes is the presence of numerous zeolites[1].

The Platinum Group. It is unnecessary to occupy space in describing the properties and technical uses of platinum and the allied metals. They are all characterized by very high density and remarkable stability, and resistance to chemical agents. The physical properties of platinum also render it almost indispensable for a vast number of electrical and other uses. It is

---

[1] Beyschlag, Vogt and Krusch, trans. Truscott, vol. II, pp. 683–92.

far from being abundant, and consequently commands a very high price. Since the greater part of the world's supply comes from Russia, the platinum industry and market are in a very disorganized state, and much research has been devoted to the discovery of substitutes for various purposes, with, in many cases, a good deal of success.

Platinum nearly always occurs as the native metal, alloyed with other members of the same group, palladium, iridium, osmium, rhodium and ruthenium, most of which also possess technical applications. Some iron is always present in the natural alloy. Only one natural compound of platinum is known, the di-arsenide, sperrylite, $PtAs_2$. The occurrence of this mineral will be described presently.

Platinum is known to occur in minute quantities in a good many widely-separated localities, but only three occurrences are of any real importance, namely, those of the Ural Mountains, Colombia and Sudbury, Ontario.

Up to the year 1914 Russia produced about 93 per cent. of the world's supply and Colombia about 5 per cent. In these two regions practically the whole output was from alluvial deposits, although the mining of platiniferous rock in the Urals was just beginning before the industrial collapse of Russia. The exploitation of platinum deposits in Russia and Colombia is therefore essentially similar to alluvial gold or tin working.

The Platinum Deposits of the Urals. The crystalline schists of the central chain of the Urals are invaded by masses of igneous rocks, including ultrabasic types, which are the source of the platinum. The dunites form oval masses with their longer axes parallel to the main lines of folding of the chain. Each mass of dunite is surrounded by a ring of pyroxenite and then a ring of gabbro, giving evidence of magmatic differentiation. In the dunites platinum, nearly always closely associated with chromite, occurs as grains and masses of early crystallization, while in the pyroxenites, when present, it was the last mineral to form[1]. It is beyond doubt that the platinum metals are here a primary constituent of an igneous magma characterized

[1] Duparc and Tikonowitch, *Le Platine et les Gîtes platinifères de l'Oural et du Monde*, Geneva, 1920.

specially by low silica and high magnesia: such rocks are dunites when fresh, but are very commonly altered to serpentine.

However, it is the alluvial deposits formed by denudation of these rocks in the Urals that are the immediate objects of attention, since it is only when naturally concentrated in gravels and sands that the platinum is found in payable quantity. The two principal producing regions are on the river Iss, near Goroblagodat, and the Nijni-Tagilsk district. Most of the platiniferous deposits are river-gravels of normal type, laid down by existing rivers, although, according to Duparc, some are related to older systems of drainage. The platinum mostly occurs as very small flattened grains, large nuggets being very rare. The biggest on record, from the Nijni-Tagilsk district, weighed 25 lb. The average yield was formerly said to be about half an ounce to the cubic yard of gravel, but the deposits recently worked are less rich than this. As usual the best pay gravel is found at the bottom and it is often necessary to remove many feet of unpayable overburden. Gold is also obtained along with the platinum. Large-scale working by dredges was introduced about 20 years ago, replacing very primitive methods, but the frozen condition of the ground for a considerable part of the year is a great obstacle to profitable working. Most of the modern machinery now appears to be in a state of decay, and it is believed that the workable alluvial deposits are approaching exhaustion.

Platinum in Colombia. The presence of platinum in the gold placers of the department of Chocó in Colombia has been known for about 150 years: in 1914 the production was about 17,500 oz., and is likely to be of much importance in the future. The chief deposits are those of the Rio San Juan and its tributary the Rio Condoto, and of the head waters of the Rio Atrato. The sands are described as brown in colour, carrying besides platinum and gold, also chromite, magnetite and ilmenite. Platinum is said to have been found *in situ* in gabbro and other more basic rocks, but little is really known of its origin[1]. It is only recently that modern methods of·exploitation have been introduced

[1] Miller and Singewald, *The Mineral Deposits of S. America*, New York, 1919, pp. 383–4.

into the country: large-scale dredging operations are now in progress.

Platinum at Sudbury, Ontario. The mineral sperrylite, the only known native compound of platinum, was discovered in the tailings from a mill at Sudbury used for dressing the copper-nickel ores of that region. The geology of the Sudbury nickel deposits is described elsewhere in detail, and it suffices here to say that sperrylite is widely disseminated in the sulphide ores, usually in association with the chalcopyrite rather than with the nickel minerals: there is also a good deal of palladium here, which appears to go rather with the nickel. The precious metals, which include also some gold, are extracted during the refining of the nickel matte and several thousand ounces per annum of platinum and palladium are recovered, especially by the Mond Nickel Co. This will probably prove to be an important source of supply in the future. Platinum has also been observed to occur in a similar nickel-bearing igneous rock at Insizwa, in Griqualand East (see p. 372).

Small quantities of platinum are also obtained from alluvial deposits in British Columbia, Australia, Borneo, Rhodesia and the United States, but the total is very small. There is always a chance of finding platinum in payable quantities in alluvial deposits in any district where peridotites and serpentines are undergoing denudation: traces have even been found in river-gravels in the Lizard district, Cornwall[1].

[1] McPherson, *Geol. Mag.* vol. LVIII, 1921, pp. 512–14.

GENERAL INDEX

absorption, 26
acid steel, 318, 342
alloys, 25, 173, 216, 224, 227, 260, 441, 455, 495
alluvial deposits, 48, 71, 85, 105, 261, 276, 278, 280, 415, 419, 438, 441, 459, 475, 495
aluminium (general), 446
alumino-silicates, 6
alunite, 142
amalgam, 212, 403, 455
antimony (general), 403
apatite dykes, 46
aplite, 34, 37, 178, 241, 366
aqueous deposits, 109
aridity, 65, 121, 233, 254, 327, 418, 431, 491
Armorican folding, 168, 301, 304, 494
arsenic (general), 407
arsenical nickel ores, 375
arsenides, 154, 407, 486
artesian water, 61, 306
asbolane, 392
assimilation, 26
aureole of metamorphism, 54
auriferous conglomerate, 91, 455, 470

baddeleyite, 440
banded ironstone, 329
banket, 471
barilla, 274
basic steel, 318, 342
bathylith, 30, 128, 176, 263, 280, 303, 459
bauxite, 361, 447
beach gravels, 437, 474, 481
bearing metals, 403
bedded ironstone, 86, 136, 341
— ores, 111
Bessemer ores, 331, 340
bismuth (general), 409
bitumen, 397, 399
black-band ironstone, 342
— sand, 435, 437, 473, 478
Blue Billy, 249
bog iron ore, 360
bonanza, 88, 492
borings, 472
boron springs, 58
boulder clay, 276, 308
breccia, 298, 368
brecciated lodes, 55, 74, 100, 270
Britannia metal, 403
brockram, 333

bronze, 260
brown haematite, 339
bunch, 88

calcination, 343
Caledonian earth movements, 189, 301, 304
calico rock, 329
campanil, 340
canga, 329
capel, 98, 266
carbona, 74, 102
carbonatization, 142, 462
carbonato, 340
carnotite, 433
cavern, 297
cementation, 50, 111
champion lodes, 184
chert, 52, 326
chilled edges, 32
china clay, 40
chirta, 340
chrome steel, 391
chromium (general), 390
chute, 88
circulation of water, 63
classification of ore deposits, 106
clay ironstone, 342, 350
cleavage, 56
climate, 65, 120, 133, 233, 254, 327, 418, 431, 491
Coal-measure ironstones, 343
cobalt (general), 376
— steel, 376
colloids, 220, 360, 448
columbite, 439, 453
complex ores, 225
concentration, 85
conglomerate, 239, 470
conjugate solutions, 24
constituents of the earth's crust, 4
contact deposits, 12, 33, 71, 84, 103, 147, 231, 239, 245, 315, 418, 422
contact pyroxenite, 131, 422
convection currents, 27, 319
copper (general), 227
— matte, 25
— shale, 258
critical temperature, 59, 202, 222
crofesima, 7
cross course, 82, 184, 295
crustal stresses, 29, 31
crust movements, 8, 29, 168, 186, 189, 301, 304, 369, 494

cryolite, 452
crystalline schists, 29, 54, 56
cupola, 184, 191, 241, 263
cupriferous pyrite, 231
curve, liquidus, 18
— solidus, 18
cyanidation, 158, 454
cycles of mineralization, 175

dedolomitization, 130
deep leads, 278, 477, 480
degree of freedom, 204
density of the earth, 2
descloizite, 432
differentiation, 14, 17, 25, 29, 41, 45, 178, 315, 324, 364, 406, 487
diffusion, 20, 33, 37
dimorphism, 162, 213
dip, 75
disseminations, 12, 70, 85, 102, 178, 239, 372
dolomite, 223
dolomitization, 142, 154
drainage tunnel, 295, 469
dredging, 479, 496
dynamic metamorphism, 55, 127

earth movements, 73, 88, 166, 174, 186, 236, 247, 261, 301, 494
— zones, 7
earthquake waves, 2
electrolysis, 452, 456
electrum, 482
eluvial deposits, 110, 279, 419
elvan, 185, 266
epigenetic deposits, 106
equilibrium, 19, 27, 30, 187, 203
eutectic, 15, 24, 33, 208, 219

fahlband, 485
fault, 55, 285, 302
— -breccia, 55, 73
ferromanganese, 381
ferrotitanium, 323, 436
ferrovanadium, 431
fissure-fillings, 71
flats, 72, 287, 297, 299, 332, 404
flat veins, 72
flucan, 75, 87, 99, 302
flux, 37, 43, 52, 94, 219, 351
folding, 51, 177, 275, 286, 308, 374, 386
forms of ore-deposits, 70
fossil ore, 356
— placers, 91, 473
fracture, zone of, 55
freezing-point, 14, 32, 34, 206
frozen ground, 480, 496
fusible metals, 210, 409
fusion, 127

gangue, 92, 95
garnierite, 363, 374
gas cavities, 37
gases, 34, 115, 216, 222
genealogical tree, 180
geological maps, 51
germanium, 172, 274, 492
geysers, 58
glacial drift, 137
glass, 13, 206
— sands, 94
gold (general), 454
gondite, 383
gouge, 75, 87, 99
gravel terraces, 478
gravity, 20, 151, 319
greenalite, 328
greenstone, 264
greisen, 138, 265, 279
ground-water, 38, 61, 182, 232, 385, 491

hade, 76
haematite ores, 321, 326, 330, 357
hard lead, 403
head, 420
high-speed steel, 376, 412, 420
hollandite, 385
horse, 78
hot springs, 58, 177, 216, 397, 399, 401, 457, 489
hydrates, 59
hydraulicking, 419, 479
hydrocarbons, 429
hydrothermal deposits, 109

igneous rocks (general), 12
ilmenite, 323
immiscible liquids, 5
infiltration, 473
inlier, 297
intermolecular compounds, 17, 225
inversion point, 207
iron (general), 313
— bacteria, 360
— in igneous rocks, 313
— pan, 360
ironstone, 314
isobars, 200
isomorphism, 17, 160, 210, 224
isotherms, 185, 200
itabirite, 329, 388, 464

jasper, 385, 397
joints, 51

kaolinization, 138, 265
killas, 262, 272

laccolith, 21, 30, 32
laccolithic differentiation, 21

lake ore, 360
lateral secretion, 153, 281, 303
laterite, 361, 383, 447, 449
laws of solution, 13
lead (general), 281
leader, 74
ledge, 82
limited miscibility, 20, 43, 150, 218, 250, 371
liquidus curve, 18, 23, 210
lithomarge, 450
lode, 12, 82
lowering of the freezing point, 15, 37, 208
luxulyanite, 265

magma, 19, 26, 41, 54, 58, 218
— basin, 31
magmatic segregation, 84, 90, 109, 147, 150, 155, 194, 205, 231, 315, 390, 462
— water, 58
magnetic separation, 85, 324, 413, 417, 439
— survey, 316
magnetite ores, 86, 316
manganese (general), 379
— ores, 380
— steel, 379
marble, 50, 54, 305
marginal concentration, 20
— facies, 33
marine placers, 475
marl, 354
marlstone, 346
marne, 354
matte, 6, 22, 25, 150, 375, 497
Mayari steel, 395
mercury (general), 396
metallogenesis, 144
metallogenetic epochs, 174
— provinces, 166, 272
— zones, 181
metallogenic elements, 145
metamorphism, 12, 29, 48, 53, 125, 147, 305, 308, 384
metasomatism, 43, 47, 52, 126, 215, 315, 330, 339
meteoric water, 60, 146, 153, 233, 304, 307, 450
meteorites, 2, 313
mineral, definition of, 10
mineralogy, 10
minette, 355, 427
mixed crystals, 10, 93, 160, 164, 210, 224
molecular compounds, 223
molybdenum (general), 420
monazite, 160, 437
Monel metal, 362
mottramite, 429
mustard gold, 467

native metals, 8, 157, 161, 212, 313, 482, 495
nebular hypothesis, 1
nickel (general), 362
— -chrome steel, 362
— -iron, 158
— ores, 362, 375
— steel, 362
nickeliferous sulphides, 364
nife, 7
nifesima, 7
norite, 364, 371
nuggets, 476, 479, 496

obsidian, 13, 58
oolitic ironstone, 345, 356, 359
open working, 83, 279, 355, 370, 418, 431
order of separation of minerals, 16
ore-channel, 310
— -shoot, 88, 185, 240, 268, 458, 463, 472
— -zones, 19, 181, 289
osmotic pressure, 67
oxidation zone, 113, 196, 233, 246
oxide segregations, 218

palladium, 497
paragenesis, 159
patronite, 430
peach, 98, 266
pegmatite, 29, 35, 41, 59, 84, 111, 147, 156, 194, 237, 309, 416, 426, 453
permeability, 63
petrogenic elements, 145
petrology, 10
phacolith, 83
phase rule, 203
phyllite, 56
pipe, 71, 171, 188, 297, 400, 410, 421, 424
pitchblende, 442
pitchstone, 13
placer, see alluvial deposits
planetesimal hypothesis, 2
platinum, 373, 393, 494
pneumatolysis, 34, 38, 43, 156, 168, 186, 261, 275, 303, 372, 416, 424
pocket, 88, 297
polymorphism, 213
porosity, 51, 63
porphyry copper, 239, 243
Portland cement, 129
precipitation, 215, 220, 253, 473
primary crust, 3
— ores, 6, 31, 95, 111, 146, 152, 182, 217, 242
primitive earth magma, 5
propylitization, 40, 140
pumice, 13

purple ore, 249
pyritization, 140, 462
pyroxenite, 365, 422

quisqueite, 430

radioactivity, 1, 43, 58
radium, 434, 442
rake, 82
rare earths, 45, 160, 436
— metals, 93
rate of cooling, 30
reaction principle, 17, 210
reascensionist theory, 66, 146, 153
red beds, 254, 257
reef, 82, 457, 462, 472
refractories, 42, 214
re-fusion, 19, 30
replacement deposits, 52, 84, 103, 111, 232
residual deposits, 110, 136, 279, 419, 450
ring ore, 100
rock, definition of, 10
— -flow, 66
— -forming minerals, 19
roscoelite, 428
rubio, 340
ruby silver ores, 165
run, 306

saddle reef, 82, 309, 462
sal, 7
salt deposits, 49
— lakes, 257
sapphire, 27
scapolitization, 47, 138, 422
schalenblende, 282
scheelite, 413, 417
schistosity, 55
schorl rock, 264
secondary enrichment, 6, 9, 113, 124, 155, 196, 231, 307, 375, 407
— ore-deposits, 9, 105, 111, 145, 277, 454, 496
sedimentary rocks, 9, 35, 48, 71, 105, 135, 232, 254
selenides, 155, 163, 455
sequence of rock-types, 31
sericitization, 139, 198
serpentine, 158, 374, 390, 393, 496
shatter-belt, 237, 267, 275, 285, 302
shear zone, 465
sheet ground, 306
shoad, 110, 415
shoot, 88, 268, 458, 460, 463, 472
silicate solutions, 33
silicification, 52, 138, 154, 198, 397, 403, 462
silicospiegel, 381
sill, 21, 30

silver (general), 481
— -cobalt ores, 154, 377, 485
— -lead ores, 483
sima, 7
skarn, 131, 417, 422, 425
slag, 3, 22, 150, 203
slate, 56
slickenside, 55, 75, 458
smelting, 25
solid solution, 10, 17, 24, 203, 210, 483
solidus curve, 18, 23
solubility, 187, 206
solution, 13, 22, 67, 116, 137, 156, 186, 202
— -cavity, 298, 302
sop, 332
special steel, 395, 411, 420, 441
sperrylite, 497
spiegeleisen, 381
springs, hot, 58, 177, 216, 397, 399, 401, 457, 489
spring waters, 57
stability of minerals, 216
steam, 30, 34, 66
stellite, 376, 391, 412, 420
stereogram, 81
stock, 70
stockwork, 12, 74, 83, 102
stoping, 26
stratification, 32, 111, 341
strike, 75
stringer, 102
sublimation, 202
sulphur, 397, 401
superficial deposits, 111
swallow-hole, 337
syngenetic deposits, 106

taconite, 328
tantalum, 439
telluride ores, 155, 158, 163, 173, 212, 225, 417, 455, 465, 482
temperature arrest, 24
— gradient, 108
ternary eutectic, 209
thermal metamorphism, 127
thorium, 436
tin (general), 260
— -gravels, 9, 276, 278, 280
titaniferous iron ore, 321, 435
titanium, 435
torbernite, 442
tourmalinization, 138, 265
travertine, 58
tungsten (general), 412
type-metal, 403

undercooling, 205
underlie, 81
uranium, 170, 442

vanadinite, 432
vanadium (general), 426
vapour pressure, 59
veinstones, 96, 190
vena, 340
viscosity, 205, 319
volcanic ash, 49
volcanoes, 13
vug, 96, 220
vulcanicity, 8, 12, 40, 59, 66, 141, 147, 166, 177, 241, 397, 401, 457, 469, 488

water, 8, 35, 57, 146, 217
— artesian, 61
— -bearing rocks, 63
— ground, 38, 61, 182, 232, 385, 491
— juvenile, 59
— surface, 60

water-table, 62, 117
— vadose, 61
weathering, 132
wire-silver, 485
wood tin, 221, 260
wulfenite, 421, 432

xenolith, 26

zeolites, 240, 485, 494
zinc (general), 281
— ores, 131, 281
zirconium, 440
zoned crystals, 19
zone of cementation, 62, 68
— — fracture, 55
— — oxidation, 68, 115, 262
— — secondary enrichment, 118, 164, 284

INDEX OF PLACE NAMES

Aberdaron, 382
Alabama, 176, 355, 447
Alaska, 172, 176, 480
Alderley Edge, 255
Algeria, 405
Allendale, 299
Almadén, 398
Alston Moor, 299
Altenberg, 83
Anaconda, 242
Anchor Mine, 278
Andes, 65, 120, 147, 273, 409, 430, 490
Anglesey, 189, 235
Annaberg, 378
Anneröd, 443
Araca, 273, 490
Arendal, 443
Arenig, 382
Argentina, 273, 416, 427, 490
Ariège, 448
Arizona, 239, 245, 421, 426
Arkansas, 448
Ashover, 297
Askham-in-Furness, 332, 337
Atolia, California, 418

Balaghat, 457
Ballandean, 423
Ballymena, 451
Bamle, 47, 366
Banbury, 346
Banka, 171, 275
Barberton, 329
Baringer Hill, Texas, 44

Bauchi, 280
Bawdwin, 310, 483
Belgian Congo, 256
Belmont, 351
Benallt, 382
Bendigo, 82, 462
Bettws-y-Coed, 289
Bigrigg, 332
Bilbao, 340
Billiton, 171, 275
Bingham Canyon, Utah, 195, 239, 244
Birmingham, Alabama, 355
Bisbee, Arizona, 239, 245
Bishop, California, 418
Black Hills, South Dakota, 44, 437, 453
Bleiberg, 195
Blencathra, 291
Blue Tier, 278
Bodmin Moor, 263, 420
Bohemia, 168, 223, 376, 408, 444, 494
Bolivia, 65, 168, 196, 223, 272, 406, 409, 414
Boltsburn, 300
Borneo, 395
Boulby, 351
Boulder County, Colorado, 417, 428
Branchville, Connecticut, 440
Brazil, 78, 91, 183, 329, 439, 463
Briey, 354
Brittany, 170
Broken Hill, N.S.W., 307, 483
— — Rhodesia, 432
Burma, 168, 275, 415
Butte, Montana, 173, 195, 241, 408

Cae Coch, 193
Caldbeck Fells, 291
California, 89, 176, 400, 418, 459, 477
Calumet, 240, 326
Camborne, 80, 183, 267, 408
Canada, 47, 128, 131, 151, 175, 325, 366, 377, 391, 408, 422, 460, 480, 485
Carbis Bay, 265
Cardiganshire, 285
Carlinhow, 351
Carnarvonshire, 193, 289, 382
Carn Brea, 263
— Marth, 263
— Menellis, 263
Carrock Fell, 21, 188
Castle-an-Dinas, 170, 416
Castleton, 297
Caucasus, 101, 387
Ceylon, 437, 440
Champion Reef, Mysore, 457
Chattanooga, Tennessee, 355, 449
Cheshire, 189, 255, 301
Chhindwara, 383
Chile, 65, 120, 233
China, 197, 404, 412
Chorolque, 272, 409
Clausthal, 98, 173, 195, 493
Cleator Moor, 333
Cleveland, 348
Cligga Head, 266
Cobalt, Ontario, 154, 173, 375, 408, 485
Coeur d'Alene, Idaho, 173
Colorado, 155, 164, 174, 417, 428, 434, 445, 453, 465, 469
Comstock Lode, 489
Coniston, 291
Cooktown, 278
Copper Queen, 246
Corbières, 448
Cornwall, 72, 83, 97, 102, 141, 154, 168, 174, 183, 190, 198, 235, 262, 382, 404, 408, 416, 420, 443, 478, 497
Crich, 297
Cripple Creek, 469
Crowgarth, 333
Crusnes, 354
Cuba, 394
Cueva de la Mora, 248
Cuyuna, 326

Dakota, South, 44, 453, 475
Dalen, 426
Dannemora, 330
Dartmoor, 263
Denbigh, 189, 295
Derbyshire, 72, 235, 237, 297, 483
Devon, 73, 83, 102, 170
Dolcoath, 183, 198, 268
Dome, 462

East Pool, 183, 270
Eaton, 346
Ecton, 188, 237
Egremont, 332
Ekersund, 323
Ely, Nevada, 239, 244
Erteli, 366
Erzgebirge, 83, 102, 168, 377, 414, 493
Eston, 348, 350
Etta Knob, South Dakota, 44, 439

Finland, 131, 360
Flaad, 366
Flint, 189, 295
Force Crag, 293
Forest of Dean, 339
Franklin Furnace, New Jersey, 131, 282, 305, 380
Fredrikshald, 46
Freiberg, Saxony, 173, 408, 493
Frodingham, 345
Furness, 331

Geirionydd, 289
Gellivare, 320
Georgia, 447
Glenarm, 451
Globe, Arizona, 244
Godolphin, 263
Gogebic, 326
Golden Stairway, 461
Goldscope, 293
Grainsgill, 291
Grängesberg, 321
Grass Valley, California, 459
Great Basin, 65, 176, 418
Greece, 375, 391, 393, 483
Greenhow, 298
Greenland, 158, 439, 452
Green River, Colorado, 433
Greenside, 293
Grinkle Park, 351
Griqualand East, 21, 151, 372, 497
Grosmont, 348
Gursli, 426

Halkyn, 295
Heemskirk, 278
Hemerdon, 83, 102, 170
Herberton, 278
Hodbarrow, 335, 339
Holwell, 348
Holywell, 295
Homestake, South Dakota, 475
Huayna Potosí, 409
Huelva, 103, 248
Hungary, 141, 147, 155, 164, 174, 469

Iceland, 58
Idaho, 173

Idria, 398
Ilmen Mountains, 440
Insizwa, 151, 179, 372, 497
Isle of Man, 190
—— Skye, 31
Italy, 58, 142, 398
Ivigtut, 439, 452

Joachimsthal, 408, 427, 443
Johanngeorgenstadt, 378
Joplin, Missouri, 306

Kambove, 256
Kangaroo Hills, 278
Karlsbad, 58
Katanga, 256, 379
Kelton Fell, 332
Kentucky, 355
Keswick, 291
Kettering, 351
Kettleness, 348
Keweenaw Peninsula, 239
Kiirunavaara (Kiruna), 316, 324
Killifreth, 269
Kilton, 351
Kingsgate, 423
Kirghiz Steppes, 258
Knaben, 425
Knockmurton Fell, 332
Kolar, 91, 457
Kragerö, 435
Kristiania, 28, 46

Lafayette, 388
Lake District, 174, 188, 235, 284, 291
— Superior, 86, 139, 238, 325, 427
Land's End, 27, 262
Larne, 451
Leadhills, 174, 294, 427
Les Baux, 447
Levant, 265
Lillestrand, 47
Little Rock, Arkansas, 448
Liverton, 351
Lizard, 262, 497
Llallagua, 272
Llanbedr, 382
Llanrwst, 289
Loftus, 351
Longwy, 354
Lorraine, 330, 343, 353, 427
Lumpsey, 351
Luossavaara, 316
Luxemburg, 343, 353, 427
Luxulyan, 265

Malay Peninsula, 168, 275, 414, 419, 449
Malmedy, 132
Mansfeld, 86, 254, 258, 427

Marquette, 326
Maryland, 391
Mayari, 394
Meinkjär, 366
Melton Mowbray, 348
Mendips, 189, 301
Menominee, 326
Merionethshire, 193
Mesabi, 326
Mexico, 147, 156, 163, 223, 406, 427
Miami, Arizona, 239
Miask, 440, 443
Michigan, 325
Michipicoten, 325
Miguel Burnier, 388
Millclose, 298
Millom, 335
Minas Geraes, 329, 441, 463
Minasragra, 430
Minera, 295
Minnesota, 325
Minsterley, 87
Mississippi Valley, 153, 199
Missouri, 67, 153, 305
Montana, 27, 173, 177, 195, 239, 408
Monte Amiata, 400
Montgomeryshire, 100, 285
Montreal Mine, 335
Moonta, 237
Morenci, 239
Moresnet, 153
Morro Velho, 78, 91, 183, 463
Mottram St Andrew, 255, 429
Mount Bischoff, 278
— Morgan, 117
— Painter, 427
Mutue Fides, 408
Mysore, 91, 183, 386, 457

Nababeep, 253
Nagpur, 383
Namaqualand, 252
Nancy, 354
Naples, 40
Narukot, 383
Nevada, 164, 239, 401, 418, 487
Nevada City, California, 89, 459
New Almaden, 402
— Caledonia, 373, 391
Newfoundland, 330, 356
New Idria, 402
— Jersey, 32, 131, 282, 305, 380
— South Wales, 142, 171, 277, 307, 410, 423, 479, 483
— York State, 32, 355
— Zealand, 58, 457
Nigeria, 168, 171, 280
Nijni Tagilsk, 393
Ningi Hills, 171
North Carolina, 441

North Wales, 193, 235, 284, 295, 382, 483
Norway, 45, 84, 103, 152, 246, 250, 364, 392, 425, 436, 443, 484
Nova Scotia, 176
Nundydroog, 457

Ontario, 154, 326, 363, 375, 378, 460, 485, 495, 497
Ookiep, 253
Ooregum, 457
Oregon, 363
Oruro, 272
Ouro Preto, 464

Park Mines, 337
Parys Mine, 235, 291
Passagem, 409, 463
Pearl Lake, 461
Peckforton Hills, 189, 255
Penokee, 326
Perranporth, 192
Peru, 65, 120, 427, 429, 490, 492
Piquery, 389
Porcupine, Ontario, 460
Portugal, 168, 247, 414
Potosí, 272
Przibram, 173

Queensland, 277, 408, 410, 423
Quyon, Quebec, 422

Radium Hill, 427
Redjang Lebong, 163
Redruth, 80, 183, 267
Republic, Washington, 163
Rhodesia, 391, 408, 432, 474
Ringarooma, 278
Ringerike, 366
Rio Tinto, 105, 155, 247, 251
Risör, 436
Roman Gravels, 289
Romsaas, 366
Röros, 152, 250, 392
Roseberry, 351
Rosedale, 350
Roseland, Virginia, 435
Roughtengill, 291
Routivare, 323
Russia, 158, 257, 387, 391, 393, 427, 437, 495

St Agnes, 263
St Austell, 263
St Dennis, 266
St Ives, 74, 102, 265, 443
St John del Rey, 78, 91, 183, 408, 463
St Michael's Mount, 263, 266
Saintonge, 448
San Dionisio, 248

Santander, 340
Saxony, 86, 168, 173, 223, 254, 258, 377, 408, 493
Schneeberg, 378
Schultz, Arizona, 426
Schwatz, 403
Scunthorpe, 345
Seboekoe, 395
Selukwe, 392
Serra de Caldas, 441
Serranía de Ronda, 158
Shelve, 185, 190, 289
Shropshire, 87, 185, 189, 255, 285, 289
Sierra de Córdoba, 416
Skelton, 351
Snailbeach, 289
Soggendal, 323
Solares, 340
Somorrostro, 340
South Africa, 21, 27, 139, 151, 168, 252, 329, 372, 392, 408, 432, 449, 470
South Australia, 237, 427
— Crofty, 183, 270
— Terras, 444
Spain, 103, 158, 168, 246, 330, 340, 398, 414, 427
Spawood, 351
Staffordshire, 237, 255, 343
Stanthorpe, 278, 423
Star of the Congo, 257
Start Point, 262
Stathern, 348
Steamboat Springs, 401
Sudbury, Ontario, 84, 151, 179, 366, 497
Sulitjelma, 152, 250
Sulphur Bank, California, 401
Sumatra, 155, 163, 275
Swaledale, 298
Sweden, 84, 131, 316, 330, 360, 427, 436
Sydvaranger, 324

Tabankulu, 21, 373
Taberg, 321, 324, 427, 436
Table Mountain, 27
Tasmania, 171, 277, 410, 423
Tasna, 409
Tavoy, 171, 415
Telemarken, 46, 426
Tennessee, 355, 448
Texas, 44, 437
Tharsis, 249
Thornthwaite, 293
Threlkeld, 291
Tiebaghi, 374, 392
Tilton, 348
Timiskaming, 379
Tintic, Utah, 408
Tonopah, Nevada, 164, 487
Transylvania, 141, 469

Travancore, 437
Trecastell, 289
Trefriw, 193, 289
Trenwith, 443
Trevalgan, 265
Trondhjem, 250
Truro, 262
Tuscany, 399
Tvedestrand, 366

Ullbank, 333
Ullcoats, 333
Uncia, 272, 406
United States, 27, 44, 58, 65, 131, 141, 147, 175, 196, 227, 238, 257, 381, 391, 421, 427, 435, 447, 469, 489
Upleatham, 351
Urals, 158, 246, 495
Utah, 142, 195, 239, 408, 418, 433

Van, 100, 287
Vanadium, 428
Var, 448
Vegetable Creek, 278
Vermilion, 326
Victoria (Australia), 82, 278, 462, 479
Villeveyrac, 448
Virginia, 435
Visagapatam, 383

Viscaya, 340
Vogelsberg, 447

Wabana, 330, 356
Wallaroo, 237
Wanlockhead, 294, 427
Wartnaby, 348
Washington (State), 163, 176
Weardale, 299
Wellingborough, 351
West Australia, 65, 155, 164, 397, 427, 449, 465, 474
Wheal Seton, 269
Whipstick, 423
Whitehaven, 332
Winster, 297
Wisconsin, 325, 355
Witwatersrand, 91, 455, 470
Wolfcleugh, 299
Wolfram, 423
Woodend, 333
Wood River, Idaho, 173

Yellowstone Park, 58
Yetholme, 425

Zeehan, 279
Zimapan, 407
Zinnwald, 83

INDEX OF AUTHORS CITED

Adams, F. D., 88
Allen, E. T., 125, 216
Andrews, E. C., 307, 425
Armas, S., 273
Arrhenius, S., 59, 60

Baragwanath, J. G., 429
Barrow, G., 29, 348, 420
Bastin, E. S., 488
Becker, G. F., 473
Bell, J. M., 461
Bertrand, M., 166
Bewick, J., 348
Billingsley, P., 177, 196, 241
Bodenbender, G., 417
Bonney, T. G., 265
Bowen, N. L., 5, 17, 19, 20
Brögger, W. C., 28, 45, 46, 160, 487
Brown, J. C., 312, 415
Buttgenbach, G., 257

Campbell, J. M., 59, 415
Carruthers, R. G., 301
Chamberlin, T. C., 2
Clarke, F. W., 4, 25, 57, 148, 446

Cole, G. A. J., 447, 451
Collins, H. F., 247, 248, 444
Compton, R. H., 374
Corstorphine, G. S., 473
Cronshaw, H. B., 190, 302
Curtis, A. H., 388

Daly, R. A., 5, 20, 26, 317
Davison, E. H., 170, 184, 190, 268, 416, 444
Davy, W. M., 196, 272
De la Beche, H., 97, 99
de Launay, L., 427
Desch, C. H., 17, 225
Dewey, H., 185, 289
Dixey, F., 451
Drabble, H., 297
Duparc, L., 495
du Toit, A. L., 21, 151, 373

Eastwood, T., 292
Eddington, A. S., 2
Elles, G. L., 76
Emmons, W. H., 107, 123, 194, 198, 243, 460, 475

Farquharson, R. A., 468
Feldtmann, F. R., 466, 468
Fermor, L. L., 386, 450
Findlay, A., 204
Finlayson, A. M., 154, 189, 304, 483
Foster, C. le N., 287

Gale, H. S., 429, 433
Geijer, P., 317, 320
Gibson, C. G., 468
Goodchild, W. H., 151, 373
Graton, L. C., 216
Gregory, J. W., 473
Grimes, J. A., 177, 196, 241
Grout, F. F., 28
Guild, F. N., 483
Gulliver, G. H., 17

Haenig, I. A., 387
Hall, T. C. F., 190, 193, 289
Harder, E. C., 329, 387
Harker, A., 5, 21, 26, 29, 31, 38, 76, 83, 130, 166
Hatch, F. H., 48, 94, 130, 228, 305, 331, 339, 352, 458, 471, 473
Hayes, A. O., 357, 360
Heron, A. M., 415
Hess, F. L., 44, 160, 413, 417, 428
Hewett, D. F., 429
Hibbard, H. D., 376
Hidden, W. E., 44, 160
Hillebrand, W. F., 428
Hobbs, W. H., 2
Holland, T. H., 449
Hore, R. E., 461
Hoyt, S. L., 173, 213, 483
Hussak, E., 464

Irving, R. D., 475

Jaeger, F. M., 223
Johansson, H., 321
Johnstone, S. J., 436
Jones, O. T., 286
— W. R., 277, 415

Kemp, J. F., 164, 195, 436, 455
Kilroe, J. R., 447
Kinahan, G. H., 452
King, L. V., 88
Kithil, K. L., 433
Kjerulf, T., 250
Knight, C. W., 175, 367
Kossmat, F., 399
Krusch, P., 444, 484
Kuntz, J., 252

Lacroix, A., 392, 406, 448
Lake, P., 166, 449
Lamplugh, G. W., 347, 349

Laney, F. B., 488
Le Conte, J., 402
Leith, C. K., 328
Lenher, V., 455
Lewis, J. V., 32
Lindgren, W., 89, 109, 175, 194, 460, 467
Lotti, B., 399
Louis, H., 300
Lundbohm, H., 317

MacAlister, D. A., 269, 444
McCaskey, H. D., 403
McPherson, G., 497
Mahony, D. J., 463
Mawson, D., 308
Mellor, E. T., 473
Merwin, H. E., 125, 216
Miller, B. L., 388, 406, 409, 430, 493, 496
— W. G., 175, 367
Mitchell, J., 294
Moore, E. S., 433
— N. P., 125, 216
Morosewicz, J., 26
Münster, H., 447
Murdoch, J., 216

Nicholas, T. C., 382
Nicou, P., 354
Nissen, A. E., 173, 213, 483

Penrose, R. A. F., 444
Petrenko, J., 225
Pirsson, L. V., 153
Pittman, E. F., 479
Pošepný, F., 61
Pratt, J. H., 391, 441
Pringle, J., 347

Ransome, F. L., 245, 467
Rastall, R. H., 48, 130, 150, 166, 179, 305, 414, 419
Rickard, T. A., 462
Ries, H., 107, 152
Rising, W. B., 402
Rogers, A. W., 474
Rosenbusch, H., 33
Rumbold, W. G., 395
Rutherford, E., 442
Rutley, F., 94

Salisbury, R. D., 2
Schaller, W. T., 417
Scott, H. K., 375, 387, 395
Scrivenor, J. B., 276
Shaler, H. S., 256
Sibly, T. F., 339
Siebenthal, C. E., 306
Simpson, E. S., 455, 468

Singewald, J. T., 324, 388, 406, 409, 430, 436, 493, 496
Sjögren, H., 317
Smith, B., 185, 193, 289, 295, 332, 338
— H. H., 254
— S., 301
Smyth, W., 287
Somers, R. E., 152
Spencer, L. J., 164, 455
Spirek, V., 400
Spurr, J. E., 194, 488
Stead, J. E., 358
Sterrett, D. B., 437
Strahan, A., 301
Stutzer, A., 317
Suess, E., 7, 58, 275

Teall, J. J. H., 130, 345
Tegengren, R. G., 405
Tikonowitch, M. N., 495
Tilley, C. E., 487
Tipper, G. H., 437
Turner, H. W., 419

Van der Veen, R. W., 341
Van Hise, C. R., 328
van Klooster, H. S., 223

Verbeek, R. D. M., 276
Vogel, J. L. F., 413
Vogt, J. H. L., 6, 22, 47, 150, 250, 318, 324, 364, 391
von Bolton, W., 439
von Cotta, B., 70

Wagner, P. A., 408
Warren, C. H., 44, 160
Warth, F. J., 447
— H., 447
Washington, H. S., 4, 145
Watson, T. L., 435
Wedd, C. B., 297, 347
Wilcockson, W. H., 150, 419
Williams, G. W., 171
Wilson, G. V., 294
Woakes, E. R., 426
Wong, W. H., 197
Woodward, H. B., 63

Young, R. B., 474
— S. W., 125, 216

Zealley, A. V., 392
Zies, E. G., 125, 216

Printed in the United States
By Bookmasters